CSR, Sustainability, Ethics & Governance

Series Editors

Samuel O. Idowu, London Metropolitan University, London, UK
René Schmidpeter, Cologne Business School, Cologne, Germany

In recent years the discussion concerning the relation between business and society has made immense strides. This has in turn led to a broad academic and practical discussion on innovative management concepts, such as Corporate Social Responsibility, Corporate Governance and Sustainability Management. This series offers a comprehensive overview of the latest theoretical and empirical research and provides sound concepts for sustainable business strategies. In order to do so, it combines the insights of leading researchers and thinkers in the fields of management theory and the social sciences – and from all over the world, thus contributing to the interdisciplinary and intercultural discussion on the role of business in society. The underlying intention of this series is to help solve the world's most challenging problems by developing new management concepts that create value for business and society alike. In order to support those managers, researchers and students who are pursuing sustainable business approaches for our common future, the series offers them access to cutting-edge management approaches.

CSR, Sustainability, Ethics & Governance is accepted by the Norwegian Register for Scientific Journals, Series and Publishers, maintained and operated by the Norwegian Social Science Data Services (NSD)

More information about this series at http://www.springer.com/series/11565

Panagiotis Dimitropoulos • Konstantinos Koronios

Corporate Environmental Responsibility, Accounting and Corporate Finance in the EU

A Quantitative Analysis Approach

Springer

Panagiotis Dimitropoulos
Department of Sport Organization
and Management
University of Peloponnese
Sparta, Greece

School of Social Sciences
Hellenic Open University
Patra, Greece

Konstantinos Koronios
Department of Accounting and Finance
University of Peloponnese
Kalamata, Messinia, Greece

ISSN 2196-7075 ISSN 2196-7083 (electronic)
CSR, Sustainability, Ethics & Governance
ISBN 978-3-030-72772-7 ISBN 978-3-030-72773-4 (eBook)
https://doi.org/10.1007/978-3-030-72773-4

This Springer imprint is published by the registered company Springer Nature Switzerland AG.
The registered company address is: Gewerbestrasse 11, 6330 Cham, Switzerland

To my wife Anna for all her love and support.

—Panagiotis Dimitropoulos

I would like to dedicate this work to my grandmother Theoni, whose dreams for me have resulted in this achievement and without her loving upbringing and nurturing; I would not have been where I am today and what I am today. A grandmother can love and give without expecting anything in return. Had it not been for my grandmother's unflinching insistence and support, my dreams of excelling would have remained mere dreams. I thank her with all my heart and I know she is up there, listening, watching over me, and sending me her blessings constantly and is my guardian angel. This one is for you grandma!

—Konstantinos Koronios

Contents

Chapter 1
Introduction to Corporate Environmental Responsibility, Accounting and Finance

The concept of corporate social and environmental responsibility was first introduced in the management and business literature by Sheldon (1924) and since then the academic and business community has offered a lot of different explanations for CER. After decades of conceptual, theoretical and empirical development, CER has been evolved gradually by being a goal of non-profit organizations and governments, to a strategic tool by corporations which has a significant impact on the managers of enterprises (Freeman 1994). Under the notion of sustainable development and globalization, the traditional corporate theory, which takes the interests of shareholders as the core of its existence, cannot longer achieve long-term sustainable development, unless it incorporates structures, tools and mechanisms in order to adhere to society's and stakeholders' needs and demands. An increasing number of stakeholders have required enterprises to undertake their corresponding CER in return for the society's acceptance and legitimacy (Pan et al. 2014). Therefore, CER is gradually becoming an inevitable trend for enterprises to achieve long-term and stable development. However, CER implementation and performance presents several discrepancies between nations and even among nation unions such as the EU, despite the fact that several efforts have been made in order to harmonize environmental regulations within the EU and the rest of the world. Nevertheless, cases such as in China, where there are still many enterprises seeking profit maximization and neglecting CER, we have instances of food pollution, illegal addition of non-food substances, the abuse of food additives and other phenomena, which are resulting in consumers' lack of confidence (Martínez-Ferrero and Frías-Aceituno 2015).

With the rapid development of new markets and the growth in developing economies, higher information transparency and the increasing awareness of the public's environmental protection has led to an unprecedented pressure towards firms to listen to those concrete concerns and adjust their behavior accordingly, even though this could mean to pay less attention to their short-term profit targets, instead of creating a conflict with the overall interests of the society. This may result into

P. Dimitropoulos, K. Koronios, *Corporate Environmental Responsibility, Accounting and Corporate Finance in the EU*, CSR, Sustainability, Ethics & Governance, https://doi.org/10.1007/978-3-030-72773-4_1

huge costs, which will affect the business development in the future. Those violations of morality or law are mainly due to the weak concept of CER in enterprises, and a lack of knowledge about how CER will affect individual business development (Surroca et al. 2010). As a profitable economic organization, it is fundamental to promote CER's positive commitment to the enterprise by giving incentives to the economy. Therefore, the goal to understand the relationship between CER and corporate accounting and finance is of great practical significance for guiding enterprises to take the initiative to incorporate CER into their operations and strategy, to help the government establish several CER incentive mechanisms, and to promote the healthy operation of the whole economy.

During the last 15 years, the EU economy has faced significant downturns and upturns, while a series of environmental problems appeared. Environmental damage has become increasingly serious, so the concept of sustainable development has gradually become a consensus. The public so far has fully realized that only under the precondition of environmental protection the economy can develop in the long-term. Along with the public's environmental protection consciousness enhancement, the enterprises, especially those heavy polluting enterprises, are necessary to accept the supervision from the public authorities, to engage with their stakeholders, take into consideration their concerns and needs and communicate their dedication to environmental protection. This dissemination of information to the public is achieved by several forms and types of environmental information disclosures, including details about their activities, policies, and behaviors related to the natural environment, their attitudes toward protecting the environment, and the utilization of natural resources (Zeng et al. 2010).

Taking a step further, the establishment of an environmental accounting system can provide a feasible mechanism of environmental responsibility supervision for industries that are prone to create environmental damage. Moreover, CER and disclosure could be deemed as a tool for more efficient resource allocation within the economy, since it will provide information about the processes of production, targeted activities for the protection of the environment, and firms' attitudes towards environmental protection. It is the requirement to enhance the level of internal control and to cooperate with the governments' macro management. This procedure will effectively put the utilization of environmental resources, emission reduction and even environmental innovation within the core of corporate governance. The firms' engagement with CER has become a pressing issue as environmental protection is highly valued by investors, lenders and other internal and external firms' stakeholders around the globe (Zeng et al. 2010). This fact creates implications of CER for corporate accounting and finance. This book tries to disentangle the various implications and impacts of CER on several issues related to accounting information quality and financial management of corporations within the EU countries.

1.1 Introduction

Corporate environmental responsibility (CER) has gained significant attention by academics, researchers and regulators during the past 25 years. As discussed above, CER development can mainly be attributed to the enhanced pressure that firms face from several stakeholders so as to adhere to society's environmental concerns (Monteiro and Aibar-Guzmán 2010). According to Camilleri (2017), the first theoretical arguments for social and environmental responsibility were built under the notions of corporate philanthropy, ethics and stewardship principles. Since then, the establishment of CER has expanded by including the vast majority of our society. Based on that stance, firms are operating within the society (utilize human and natural resources) in order to create value for it. The shareholders are no longer the only legitimate contractors of the firms, but other external parties with significant interests have emerged. According to McWilliams and Siegel (2001), Porter and Kramer (2007), CER has been recognized as the economic justification of doing what is good and creates significant opportunities and competitive advantages. Practically, corporations' success is dependent on how efficiently they address several stakeholders concerns including social and environmental issues. This has led several corporations to devote themselves into responsible corporate activities (investments, donations, cultural and heritage protection, relationships with the communities and environment protection). They also devote significant efforts to communicate those activities to the public by publishing a large amount of voluntary CER reports and information (Dienes et al. 2016; Camilleri 2017).

Corporate environmental and social responsibility has originated as a concept of managerial responsibilities towards employees, investors, customers and the public at large (Abrams 1951), but Carroll (1979) was the first to explicitly assert the obligation of firms to engage in economic, legal, social and ethical activities. He described the economic responsibility of the firms (towards their shareholders) as the base of the pyramid, while legal responsibility is referring to the law abiding obligation of firms. The ethical and social responsibility of the firms' refers to activities and strategies directed towards the society, which go beyond the regulatory or legal prerequisites and obligations. Since then, the issue of CER has exploded into a global concern after several environmental issues (climate change, environmental resources depletion etc.) and corporate scandals have burst over the world, affecting several sectors of the economy and the lives of million of people worldwide.

CER has been conceptualized from different theories. According to Suto and Takehara (2018) and Camilleri (2017), stakeholder theory is the main framework under which CER is been developed and assessed. Also, the resource-based view (Russo and Fouts 1997; Hall 1992; Roberts 1992) is another theory indicating that CER activities utilize tangible and intangible corporate resources and the creation of sustainable stakeholder relationships requires the creation of intangible assets. Moreover, legitimacy theory of the firm (Tang and Luo 2016) also contributed to the theoretical explanation of CER motives and activities by considering CER as the result of social pressure that firms face regarding their environmental performance.

Corporate environmental performance is determined by strategies, policies and countable results on the responsiveness of firms to social and environmental concerns (Frederick 1986; Camilleri 2017). So CER performance includes various processes such as environmental innovation, emission reduction, efficient use of resources and the implementation of environmental friendly processes and policies. According to Camilleri (2017), environmentally responsible firms could improve their image to the public, differentiate themselves from competition and even gain economic advantages. This fact provide a significant incentive for firms to increase their CER related activities and reports, so as to gain legitimacy for their operations and decisions and consequently to mitigate the risk which arises from non-compliance with stakeholders' concerns and interests (Huang and Watson 2015).

Nevertheless, we must not exclude the dimension of ethical business conduct inherent on CER activities, since business ethics was one of the first conceptual orientations of CER. According to Goodpaster (1991), corporations should pay significant attention to the ethical concerns and stances of their stakeholders. For the early stages of its development, corporate social and environmental responsibility was envisaged through the lens of ethical corporate behavior dictating that firms must establish official mechanisms of communication with stakeholders. As a result, some firms were becoming more proficient on disclosing their environmental performance, while steadily this trend evolved as one of the most significant mechanisms for information dissemination to the public. Such information could improve the decision environment of the firms' leading to more informed customers, investors, lenders, creditors and other related parties, supporting efficient economic decision making and resource allocation within the economy.

Under this framework, the book tries to extend the existing empirical literature on the accounting and corporate finance decisions by examining the determinant factors of CER, the impact of CER on financial management decisions including financial performance, corporate risk, working capital management, dividend policy, innovation, capital structure and accounting information quality. The book will be focused on the EU member countries, since they display several communalities regarding environmental laws and regulations while simultaneously they present several differences regarding their structure of their economies, legal frameworks and levels of economic development. The rest of the chapter is organized as follows: In the following section we provide a brief discussion on corporate sustainability followed by a chapter on the European policies and regulatory instruments implemented within the EU countries. The fourth section will analyze the environmentally and socially responsible investments as a mechanism of CER which is closely connected to corporate accounting and finance. The fifth section will establish a connection of CER with accounting and finance and will provide the main research questions that will be addressed in the following chapters of the book, along with the book's contribution to existing literature. The last section will conclude the chapter.

1.2 Corporate Sustainability and CER

Corporate sustainability has emerged within the corporate world since 1987 and today has been evolved as more crucial than ever since it encompasses several dimensions of corporate activity including environmental responsibility. A corporate activity is considered as sustainable if it's socially and environmentally justified, is ethically conducted, creates economic value and of course abides to existing laws and regulations (Brockett and Rezaee 2012). Over the years sustainability activities and reporting has grown exponentially and the role of CEOs, employees, and other stakeholders have evolved them to key players on sustainability activities. Also, environmental awareness is rising due to the importance of corporate resources and the scarcity evidenced on them. The whole concept of sustainability originates on the stakeholder theory which argues that contemporary organizations and corporations have obligations towards several constituents such as customers, employees, creditors, investors and of course shareholders. Corporate sustainability proposes the consideration of stakeholders' interests on corporate decisions. According to Brockett and Rezaee (2012), the so called "Benefit Corporations" have emerged as a corporation lying between the well-known for-profit firms and the not-for-profit organizations, since their goal is to perform certain social and environmental goals while covering their operating costs and achieving a sustainable level of profit, which is not re-distributed among the owners (as dividends for example) but is re-invested in the firm. Consequently, board of directors on benefit corporations incorporate stakeholders' interest and needs on their decision making process.

Nevertheless, the consideration of stakeholders' interest by the board of directors has been extended to the majority of for-profit corporations where several CEOs and directors take under serious consideration the repercussions of their decision on local communities and the environment. This trend has evolved into a crucial dimension of corporate activity monitored by investors, stock markets, funds and auditing corporations. As the various accounting and auditing standards had firstly were developed for shaping the financial reporting environments of firms, along the way several standards were developed to recognize, measure and report to the public the ethical, social and environmental activities of firms (the GRI standard is an example). The scope of those standards was to assist firms disseminate accurate, reliable and transparent information that could be proved useful to stakeholders in order to assess the prospects and risks of the firm. Thus, firms realized that sustainability and environmental reporting was a valuable addition to the financial statements so as to mitigate investors' concerns regarding their future economic prospects and risks.

Despite the fact that the primary goal of businesses is to enhance their financial performance (profitability) and safeguard their shareholders' investments, they also have to deal with environmental and social issues in order to add value to both their shareholders and stakeholders. Corporate sustainability provides the background for long-term competitive advantages, profits and the wellbeing of the society. According to the International Federation of Accountants (IFAC 2011), corporations nowadays have to build a sustainability framework which must focus on the achievement of long-term goals and performance (instead on focusing on short-term financial oriented goals), engage internal management and investors on directing their activities towards sustainable performance, and finally incorporate financial and non-financial key performance indicators on business decisions and provide assurance to the disclosed information on sustainable performance. There propositions have been made under the notion that global investors and investment funds have shown immense interest and have increased their demand for such types of information because they incorporate them on the investment decisions.

Corporate environmental responsibility (CER) has evolved as the most important dimension of sustainable practices within corporations and can be viewed as a tool for more efficient risk management. The reason for that is because CER can assist investors and other stakeholders to identify opportunities and risks connected to corporate operations. So transparent and reliable information disclosure on CER creates opportunities and allows the identification and improvement of operational deficiencies. This in turn can mitigate any reputational and financial risks that could have a devastating impact on future financial performance and viability of corporations and the stability of the national and world economies. That is the main reason that several national and supranational organizations have drafted rules and regulations that aim to promote environmental and sustainability reporting internationally.

1.3 International and European Policies and Regulations on CER

The development of international CER awareness and engagement was initially considered by the side of corporations' expansion within the international trade legislation. Several international organizations including the World Trade Organization (WTO) and the Organization for Economic Cooperation and Development (OECD) have issued over the years several standards trying to regulate the operation of multinational enterprises at first and then the national corporations of their member states. For example the OECD (1999) principles introduced the issues of shareholder rights protection (in order to maintain investor confidence and the undisruptive flow of capital within the capital markets) and the important role of stakeholders on firms' corporate governance structure and decisions (balancing the interests of various stakeholders such as customers, employees, local communities, creditors, lenders, etc.). Despite the fact that those guidelines were providing sufficient directions to firms to engage in appropriate CER activities via publications, workshops and other channels, those principles were not exhaustive or suitable for every country and market. Moreover, the guidelines of CER reporting were voluntary in nature (not being mandated by each member's country national legislation) creating differences between counties in term of the tools and standards that have been developed and implemented.

Additionally, several other global environmental initiatives and regulations were developed since the 2000s including the Kyoto Protocol (which introduced specific mechanisms on emission reductions with important ramifications for corporate activities and investments), the EU emissions trading system (EU ETS) and the carbon reduction commitment (CRC), the World Bank carbon finance units, the UN environmental program (UNEP) and other. Also several national regulations were put into action in the US (the Environmental reporting Agency—EPA; the California assembly bill No 32), Denmark (the Copenhagen Accord in 2009), Canada (the Montreal protocol) and various other initiative in the United Kingdom, Japan and the EU.

Since the last decade, the role of corporations for both national and international markets has raised the interest and the need for more transparency and accountability of their operations and decisions. The CER guidelines proposed by international non-governmental organizations were aiming to hold corporations accountable and raise stakeholders' awareness on that matter (something that national laws have not been very efficient at that point). This accountability issue assisted firms to take into consideration the interests of their stakeholders by incorporating predefined norms which were voluntary in nature, regarding several social and environmental issues (Rasche et al. 2008; Rasche 2010; Camilleri 2017). Consequently, firms were incentivized to engage and communicate their social and environmentally responsible activities to the public and their stakeholders specifically. Several examples of social and environmental responsibility standards include the ISO 14001 and OHSAS 18000 which aim at improving the quality and safety of products and the GRI project that aims at guiding corporations to report on the environmental sustainability of their operations. Several times, those standards were covering a significant gap on the national legislation regarding governance and accountability and could be viewed as complementary tools for firms to gain legitimacy from the local societies and their stakeholders (Michelon and Parbonetti 2012; Camilleri 2017).

It was not long enough when national governments took decisive action in order to draft and circulate CER specific frameworks which can guide firms' behavior. The EU Lisbon Treaty in 2000 and the Gothenburg Sustainability Strategy in 2001 paved the ground of corporate responsibility and sustainability by introducing CER as a strategic tool for the transformation of the European economies into dynamic knowledge-based economies. The goal was to take advantage of local capabilities and strategic advantages of nation states and enhance their economic growth within a sustainable manner, increase employment levels and social consistency. Moreover, in 2006 and 2007 the EU Commission drafted and passed the "Corporate Social Responsibility: National public policies in the European Union" which set the agenda for social and environmental responsibility aiming at enhancing stakeholders' trust in businesses, improving dissemination of CER related information and visibility of CER, improve corporate disclosures on social and environmental activities and even create market rewards for socially and environmentally responsible firms.

Environmental protection was given top priority and was characterized as an important pillar of economic development and social cohesion (EU 2014; Camilleri 2017). Moreover, the EU parliament amended the directive 2013/34/EU regarding the published financial information of listed corporations and their parent companies, and required listed firms to mandatorily disclose non-financial type of social and environmental information on their annual reports on a “comply or explain” basis where firms have to be certain if they complied with the regulation or explain in detail why not (EU 2014; Camilleri 2015, 2017). These regulations introduced into the public arena issues such as public health, climate change, sustainable reporting and efficient use of natural resources which were not adequately addressed on previous regulations. All the above mentioned interventions aimed at improving the information environment of CER activities, increase public awareness on the matter and establish CER engagement and performance as an indispensable corporate activity with significant implications for corporate resource allocation, investment decisions and corporate accounting and finance.

1.4 Social and Environmental Responsible Investments

Social and environmental responsible investments (SERI) refer to the incorporation of social and environmental goals or criteria within investment decisions. The notion behind SERI is to achieve both financial (economic returns) and non-financial goals related to environmental protection, emissions reduction community engagement, employee relations and human rights protection. Also SERI incorporates investments which aim to reduce negative externalities to the environment and local communities (Camilleri 2017). Since 2003 the European Sustainability Investment Forum (EUROSIF) publishes annual reports highlighting the development of the SERI activities, practices and trends in the EU, which have been substantiated over the years as a dynamic segment of the international capital markets. According to the EUROSIF (2018) report, sustainability and environmental investments focused mainly on, renewable energy, climate change and water and natural resources management and seem to remain stable relative to the previous 2 years 2017 and 2016. After SERIs exponential growth experienced in 2016, the trend was rather stabilized in 2017, endorsing the investors’ interest towards a specific sustainable type of investment. SERIs overall have marked a vigorous growth in the past year with a positive cumulative return of 20% over the last 8 years, reaching almost 600 billion euros in market value, confirming that SERI remains as a valid investment choice among market participants.

Impact investing is another type of SERI strategy where investors seek for businesses that their investments combine financial and social returns, aiming at societal wellbeing through social or environmental projects (Camilleri 2017). According to EUROSIF (2018) report, impact investing included several types of investments such as community development, renewable and clean energy projects, microfinance and other forms of investments with a clear environmental and social footprint. Moreover, impact investing in the EU has confirmed its positive upward trend in 2018 by keeping its growth rate modest relative to 2017. The growth rate of impact investing has reached up to 52% during the past 6 years and exemplifies many assurances and possibilities linked to sustainable development. Impact investing incorporates several Sustainable Development Goals (SDGs) as set by EU, a fact that gives promises for further growth in the years to come. Also, positive investing is an additional SERI strategy that allows investors to express their values on environmental issues without deteriorating financial returns. According to Camilleri (2017), positive investing involves investments in environmentally and sustainable projects which improve firms' long tern potential and at the same time create value in an environmental sense.

Furthermore, community investing is another type of socially responsible investments within local or community organizations (Benjamin et al. 2004). Such types of investments aim at battling poverty, inequality, improve economic development and create other positive externalities within local communities. According to Camilleri (2017), community investing aims at providing financing into low-income individuals with significant difficulties of raising funds through regular financing channels. Also, community investing could involve additional support (beyond pure financing) in the form of business training, entrepreneurship and consulting in order to guarantee the success of financing and the required returns to the investors. All the above mentioned types of SERI assumes the active engagement of stakeholders with business managers so as to support and complement changes within corporate policies and activities. Environmental investors several times cooperate closely with managers in order to navigate the firm towards a sustainable corporate financial performance over time, while at the same time improve the status of various other stakeholders like customers, creditors, suppliers, employees and shareholders among others.

The development of SERI over the years can be attributed to several factors including the technological advancements in the fields of computer science and communications, were investors now are in better position to receive acute information on corporate activities and behaviors and reach decisions in a faster pace relative to the past. That along with the increase corporate disclosures on several social and environmental issues provide a thorough picture on firms' economic condition which assist investors to assess firms' prospects and risks. Secondly, it is becoming clearer over the years that SERI is not a simple waste of corporate resources (as has been viewed several decades ago) but it can create value for shareholders. SERI can create value for corporations through three different ways. At first, the adoption of environmentally responsible practices can assist firms to be ready to deal with governmental regulations, policies and taxation that some time seem inevitable, especially in developed countries. Thus, firms that have established SERI policies already are in position to conform to regulatory pressures and so gain an advantage relative to their competitors. Secondly, SERI practices are viewed positively by current and future stakeholders, providing an advantage over competition. Thirdly, SERI investments and activities inspire firms' human resources, leading to longer tenure of staff contributing positively to corporate financial performance and viability on the long run (Brockett and Rezaee 2012).

Practically, investors (and especially institutional investors) who have a long-term investment horizon can benefit from environmental (and other non-financial) disclosures and have shown significant interest on such types of corporate disclosures. SERI disclosures can have an effect on firm's performance and investment portfolios and are incorporated on operating and investment decisions. According to Brockett and Rezaee (2012), investors utilize such disclosures on SERI for identifying and assessing risk-return potential, customization of investment products and portfolios, evaluation of managerial quality and for stakeholder signaling and dissemination of information. So it has become clear over the years that shareholders do not have to lose part of their desired returns when investing in environmentally responsible projects. Since SERI creates a win-win situation for both firms and societies, investors realize now more than ever that responsibility is compatible with prosperity and they can add value to their portfolios by investing in environmental activities (Porter and Kramer 2011; Camilleri 2017). Consequently, environmental responsible investments have implications for corporate finance and accounting because investors and analysts have to analyze corporate strategies, policies and effects of such investments on the firms' profitability potential beyond the traditional tools of financial analysis.

1.5 CER Implications for Corporate Accounting and Finance

Despite the fact that CER seems to be contained and mostly explained within firms' ethical and responsible behavior (thus involving the discretionary behavior of management) there are several external sources that seem to dictate or exert pressure towards firms to engage in CER. The initiative of international organizations and European institutions discussed in the previous section seem to validate this opinion. This is more evident in the environmental accounting and finance literature were according to Johnstone (2018), the launch of practices and policies regarding CER are the outcome of external stakeholders' pressure. Under this framework, corporations exert efforts to improve their environmental footprint and performance while simultaneously enhancing their image, trust and legitimacy towards their stakeholders (Hassan and Ibrahim 2012). The association between the operational and strategic dimensions of CER which originate both from inside and outside of the firm has important implications for accounting and finance research. Since accounting refers to internal operational decisions and control systems of the firm to sustain and enhance performance, CER stems from both internal and external corporate sources. This fact has direct implications for the allocation of corporate resources with an impact on anticipated investment returns and shareholders wealth.

Moreover, the Eco-Management and Audit Scheme (EMAS) along with the EU Accounts Modernization Directive 2003/51 are two distinct legislatures which validate the argument regarding the external pressure towards firms to engage in CER. The EMAS was established in 1995 and was amended in 2009 and is an environmental reporting tool which scope is to direct firms to enhance their environmental performance. This tool includes several performance indicators which have to be reported in environmental or social responsibility-type of disclosures in order to complement the existing financial statements and reports published by firms. In addition, the EU modernization directive dictates that European firms must disclose both financial and non-financial key performance indicators which are pertinent to each specific sector. These indicators could be related to social and environmental activities, employee relations and other forms of environmental responsibility. Since 2009 all national legislations of each EU member country has incorporated that directive and each government has undertaken specific initiatives as to provide further guidance and assistance for firms to comply with the new reporting requirements (Camilleri 2017). Thus, the ultimate goal of the EU was to improve the corporate reporting environment by increasing the amount of information relating to corporate activities.

CER disclosures and performance are expected to have value relevance for internal and external users of this information such as suppliers, customers, governmental organizations, creditors and lenders, shareholders and employees. All those parties can reach to more informed decisions utilizing such environmental information. Also, CER related information can be useful for risk management (identifying risks and opportunities), correcting operational inefficiencies and dealing with reputational and economic hazards which in turn diminish financial performance. Practically, environmentally irresponsible firms can face troubling situations including higher risk of regulatory interventions and penalties, seize of operations, loss of customers' confident and support (with direct impact on revenues and profits), higher cost of capital and debt, low analyst coverage and a reduction on investors' confidence with direct repercussions on market valuation.

The disclosure of environmental non-financial information can contribute to enhanced monitoring on behalf of stakeholders', better allocation of scarce economic and financial resources within the firms, and to more informed and value relevant financial decisions by investors. Also, CER related disclosures tend to make corporations more accountable towards their stakeholders and at the same time can exert financial and commercial benefits. Recent studies argue that businesses can increase their profitability and reduce their operating costs by transmitting environmental responsibility actions to its customers, as an important corporate value. This claim is based on two considerations: First, firms can experience significant cost savings since many corporate activities performed via environmentally safe actions are less costly than doing things the "old" way. Secondly, the communication of environmentally responsible activities to customers can improve firm's corporate image and enhance the attractiveness of firm's product. These facts may lead to competitive advantages which are practically translated into higher financial performance and viability. These arguments have been verified empirically in the past documenting that environmentally proactive firms have greater profitability due to reduced costs for compliance to environmental regulations and enhanced operational efficiencies. Moreover, organizations which have recognized that their operations have a significant negative impact on the environment, are associated with increased motivation to engage in environmentally responsible actions so as to build stronger relationships with key stakeholders, saving money and even gaining strategic or competitive advantages.

Consequently, the engagement on CER related activities and disclosures creates direct implications for financial decision making, the quality of published financial information and decision making ability of practitioners since environmental and sustainable behaviors can bring important advances in operational efficiency, innovation, economic performance and competitiveness. Consequently, this book aims to provide direct evidence on the association between CER performance and various issues on corporate accounting and finance, contributing to the ongoing debate regarding the benefits and detriments of environmental activities and disclosures.

1.6 Scope of the Book and Research Questions

Following the above mentioned discussion, Corporate Environmental Responsibility (CER) has received increased attention from academics, executives and international organizations over the last decades, especially due to their significant impact on the operation of the organizations. Several studies assert that every firm—regardless of the business sector—has some primary (employees and customers) and secondary (communities, local governments) stakeholders who both assert increased pressure to the firms, as arising out of a growing concern for human rights, labor standards and environmental concerns. Empirical evidence so far suggest that positive stakeholder relationships can reduce the difficulty when dealing with their sub-groups such as employees, customers and the community. Despite the increased interest of CER on various business sectors the concept has recently entered the accounting and finance area, since many corporations operating product facilities or offering services, have evolved as a growing economic force within the national and international markets, and are more intensively engaging in socially and environmentally responsible activities.

The business environment consists one of the most turbulent parts of the social function characterized by a diverse variety of involved organizations. This variety ranges from profit oriented organizations, central public corporations and even non for-profit services organizations. According to many recent studies, CER has evolved as an important focus area of large listed corporations due to the increased recognition of their ability to deal with crucial social issues. Moreover, several arguments regarding the popularity and global reach of listed organizations and their appeal on the local and European society, makes them more effective promoters of environmental awareness, cultural understanding, social integration and social responsibility. This argument has been verified empirically by the fact that managers within listed corporations, can enhance their economic prospects through the achievement of social and environmental objectives, by improving sectors of activity pointed by stakeholders or even materializing innovating ideas and goals.

Being more specific, environmental responsibility has received increased attention internationally, due to the global awareness that took place the last decade about the planet's global warming and pollution. This fact urged corporations to shift their attention and actions towards this issue, since it was strongly connected with social cohesion and approval of their corporate activities. The stock market listed for-profit firms, are corporations whose main business revolves in economic activities that are not related within country borders and most of the times they make essential use of the natural environment. It is a common phenomenon that several times managers within such corporations have to confront with environmental and social issues in order to achieve competitive advantages. Therefore, under this framework, the involvement in environmental protection (or responsibility as it is widely known) activities assist firms to communicate their environmental awareness to the public and customers (and other stakeholders), enhance their image and stakeholder's trust, leading to various advantages from improved reputation to enhanced financial performance, reduction of risk and other related economic benefits.

Consequently, the book will build on the stakeholder and institutional theories stating that CER is a significant dimension of corporate responsibility and will address all key implications of CER related to corporate accounting and finance and specifically on financial performance, capital structure, firm risk, innovating activities, and the quality of accounting reporting. The aim of this book will be to provide answers on several research questions referring to CER and corporate finance and accounting by utilizing data from a large sample of EU listed and unlisted firms for the period 2003–2018, thus providing thorough and consistent inferences for future research and policy decisions. This characteristic makes the contribution of this book to the existing literature fairly distinguishable regarding CER and its embeddedness into business and financial decisions. Finally, one more contribution of the book will be that it adds to the ongoing debate on CER by providing a thorough empirical analysis yielding results with supporting arguments that firms must develop more strategic forms of environmental involvement that are able to satisfy stakeholders and deliver financial returns to the firm. Furthermore, the analysis on this book broadens the scope of CER from being solely altruistic, to a more strategic focus which can realize direct economic benefits tied to corporate finance and accounting. Consequently, the main research questions that we will try to provide answers to are summarized as follows:

1. Which are the main financial determinants of CER performance?
2. How CER performance contributes to corporate innovative activities and how it affects firms' decisions to invest in R&D, and the impact of that decision on their financial performance?
3. What is the association between CER and corporate capital structure decisions?
4. How CER performance is associated to corporate systematic risk?

5. What is the impact of CER performance on working capital management decisions, and specifically on cash holdings and cash dividend payments?
6. How CER performance contributes to the value relevance of accounting information, the persistence and predictability of earnings and the firms' stock prices? Is CER performance useful for investors and analysts to project future earnings figures?
7. What is the contribution of CER engagement and performance in mitigating earnings management behavior? Is there any association between CER activities and various forms of accounting manipulation including accruals and income smoothing, and how CER activities contribute to a more responsible and efficient financial reporting.
8. How CER performance impacts on the corporate decisions regarding accounting conservatism?

1.7 Conclusion

Corporate environmental responsibility (CER) has received increased attention from academics, executives and international organizations over the last decades, especially due to their significant impact on the operation of the organizations. Several studies assert that every firm—regardless of the business sector—has some primary (employees and customers) and secondary (communities, local governments) stakeholders who both assert increased pressure to the firms, as arising out of a growing concern for human rights, labor standards and environmental concerns. CER has been developed as an important strategic mechanism through which organizations are trying to differentiate themselves from competition.

CER includes policies, practices and activities which are incorporated within business operations, decision-making processes and organization structures. Being more specific, environmental responsibility has received increased attention internationally, due to the global awareness that took place the last decade about the planet's global warming and pollution. This fact urged organizations to shift their attention and actions towards this issue, since it was strongly connected with social cohesion and approval of their corporate—organizational activities. Therefore, under this framework, the involvement in environmental protection activities in general, assist organizations to communicate their environmental awareness to the public and customers (and other stakeholders), enhances its image and stakeholder's trust, leading to various advantages from improved reputation to enhanced financial performance, reduction of risk and other related economic and organizational benefits.

The current chapter presented a brief theoretical background on CER performance and activities and the main regulatory interventions that have been made from international and European institutions over the last two decades. These interventions and the increased public awareness on the issue of environmental responsibility created a wave of socially and environmentally responsible investments. It is even more comprehensible today than ever before that environmental responsibility does not lead to a waste of scarce corporate resources but on the contrary it can create value for both shareholders and stakeholders, leading to a win-win situation for all related parties. Finally, the chapter discussed the implications of CER for research in corporate accounting and finance and presented the main research questions that will be addressed in the following chapters of the book.

The book is organized in three main parts. The first part provides the main EU regulatory interventions on the matter of CER disclosures and policies along with descriptive statistical data on CER performance per country over the sample years, while also offers a detailed literature review of theoretical and empirical studies on the topic. The second part includes the empirical analysis of CER performance on corporate finance issues (financial performance, capital structure, risk, innovation, cash holdings and dividend policy decisions), and the past part of the book is dedicated on the association between CER performance and accounting reporting quality (value relevance of earnings, earnings management and accounting conservatism).

References

Abrams, F.W. 1951. Management's responsibilities in a complex world. *Harvard Business Review* 29: 29–34.

Benjamin, L., J.S. Rubin, and S. Zielenbach. 2004. Community development financial institutions: Current issues and future prospects. *Journal of Urban Affairs* 26 (2): 177–195.

Brockett, A., and Z. Rezaee. 2012. *Corporate sustainability: Integrating performance and reporting*. Hoboken, NJ: Wiley.

Camilleri, M.A. 2015. Environmental, social and governance disclosures in Europe. *Sustainability Accounting, Management and Policy Journal* 6 (2): 224–242.

———. 2017. Corporate sustainability. In *Social responsibility and environmental management: An introduction to theory and practice with case studies*. Cham, Switzerland: Springer.

Carroll, A.B. 1979. A three-dimensional conceptual model of corporate social performance. *Academy of Management Review* 4: 497–505.

Dienes, D., R. Sassen, and J. Fisher. 2016. What are the drivers of sustainability reporting? A systematic review. *Sustainability Accounting, Management and Policy Journal* 7 (2): 154–189.

EU. 2014. Non-financial reporting. Retrieved January 2020, from http://ec.europa.eu/internal_market/accounting/non-financial_reporting/index_en.htm

EUROSIF. 2018. European SRI study 2018. European sustainable investment forum, Brussels. Retrieved March 16 2020, from http://www.eurosif.org/wp-content/uploads/2018/11/European-SRI-2018-Study.pdf

Frederick, W. 1986. Toward CSR: Why ethical analysis is indispensable and unavoidable in corporate affairs. *California Management Review* 28: 26–141.

Freeman, R.E. 1994. The politics of stakeholder theory: Some future directions. *Business Ethics Quarterly* 4: 409–421.

Goodpaster, K.E. 1991. Business ethics and stakeholder analysis. *Business Ethics Quarterly* 1 (1): 53–73.

Hall, R. 1992. The strategic analysis of intangible resources. *Strategic Management Journal* 13 (2): 135–144.

Hassan, A., and E. Ibrahim. 2012. Corporate environmental information disclosure: Factors influencing companies' success in attaining environmental awards. *Corporate Social Responsibility and Environmental Management* 19 (1): 32–46.

Huang, X.-B., and L. Watson. 2015. Corporate social responsibility research in accounting. *Journal of Accounting Literature* 34 (1): 1–16.

International Federation of Accountants (IFAC). 2011. Sustainability Framework 2.0. Retrieved from https://www.ifac.org/about-ifac/professional-accountants-business/publications/ifac-sustainability-framework-20

Johnstone, L. 2018. Environmental management decisions in CSR-based accounting research. *Corporate Social Responsibility and Environmental Management* 25 (6): 1212–1222.

Martínez-Ferrero, J., and J.V. Frías-Aceituno. 2015. Relationship between sustainable development and financial performance: International empirical research. *Business Strategy and the Environment* 24 (1): 20–39.

McWilliams, A., and D. Siegel. 2001. Corporate social responsibility: A theory of the firm perspective. *Academy of Management Review* 26: 117–127.

Michelon, G., and A. Parbonetti. 2012. The effect of corporate governance on sustainability disclosure. *Journal of Management and Governance* 16 (3): 477–509.

Monteiro, S., and B. Aibar-Guzmán. 2010. Determinants of environmental disclosure in the annual reports of large companies operating in Portugal. *Corporate Social Responsibility and Environmental Management* 17: 185–204.

OECD. 1999. Basic principles of corporate governance. Organisation for Economic Cooperation and Development, Directorate for Financial and Enterprise Affairs. Retrieved January 10, 2015, from http://www.oecd.org/document/49/0,3343,en_2649_34813_31530865_1_1_1_1,00.html

Pan, X., J. Sha, H. Zhang, and W. Ke. 2014. Relationship between corporate social responsibility and financial performance in the mineral industry: Evidence from Chinese mineral firms. *Sustainability* 6 (7): 4077–4101.

Porter, M.E., and M.R. Kramer. 2007. Strategy and society: The link between competitive advantage and corporate social responsibility. *Harvard Business Review* 84 (12): 78–92.

Porter, M.E., and M.R. Kramer. 2011. Creating shared value. *Harvard Business Review* 89 (1–2): 62–77.

Rasche, A. 2010. Collaborative governance 2.0. *Corporate Governance* 10 (4): 500–511.

Rasche, A., D. Baur, M. Van Huijstee, S. Ladek, and J. Naidu. 2008. Corporations as political actors—A report on the first Swiss Master Class in corporate social responsibility. *Journal of Business Ethics* 80 (2): 151–173.

Roberts, R.W. 1992. Determinants of corporate social responsibility disclosure: An application of stakeholder theory. *Accounting, Organizations and Society* 17 (6): 595–612.

Russo, M., and P. Fouts. 1997. A resource-based perspective on corporate environmental performance. *Academy of Management Journal* 40 (3): 54–559.

Sheldon, O. 1924. *The philosophy of management*. London: Sir Isaac Pitman and Sons Ltd.

Surroca, J., J.A. Tribó, and S. Waddock. 2010. Corporate responsibility and financial performance: The role of intangible resources. *Strategic Management Journal* 31 (5): 463–490.

Suto, M., and H. Takehara. 2018. *Corporate social responsibility and corporate finance in Japan*. Singapore: Springer Nature.

Tang, Q., and L. Luo. 2016. Corporate ecological transparency: Theories and empirical evidence. *Asian Review of Accounting* 24 (4): 498–524.

Zeng, S.X., X.D. Xu, Z.Y. Dong, and V.W.Y. Tam. 2010. Towards corporate environmental information disclosure: An empirical study in China. *Journal of Cleaner Production* 18 (12): 1142–1148.

Chapter 2
Corporate Environmental Responsibility in the EU

Following the discussion on Chap. 1, corporate environmental responsibility (CER) has been widespread among European firms during the last 15 years. This shift of corporate behavior towards CER and other related activities is the outcome of significant regulatory interventions made by EU member countries during the course of the previous four decades. Starting from the Paris EEC (European Economic Community) convention, where the issue of environmental protection was initially included in the agenda of the member states, until more recent regulations have ultimately shaped the European legal framework of environmental protection. This framework sets and even provides directions to corporations in order to reduce their environmental footprint, while promoting sustainable economic goals. The scope of this chapter is to provide to the reader a comprehensive discussion on the evolution of regulatory interventions within the EU and to offer descriptive evidence on the how these regulations have affected firms' decisions and CER performance within 24 EU countries (with available CER data) and across various sectors of economic activity.

2.1 Introduction

The founding Treaty establishing the European Economic Community (Treaty of Rome 1957) did not aim to protect the environment. It limited itself to formulating a framework for economic cooperation between Member States, as it was accepted that its intervention in matters of political or social charge would lead to the failure of the new institution of transnational cooperation, as it would cause distrust in candidate Member States (Dagtoglou 1985). Over time, it has become clear that the creation of a common market and economic development cannot be pursued in the absence of environmental policy and without a clear social and ecological orientation.

P. Dimitropoulos, K. Koronios, *Corporate Environmental Responsibility, Accounting and Corporate Finance in the EU*, CSR, Sustainability, Ethics & Governance, https://doi.org/10.1007/978-3-030-72773-4_2

The protection of the environment is an objective of the Community, for the first time, at the Summit held in Paris in October 1972 (Paris European Council 1972) (Koppen 1988). The declaration by the Heads of State of the Member States at the aforementioned summit on the need for a specific Community policy on the environment has led to the adoption of five Community action programs, which are, however, declarations of principles and objectives without legal binding (Baker 2012).

Subsequently, the protection of the environment is reduced to an independent Community action with the Single European Act of 1986, with which the articles 130Π, 130P and 130Σ were added to the EEC Treaty (Ehlermann 1987). The Single European Act of 1987 introduced a new "environmental chapter", which was the first legal basis for a common environmental policy, aimed at preserving the quality of the environment, protecting human health and ensuring the rational use of natural resources (Kramer 1992).

The provisions of Article 130Π specify the objects, general principles and limits of Community environmental action. They not only provide the legal basis for action by the Community institutions, but also contain the fundamental principles that should govern their action. These principles are not merely guidelines or programs, but are legally binding, in the sense that the acts of the Community institutions and the Member States must comply with them. Article 130Π also defines the procedure for taking measures. Jurisdiction is vested in the Council, which decides unanimously, while Article 130Σ allows Member States to maintain enhanced protection measures (Glaesner 1986; Kramer 1987; Swann 2017). The basic principles governing the pursuit of Community environmental policy are:

(a) The principle of integration, according to which Community policy, in all its spheres of competence, must be formulated taking into account environmental protection requirements. This requirement, due to the specificity of environmental goods, which can be affected by any kind of human activity, allows the protection of the environment to penetrate and co-shape Community policy in all areas. It follows from this principle that environmental protection measures can be taken not only under Articles 130Π, but also in the context of the exercise of other responsibilities (Weale and Williams 1992; Hooghe and Marks 1999).
(b) The principle of subsidiarity. According to article 130Π: "the Community acts in the field of environment, provided that the objectives referred to in par. 1 can be better done at Community level than at the level of individual Member States ". In combination with the possibility of derogation of the Member States from the Community regulations, based on art. 130T, the principle of subsidiarity seems to place Community environmental law second to national law and thus to prevent it from evolving into a complete regulatory system with uniform application in the Community. It should be noted, however, that tackling environmental problems at national and local level is often preferable, as it allows the specifics of each case to be taken into account and enables greater citizen participation. In any case, the institutions of the Community can be freed from

the limitations of the principle of subsidiarity by adopting provisions providing better quality protection (Toth 1992; Portuese 2010; Doussis et al. 2019).

Subsequent revisions of the Treaties have strengthened the Community's commitment to environmental protection and the role of the European Parliament in the development of environmental policy. In particular:

The Maastricht Treaty (1993) established the environment as an official EU policy area. For the first time, environmental protection is being expanded, improved and explicitly integrated into the principles and purposes of the Community (art. 2 and 3 EEC Treaty). With the articles 130P et seq. Which replaced the art. 130Π et seq. of the Single European Act (SEA), Community "action" on the environment is promoted in Community policy. Environmental protection needs are not just a "component" (art. 130Π SEA) of other Community policies but "must be taken into account in the formulation of Community policies" (principle of integration, art. 130Π EEC Treaty). In addition, this Treaty introduced the co-decision procedure and made the general rule in the Council the qualified majority instead of the unanimity rule, thus strengthening the role of the European Parliament in shaping the measures. Of course, the possibility for Member States to derogate from harmonization measures is maintained, but it is clarified that the introduction of safeguard clauses should be aimed at better protection of the environment (Wilkinson 1992; Sheldon 1998).

The Amsterdam Treaty (1999) introduced the obligation to integrate environmental protection into all EU sectoral policies, with a view to promoting sustainable development. Protecting the environment is, for the first time, a Community objective independent of economic growth (Jordan 1998).

The Treaty of Nice, which entered into force on 1 February 2003, did not bring about any substantial changes to primary Community environmental law. These focus exclusively on article 175 par. 2 on the subjects for which unanimity is required for the drafting of secondary Community law. In addition, efforts have been made to facilitate the development of enhanced cooperation between Member States in areas, including the environment, without this route being activated to date (Katz 2003).

The trend of halting the development of primary Community environmental law is also evident in the Charter of Fundamental Rights of the European Union (CFR), signed in Nice in December 2000. The Charter contains all the rights recognized by the Community legal order. It is addressed to the institutions and Bodies of the European Union, as well as to the Member States, but only to the extent that they apply Union law. The environment as a legal property worthy of protection is included in Article 37 of the Charter, which has been included in the chapter entitled "Solidarity". According to this article: "a high level of environmental protection and the improvement of its quality must be integrated into the policies of the Union and ensured in accordance with the principle of sustainable development". With the above wording, an attempt is made to summarize the fundamental positions of the Treaty establishing the European Community on environmental protection, without, however, measuring the excess, ie the recognition of a right to the environment

(Morgera and Durán 2014; Kaiser 2016). Subsequently, other provisions of the Charter demonstrate greater interest for the citizen who wishes to claim the protection of the environment through the Community legal order by submitting complaints to the European institutions, such as those that explicitly provide for the right to "good administration", i.e. the right of every person "in the impartial, fair and within a reasonable time examination of his cases . . .", the right "to a prior hearing . . .", the obligation of the public bodies to state reasons, as well as the rights of access to documents, to appeal to the Ombudsman against conduct mismanagement by the Community institutions and reporting to the European Parliament (Von Bogdandy 2000; Kanska 2004). Nonetheless, despite initial expectations, the Charter remains without clear legal binding—and certainly without constitutional scope (Bousta 2013). This is expected to change if efforts to draft and introduce a European Constitution finally succeed. However, the constitutional attempt that took place at the beginning of the twenty-first century (2002–2004) and which resulted in a final draft Constitution, which was never ratified by the 25 EU Member States, probably wronged the ecological aspect of the Community and in this light, its failure cannot be seen as a missed opportunity for Community environmental law, but rather as an opportunity to draw up a new plan that is more environmentally friendly than the previous one (Eriksen et al. 2004).

The Treaty of Lisbon (2007), signed on 13.12.2007 and entered into force on 1.12.2009, abolished the European Community and brought about significant changes in the structure and functioning of the European Union. In terms of environmental protection, the most important innovation of the Lisbon Treaty is that it has included in the objectives of the European Union's environmental policy the "fight against climate change", as well as sustainable development in relations with third countries. Legal personality now allows the EU to conclude international agreements (Vedder 2010).

Pursuant to Articles 11 and 191 to 193 of the Treaty on the Functioning of the European Union (TFEU) (renamed the TEU under the Lisbon Treaty), the EU has the power to act in all areas of environmental policy, such as the air and water pollution, waste management and climate change. Its scope is limited by the principle of subsidiarity and mandatory unanimity in the Council in the areas of budgetary issues, urban and rural spatial planning, land use, quantitative management of water resources, options concerning energy sources and the structure of energy supply (De Sadeleer 2012; Kozien and Kozien 2018).

2.2 The Guiding Principles of EU Environmental Law

In addition to the general principles of integration and subsidiarity (see above), which define the scope of Community environmental policy and delimit the regulatory competence of the Community institutions, there are certain principles derived mainly from para. 2 of Article 191 TFEU and refer to the content of Community regulations relating to the environment. These are the general principles from which the Community legislature will draw the guidelines for the proper handling of the legal interest of the environment as a regulated object of Community law (Proelss 2016; Kramer 2018; Aragão 2018).

Specifically, article 191 par. 2 TFEU states that: "Union policy on the environment shall aim at a high level of protection taking into account the diversity of situations in the various regions of the Union. It shall be based on the precautionary principle and on the principles that preventive action should be taken, that environmental damage should as a priority be rectified at source and that the "polluter should pay''. In this context, harmonization measures answering environmental protection requirements shall include, where appropriate, a safeguard clause allowing Member States to take provisional measures, for non-economic environmental reasons, subject to a procedure of inspection by the Union".

Despite their lack of explicit definition and general wording, these principles constitute complete and binding rules of law with unquestionable legal force, which produce legal obligations and develop binding on the Union authorities, which are called upon to adopt measures compatible with them. As is explicitly stated in the case-law of the Court of Justice of the European Union, these environmental authorities have no immediate effect, as they cannot directly bind or invoke national authorities in the absence of relevant national legislation. In any case, their presence in a number of Directives, regarding the formulation and implementation of the EU environmental policy, make them a key interpretive guide in the transfer and implementation of EU regulations in the national legal orders (Proelss 2016; De Sadeleer 2020).

2.2.1 The Requirement of a High Level of Environmental Protection

The principle of the "requirement of a high level of environmental protection" was introduced by the Maastricht Treaty, while the Amsterdam Treaty in 1997 reduced it to a fundamental principle of Community law. Until then, the level of quality to be achieved by the Community institutions in the exercise of their regulatory competence was indirectly inferred from the principle of subsidiarity, which requires the introduction of Community rules establishing a higher level of protection than is achieved by regulations from the legal orders of the Member States.

The requirement of high quality environmental protection is difficult, due to its vagueness, to develop binding force, without proper regulatory and case-law specialization. It is argued, however, that it introduces a form of "Union environmental acquis" that prevents environmental degradation (Wasmeier 2001; De Sadeleer 2009, 2013).

2.2.2 The Principle of Precaution and Prevention

The nature of environmental goods, their vulnerability and the often insurmountable difficulties that arise in trying to repair the relevant damage, make it imperative by nature to take precautionary action as a necessary condition for their effective protection.

The principles of "prevention" and "precaution" (precautionary action) were established by Article 191 of the Treaty on the Functioning of the European Union (TFEU) and by the 1992 Rio Declaration, which, inter alia, states that "the protection of approach as a precautionary measure" (Bodansky 1994; Wibisana 2006; Handl 2012).

The distinction between prevention and precaution concerns that the former, in principle, allows only negative measures to be taken, such as a ban on activities dangerous to the environment, while the latter is the legal basis for taking mainly positive measures to protect the environment, significant interference for environmental protection and ultimately the undertaking of greater efforts and expenditures by public bodies (Rezza 2006).

The precautionary principle came into being when it was applied to emerging issues at Community level, such as bovine spongiform encephalopathy, electromagnetic radiation generated by mobile antennas or other fields, as well as genetically modified products. These cases do not fall within the traditional scope of preventive action, as the scientific data on them are usually insufficient or raise concerns about their dangerous effects on the environment and health in relation to the level of protection provided. The authority in question was therefore worded in particular with the following content, in order to legitimize the addressing of such cases.

The precautionary principle is a risk management tool that can be activated if there is scientific uncertainty as to an alleged risk to human health or the environment arising from a specific action or policy. In particular, recourse to the precautionary principle presupposes, first, the identification of the potential problem and the scientific assessment of the risks arising from that problem, which cannot be scientifically determined with sufficient certainty. The appropriate measures are then selected to deal with this potential risk, an option which is primarily political in nature and not legal and depends on the socially acceptable level of risk. The relevant decision must be based on scientifically substantiated evidence of a hazard (Cameron and Abouchar 1991; Fisher et al. 2006). For example, in case of doubt as to the possible harmful effects of a product, it may be ordered to stop the distribution of that product or to withdraw from the market if, after an objective scientific evaluation, uncertainty remains. In any case, the measures taken should be based on existing scientific evidence and reviewed periodically according to the progress of scientific and technical research. In addition, they should be non-discriminatory, proportionate and reviewed as soon as more scientific evidence becomes available.

The precautionary principle, as a fundamental component of European environmental law, is also applied by the national judge, with binding force. The importance attached to the principle of precaution and prevention under Community law is obvious, as most Directives and Regulations introduce precautionary measures either in the form of the introduction of precautionary measures of environmental impact or precautionary measures adapted to the specific characteristics of certain environmental goods or finally, in the form of information and participation of citizens in environmental issues (Douma 2000).

The following Instructions are indicative:

- Directive 2008/1/EC of the European Parliament and of the Council of 15 January 2008 concerning integrated pollution prevention and control.
- Directive 2008/50/EC of the European Parliament and of the Council of 21 May 2008 on ambient air quality and cleaner air for Europe OJ L 152, 11.6.2008.
- Directive 2001/18/EC of the European Parliament and of the Council of 12 March 2001 on the deliberate release into the environment of genetically modified organisms and repealing Council Directive 90/220/EEC.
- Directive 2001/42/EC of the European Parliament and of the Council of 27 June 2001 on the assessment of the effects of certain plans and programs on the environment.
- Directive 2000/60/EC of the European Parliament and of the Council of 23 October 2000 establishing a framework for Community action in the field of water policy.

2.2.3 The Principle of Sustainability

The "principle of sustainability"—in other words, the principle of sustainable development—is enshrined in Articles 11 and 191 of the Treaty on the Functioning of the European Union. It has its roots in eighteenth century German forest science and, as a general environmental principle, is also found in the United Nations Declaration of the Stockholm Environment (1972), in the World Commission on Environment's Brundtland Report (1987), in the United Nations Declaration on Environment and Development in Rio (1992) and in the Johannesburg Declaration on Sustainable Development (2002) (Handl 2012). With the revision of article 24, in 2001, it was explicitly enshrined in the Greek Constitution and.

A key element of the "sustainability principle" is that all human intervention in the environment must be done in a way that ensures a balanced relationship between the exploitation of natural resources and the evolution of ecosystems for the sake of future generations. Its goal is therefore a model of economic growth that will ensure ecological balance in perpetuity (Emas 2015; Bosselmann 2016).

The most important specialization of the principle of sustainability in the Community legal order is the introduction of the principle of integration into the Community Treaties, according to which, in the formulation and implementation of Community policies and actions, such as agriculture, energy, transport and tourism parameters related to the protection of the environment and the management of natural resources should be taken into account. The integration seeks in particular the functional integration of the protection and management of natural resources in the design and implementation of almost all individual policies, so that a gradual ecological transformation of economic processes can take place (Liberatore 1997; Barral and Dupuy 2015).

With regard to the substantive content of the principle of sustainable development in the Community legal order, there is an approach according to which it is possible to replace natural capital with artificial capital marginally, while irreparable damage to the environment must be avoided. Key tools for this type of development are "clean", environmentally friendly technologies and services. The adoption of the principle of sustainability in European Union law has been implemented mainly with Directives 43/1992 on natural habitats and 403/1979 on birds.

2.2.4 *The Principle of Remediation of Environmental Infringements as a Matter of Priority at Source*

The principle of "remediation of environmental disasters as a matter of priority at source" is of particular practical importance, as ecological damage is not often limited to a certain area, but is diffuse, as in the case of cross-border pollution or waste transport. In the context of secondary Community law, this principle appears, first, in the text of Directive 84/631/EEC on the transboundary movements of hazardous waste, treating waste as a commodity and setting as a rule their free transport, with the exception of possibility of its limitation by the Member States. This Directive has been replaced by Regulation (EC) No 1013/2006 of the European Parliament and of the Council of 14 June 2006 on the transport of waste concerning the monitoring and control of the transport of waste within the Community, by which free waste transport became the exception, and waste management was determined to be governed by the "principle of disposal at source", as partly analyzed in the principles of "proximity" and "self-sufficiency" (Jans 1999).

The principle of self-sufficiency requires the establishment of a waste installation and disposal network in the Member States, which will allow the Community to become self-sufficient in the field of waste disposal and the Member States to pursue this goal separately, taking into account geographical conditions or the need for special facilities for certain categories of waste (Jans 1994). The principle of proximity requires the formation of a network that allows the disposal of waste at the nearest possible facility with a parallel concern for the protection of the environment and public health (Reese 2018).

The Court of Justice of the European Communities (CJEC) in its relevant case-law has widely applied the "principle of remediation of environmental harms at source", balancing it with the principle of free distribution of goods (Article 28 et seq. EEC Treaty). Disintegration of the environmental protection case-law of the CJEC is the decision of 25.6.1998 (CJEC C-203/1996), which, abandoning the unified treatment of waste, considered that the Community legislation on waste did not enforce its principles of self-sufficiency and proximity to waste destined for economic recovery.

2.2.5 The "Polluter Pays" Principle

The Community "polluter pays" principle binds the Community legislature and the Community institutions at all levels and obliges them to take all necessary measures to transform the above dictate into an effective, preventive and remedial means. At the same time, as a general principle of Community law, it takes precedence over any applicable provision of national law, irrespective of the time of its adoption, in so far as the latter contradicts or conflicts with the purpose pursued by that principle (Wibisana 2006).

The "polluter pays" principle was first articulated in 1972 in an OECD Recommendation, in the context of a purely economic rule, according to which "The polluter bears the costs of implementing the measures decided by public authorities to ensure maintaining the environment in an acceptable state" (Grossman 2006).

In European Union law, the "polluter pays" principle, long before its incorporation into the official text of the Treaty, is included in the following reference to Council Recommendation 75/436: "Natural person or legal entity, whether governed by public or private law, to cause the contamination, they must bear the costs of those measures necessary to eliminate or reduce the contamination, in compliance with the specifications or the corresponding regulations, which establish quality objectives or in the absence thereof in compliance with the specifications or corresponding measures put forward by public authorities" (Munir 2013).

The legal and political aim of the principle is to urge entrepreneurs and potentially polluters to reduce pollution levels due to their own activity, and to be more careful when taking actions that are dangerous to the environment. They are also provided with the necessary incentive to step up research in search of "clean technologies" and less polluting products, as well as for a more rational use of natural resources.

In 1992, the "polluter pays" principle was incorporated into the 16th Principle of the Rio Declaration on Environment and Development. It has since been a fundamental interpretative rule of the Union's environmental policy. It was mainly the basis for establishing the obligation of entrepreneurs to integrate the cost of pollution into the total cost of a product or service they provide. In this way, companies focus on adopting environmentally friendly production methods and technologies (Gaines 1991).

At the level of an economic analysis of environmental protection, the way in which this principle is applied is, however, not without controversy (Gergen 1994; Csák 2011; Lindhout and Van den Broek 2014). The "polluter pays" principle has been criticized for, due to the pervasive nature of ecological damage, often leading to the inability to find the culprit or the contribution rate of more culprits to the damage. In international practice, the rule that the culprit of the pollution is to bear the cost of remediation usually ends up being borne by the producer, at a time when the environmental risk also comes from consumers. Although the economic burden of the industrial producer may be shifted to the consumers when determining the price of the product, this solution ends up favoring the polluter over the conscientious consumer. Furthermore, the application of this rule is not obvious in cases of "chain" or "cumulative" pollution, i.e. that many natural or legal persons have contributed jointly or successively to the violations of the environment.

Also, the grammatical wording of the principle does not inadvertently lead to the line between the individual damages that the polluter has to make up for and those that fall outside his area of responsibility, so it is not clear what the polluter is required to pay. This is if we take into account that an environmental catastrophe has many and varied, direct or reflective, present or future, occurring or probable, depending on its severity, negative effects and extensions. In fact, it has been argued that overemphasis on the economic dimension of pollution, according to this principle, may even lead to a deterioration of its ecological dimension (Gergen 1994; Csák 2011; Lindhout and Van den Broek 2014).

At Court of Justice of the European Communities (CJEC) case-law level, the judgment of the CJEC of 29 April 1999 C-293/1997 Standley etc. was significant, concerning the compatibility of Directive 91/676/EC on the protection of waters against nitrate pollution by overriding rules of Community law. The matter was brought before the Court with a preliminary referral from the British High Court of Justice.

The Court expressed (paragraph 52), inter alia, the view that: "Directive 91/676/EEC does not mean that farmers with agricultural holding must bear the burden of eliminating a pollution to which they did not contribute. It is for the Member States to take into account, when implementing the Directive, the other causes of the pollution and, in the light of the circumstances, not to impose on the holders of agricultural holdings unnecessary burdens for the spread of pollution. In this light, the 'polluter pays' principle is a manifestation of the principle of proportionality." Therefore, the Court, interpreting the "polluter pays principle", ruled that in the event that the national system of protection of the environment from nitrate water pollution placed on agricultural holdings the sole burden of restoring the natural balance from damage caused by other reasons, the "polluter pays" principle would be violated. Furthermore, the Advocate General, in his Opinion of 8 October 1998, mentions another dimension of the "polluter pays" principle, the precautionary one, that the application of the principle does not presuppose the occurrence of environmental damage but allows the financial burden of activities those who have the greatest risk of causing the damage concerned.

The possibilities of practical application of the "polluter pays" principle are varied and extend to the area of both public and private law (Vícha 2011; Luppi et al. 2012).

In the context of public law, special taxation of polluting activities is possible. The issue of taxation of carriers on environmentally damaging activities is particularly relevant. It has significant advantages, as it forces companies to adapt their activities to environmental protection requirements. The imposition of the tax does not depend on the often difficult proof of causing the environmental damage, but on the inclusion or not of the polluter's activity in the quality parameters provided by the relevant legislation. It is a fact, however, that the Community has not taken any steps in this direction, nor does it favor their adoption by the Member States, considering that these are State interventions which endanger the freedom of competition.

In the field of private law, the "polluter pays" principle favors the introduction of liability for endangering the environmentally hazardous works, activities or facilities. This is the direction of the Directive 2004/35/EC on civil liability for environmental damage, which aims to prevent or otherwise repair environmental damage in protected species and natural habitats, in water and soil. Entities engaged in certain occupational activities, such as the transport of hazardous substances or activities involving discharges into water, must take precautionary measures in the event of an imminent threat to the environment. If damage has already been occurred, they must take the appropriate steps to repair it and pay the cost. The application scope of the directive has been extended three times to include the management of extractive waste, the operation of geological storage sites and the safety of offshore oil and gas extraction activities (Kramer 2005).

2.2.6 The Principle of Citizen Participation: The Right to Free Information on Environmental Issues

The principle of participation is a special manifestation of the general principle of environmental protection of Article 191 TFEU, since a necessary condition for an integrated environmental policy is the participation of the citizen: Informing the citizen on environmental issues and his participation in decision-making with the possibility of seeking full judicial protection create a two-way relationship of information exchange with the state (Proelss 2016). This relationship is obviously necessary for the environmental protection in Europe of the market economy and rapid technological progress (Luttenberger and Rukavina 2004; Renn et al. 2013).

In particular, the principle of participation highlights the role of the citizen in the field of environmental protection, covering even partially the gap that exists in the Community legal order due to the absence of an (individual or collective) right to environmental protection. On the other hand, it is perhaps the most crucial tool for enforcing Community legitimacy in view of the fact that the European Union does not have adequate control mechanisms for on-the-spot investigation and intervention.

Given the scattered provisions in various Directives concerning information on environmental matters, this principle has found wide application in Directive 90/313/EEC of 7.6.1990, on free information on environmental matters. This Directive, in conjunction with Regulation (EC) No 401/2009 of the European Parliament and of the Council of 23 April 2009 on the European Environment Agency and the European Environment Information and Observation Network (OJ L

126/21.5.2009), formulate, at European level, a general institutional framework, with the aim of collecting and exchanging data and informing and briefing not only at transnational but also at citizen level, so that the Community, the Member States, the organizations and the European citizens to be better informed and to be able to make more effective use of their right to participate and contribute to European environmental events.

2.3 EU Actions in the Framework of CSR and CER

Undoubtedly, the activities of all companies, regardless of size, can have an impact on the natural environment (Crawley and Aho 1999). The means used by companies to deal with these environmental impacts are multiple and they include, among others: environmental management systems, Eco design techniques, cleaner production technologies, product life cycle monitoring, etc.

To this end, the institution of corporate environmental responsibility is defined by policies and actions of the European Union and its institutions, with the aim of coordinating the environmental behavior of a company as much as possible, increasing political visibility, and rewarding European companies for their actions in this area and the encouragement to take greater initiatives, always in implementation of the guiding principles of EU environmental law, which have been set out in detail.

In recent decades, the public interest in the concept of sustainable development has intensified, which, in addition to balancing economic growth and social integration, prioritizes the preservation of a healthy natural environment for future generations, and is also of particular interest in management waste prevention, pollution prevention and public information activities (Moon 2007; Elliott 2012; Emas 2015; Verma 2019).

In particular, the term "sustainable development" first appeared in 1987 in the Brundland Report. It covers how organizations today seek economic growth while protecting the resources for tomorrow's growth. This is an issue of increasing importance and concern for the responsible organizations. Businesses today take into account their impact on the environment and all their business planning, management and efficiency measurements. "Sustainability" refers to a company's actions to reduce its negative impact, for example the existence of management systems to reduce emissions and the efficient use of natural resources.

On the other hand, "green actions" refer to actions that the company develops inside and outside its organization to increase knowledge of environmental issues. Information plays a key role in prevention and helps in efforts to protect the environment (Schaltegger and Burritt 2005; Esty and Winston 2009).

Since 1973, the European Commission has been issuing multiannual environmental action plans setting out future legislative proposals and objectives for EU environmental policy. In 2013, the Council and Parliament approved the 7th Environmental Action Plan for 2020, entitled "Prosperity within the limits of our planet". Based on a number of strategic initiatives, the program sets out nine key objectives, including: nature protection; greater ecosystem resilience; sustainable economic growth with resource efficiency and low carbon emissions; and the fight against the environment health threats. The program also emphasizes the need for better implementation of EU environmental legislation, advanced science, investment and the integration of environmental aspects into other policies (Colombo et al. 2019).

Furthermore, in 2001 the European Union introduced its Sustainable Development Strategy (SDS), which complemented the previous Lisbon Strategy for Promoting Growth and Jobs with an Environmental Dimension. The revised EU Sustainable Development Strategy, renewed in 2006 to combine the internal with the international dimensions of sustainable development, aims to continuously improve the quality of life by promoting prosperity, environmental protection and social cohesion. The Europe 2020 growth strategy is in line with these objectives and aims at "smart, sustainable and inclusive growth". As part of this strategy, the "flagship initiative for a resource-efficient Europe" paves the way for sustainable growth and promotes the transition to an economy of efficient use of natural resources and low carbon emissions. In addition, in 2011 the EU committed itself to halting biodiversity loss and the degradation of ecosystem services by 2020 (EU Biodiversity Strategy) (Giljum et al. 2005; Nash 2009). In this context of EU action, the implementation of responsible entrepreneurship in the pillar environment involves a number of interventions and proactive actions on a wide range of issues. Indicatively, it is selected to present actions and legislative interventions for the following issues, which relate and interact directly with the activities of all companies:

Climate Change

Climate change has begun to affect all parts of the world and all sectors of society, threatening global growth and undermining the foundations of the global economy. Global warming has led and will lead to even more extreme weather phenomena (such as floods, droughts, heavy rainfall and heat waves), forest fires, water scarcity problems, glaciers melting and rising sea levels, changes in distribution or even extinction of various species of fauna and flora, plant diseases and pest infestations, lack of food and drinking water and population migration to avoid these dangers. According to scientific data, the risks of irreversible and catastrophic changes will increase significantly if the planet's temperature rises more than 2 °C—or even 1.5 °C—compared to pre-industrial era levels.

In the face of this reality, the mobilization of the European community is increasingly intensified in the context of both foreseeable and regulatory requirements that require companies to face and manage their environmental risks with even more active policies (Gupta and Grubb 2000).

Through the Climate and Energy Framework for 2030, which also reflects its commitment under the Paris Agreement (2015), the EU has committed itself to achieving the following objectives by 2030: reducing greenhouse gas emissions by at least by 40% below 1990 levels, an improvement in energy efficiency by 32.5% and an increase in the share of renewable energy sources to 32% of final consumption. The framework for 2030 is a continuation of the "20-20-20 targets" set in 2007 by EU leaders for 2020: reduction of greenhouse gas emissions by 20%, increase of the share of renewable energy sources in final consumption energy by 20% and a reduction in total EU primary energy consumption by 20% (all percentages compared to 1990). All of these objectives have taken the form of binding legislative measures.

Specifically, not exhaustively, it is mentioned: the EU Emissions Trading Scheme (ETS), the first and largest international coal market to date, which is a key EU policy instrument in the fight against climate change. It is based on the principle of "cap and trade": a "cap" is set for the total greenhouse gas emissions that can result from more than 11,000 plants (factories, power stations, etc.) included in the system. Each installation buys or receives "emission allowances" put up for auction by the Member States. Emissions from sectors not covered by the ETS, such as road transport, waste, agriculture and building construction, are subject to binding annual greenhouse gas emission reduction targets for each Member State, as set out in the Allocation Regulation of efforts (Oberthür 2019; Kulovesi and Oberthür 2020).

In addition, Directive (EU) 2018/2001 on renewable energy sources aims to ensure that by 2030 renewable energy sources, such as biomass, wind, hydroelectric and solar energy, will cover at least 32% of the EU total energy consumption in terms of electricity generation, transport, heating and cooling. Each Member State adopts its own national action plan for renewable energy sources, including sectoral targets (Holzleitner et al. 2020).

As a result, several companies have already set specific targets for reducing greenhouse gas emissions and are working with other companies and governments to improve their environmental performance on an annual basis, choosing to be part of the solution.

Air Pollution

Air pollution, in addition to the health problems it causes to humans, also adversely affects water and soil quality and damages ecosystems due to eutrophication (excess nitrogen pollution) and acid rain. As a result, agriculture and forests are affected, as are materials and buildings. Air pollution comes from many sources, but mainly from industry, transport, energy production and agriculture.

Directive 2008/50/EC on air quality aims to reduce air pollution to levels that minimize the harmful effects on human health or the environment. To this end, it sets out measures to lay down and set targets for ambient air quality (ie limits that must not be violated anywhere in the EU) in relation to basic air pollutants (sulfur dioxide, nitrogen dioxide, oxides, (fine) suspended particles, lead, benzene, carbon monoxide and ozone). Member States must designate zones and urban complexes in order to assess and manage air quality, monitor long-term trends and provide information to the public. Where air quality is good, it should be maintained; where limit values are exceeded, action should be taken (UNION 2008).

In addition, Directive 2010/75/EU on industrial emissions covers highly polluting industrial activities which are responsible for a significant proportion of pollution in Europe. Consolidates and combines all relevant directives (waste incineration, volatile organic compounds, large incineration plants, integrated pollution prevention and control, etc.) into a coherent piece of legislation, in order to facilitate the implementation of legislation and the minimization of pollution that comes from various industrial sources. The Industrial Emissions Directive sets out the requirements to be met by all industrial installations, contains a series of measures to prevent water, air and soil pollution, and provides a basis for the granting of operating licenses or the establishment of industrial installations (Makovetska and Omelyanenko 2014).

Resource Efficiency and Circular Economy

All products have a natural base. The EU economy is heavily dependent on natural resources. However, the usual ways of using resources, both in the past and today, have led to high levels of pollution, environmental degradation and depletion of natural resources.

EU waste policy has a long history and has always focused on more environmentally sustainable waste management (Fischer 2011). The "Roadmap for a resource-efficient Europe" and the circular economy package are expected to reverse this trend, making the EU economy sustainable by 2050. The four new waste directives, as part of the new circular economy package, introduce new waste management targets for waste prevention, reuse, recycling and landfill.

An example is the Waste Framework Directive (2008/98/EC), which is a continuation of the Thematic Strategy on the Prevention of Waste Generation and Recycling (COM (2005) 0666), repealing the previous Waste Framework Directive (75/442/EEC, codified as 2006/12/EC), the Hazardous Waste Directive (91/689/EEC) and the Waste Oils Directive (75/439/EEC). Its aim was to reform and simplify EU policy by establishing a new framework and setting targets, with a focus on prevention. It has established basic concepts and definitions in the field of waste management, including the definitions of waste, recycling and recovery (Gharfalkar et al. 2015).

Chemicals and Pesticides

EU legislation on chemicals and pesticides aims to protect human health and the environment and to prevent barriers to trade. It consists of rules for the placing on the market and use of specific categories of chemicals, a set of harmonized restrictions on the placing on the market and use of certain dangerous substances and preparations, as well as rules governing serious accidents related to dangerous substances and exports of these substances. "Pesticides" means substances used to control, eradicate and prevent organisms considered to be harmful (Storck et al. 2017).

The most important achievement at EU level is REACH (EC) Regulation no. 1907/2006, which regulates the registration, evaluation and licensing of such dangerous substances, as well as the restrictions that apply to them. The aim of REACH is to better protect people and the environment from the potential dangers of chemicals and to promote sustainable development. REACH established a single system for all chemicals and abolished the distinction between "new" (entering the market after 1981) and "existing" (approved before 1981) chemicals, while transferring the burden of proof for risk assessment of substances by public authorities in industries. In addition, it calls for the replacement of the most dangerous chemicals with appropriate alternatives (Williams et al. 2009).

Furthermore, Regulation (EC) No 1272/2008 on the classification, labeling and packaging of substances and mixtures, approved in 2008, aims to align the EU system with the United Nations Global Harmonized System (GHS), given that, in order to improve the level of protection of human health and the environment, the same identification criteria and the same labeling should be used to describe chemical hazards throughout the EU and worldwide (Kim et al. 2015).

In addition, the Seveso Directive (82/501/EEC), named after an Italian city which was contaminated by dioxin emissions due to an accident at a nearby industrial plant in 1976, aims to prevent serious accidents, such as fires and explosions, and to reduce their consequences, if they occur, by requiring safety reports, emergency plans and making information available to the public (Marchi 1991). Subsequently, in 1996, the Seveso II Directive (96/82/EC) on the risk of major accidents involving dangerous substances introduced new standards for safety management systems, emergency planning and land use planning and strengthened the provisions on inspections carried out by the Member States (O'Mahony et al. 2008). In the light of a number of serious industrial accidents (in Toulouse in France, Baia Mare in Romania and Enschede in the Netherlands), and on the basis of studies on carcinogens and hazardous substances, the scope of the Seveso II Directive has been further extended by Directive 2003/105/EC. This Directive obliges Member States to provide a detailed risk assessment based on possible accident scenarios and to cover the risks arising from storage and processing activities during the extraction and storage of pyrotechnic and explosive substances, ammonium nitrate and fertilizers based on ammonium nitrate. The Seveso III Directive (2012/18/EU) was approved by Parliament and the Council and subsequently published in July 2012. The Directive takes into account the new international classifications of substances agreed at United Nations level, which allow for better risk assessment and management of substances (Vasovic et al. 2016; Pilone et al. 2017).

Specifically for pesticides, a package of measures was approved in 2009, consisting of: Directive 2009/128/EC on the rational use of pesticides, aimed at reducing the risks to the environment and health, with a view to while maintaining crop productivity and improving controls on the use and distribution of pesticides; Regulation (EC) No 1107/2009 on the placing of PPPs on the market; Regulation (EC) No 1185/2009 on statistics on plant protection products, which lays down rules for the collection of information on the annual quantities of pesticides placed on the market and used in each Member State. Directive 2009/128/EC required Member States to adopt national action plans in order to set quantitative targets, measures and timetables in order to reduce the risks and consequences of the use of pesticides on human health and the environment (Karabelas et al. 2009).

Sustainable Consumption and Production

Sustainable development is one of the European Union's key objectives (Emas 2015; Bosselmann 2016). As we face a global shortage of natural resources, the main challenge for producers and consumers is now "more results with less resources". To meet this challenge in the current context of climate change and the growing need for energy, the EU has introduced a number of policies and initiatives aimed at sustainable consumption and production. They aim to improve the overall environmental performance of products throughout their life cycle, stimulate demand for better products and improved production technologies, and thus help consumers make informed choices (Scholl et al. 2010).

In July 2008, the Commission proposed a package of actions and proposals on sustainable consumption and production and sustainable industrial policy (SIP/BIP) (COM (2008) 0397), aimed at improving the environmental performance of products throughout their life cycle, to increase consumer awareness and demand for sustainable goods and sustainable production technologies, to promote innovation in EU industry and to meet international aspects. These proposals are based on and complement existing EU policies, such as the Integrated Product Policy (IPP), which was the first to formally introduce life cycle analysis (LCT/LCA) into European policies. The aim of the LCA is to identify potential improvements in products and services that would reduce the environmental impact and use of natural resources at all stages of the product or service life cycle (raw materials/supply chains/product use/end of life: disposal results and possibilities for reuse or recycling). The BCP Action Plan has resulted in initiatives in the following areas: extension of the Ecodesign Directive, revision of the Ecolabel Regulation, revision of the ECO Regulation, adoption of legislation on ecological public procurement, the roadmap for the efficient use of resources, and the action plan for eco-innovation (Pogutz and Micale 2011).

These instruments are an integral part of the renewed EU Strategy for Sustainable Development (EU SDS), the revision of which in 2009 strengthened the EU's long-standing commitment to the challenges of sustainable development, while recognizing the importance of strengthening the cooperation with partners outside the EU, for example through the UN Marrakesh Process. In particular, the following are mentioned, for example, for certain instruments directly related to the operation of enterprises (Nash 2009).

Labeling provides key information that enables consumers to make informed choices. The European Eco-label is an optional system introduced in 1992 to encourage businesses to market products and services that meet certain environmental criteria. The criteria are defined and revised by the EU Eco-Labeling Council (EUSEC), which is also responsible for the relevant evaluation and control requirements. They shall be published in the Official Journal of the European Union. The eco-labeled products and services bear a floral logo, enabling consumers—including public and private buyers—to easily recognize them. The label has so far been awarded to cleaning products, electrical appliances, paper products, clothing, household and gardening products, lubricants, as well as services such as tourist accommodation. The criteria for the award of an eco-label are not based on a single factor but on studies that analyze the impact of a product or service on the environment throughout its life cycle; (EC no. 66/2010) in 2008 aimed to promote the use of the optional eco-label system, making the implementation of the rules less costly and bureaucratic (Capitano et al. 2014; De Chiara 2016).

The Ecodesign Directive ensures the technical improvement of products. Directive 2005/32/EC establishes a framework for the creation of ecodesign requirements for energy consuming products (EuPs) by amending Directives 92/42/EEC, 96/57/EC and 2000/55/EC on requirements energy efficiency for products such as boilers, computers or televisions. The revision of the 2005 Directive (Directive 2009/125/EC) that took place in 2009 extended its scope to energy-related products in addition to energy-consuming products; these are products that do not consume energy during operation, but indirectly affect energy consumption, such as water-using appliances, windows and insulation materials (Huulgaard et al. 2013).

Water Conservation and Management

Water is not a commodity, but a common good, and it is a finite resource that must be protected and used in a sustainable way, both in terms of quality and quantity. It is, however, under pressure from a variety of uses in various sectors, such as agriculture, tourism, transport and energy. Its protection and management go beyond national borders. The EU Water Framework-Directive (2000/60/EC) establishes a framework for the protection and restoration of clean water in the EU and ensures its long-term sustainable use (Directive 2003). To be supplemented by more specific legislation, such as the Drinking Water Directive (98/83/EC) or Bathing Water (2006/7/EC), the Floods Directive (2007/60/EC) and the Framework-Directive on maritime strategy (2008/56/EC), as well as international agreements.

Typically mentioned in this case is the Urban Wastewater Treatment Directive (91/271/EEC), which aims to protect the environment from the negative effects of urban and industrial waste water disposal. The directive sets minimum standards and timetables for the collection, treatment and disposal of municipal waste water, introduces control mechanisms for sewage discharge and requires the phasing out of sewage discharge into the sea (Calleja et al. 2000).

2.4 Civil Liability of Companies Under EU Environmental Law and the "Polluter Pays" Principle

Today in the Community there are numerous sites that have been polluted, which poses serious health risks, while at the same time in recent decades there has been a dramatic acceleration of biodiversity loss. The extent of environmental damage, the escalation of ecological damage and the growing risk of environmental damage, imposed by the intensification of production and the over-depletion of natural resources, make the provision and implementation of a comprehensive and effective regulatory, normative in general framework for environmental liability a basic and necessary parameter for strengthening environmental protection and the sustainable evolution of life on Earth. The prevention and remediation, as far as possible, of environmental damage also contributes to the implementation of the objectives and principles of Community environmental policy, as set out in the Treaty and summarized above.

The European Union went through a course of almost 10 years until ending up voting Directive 2004/35/EC on environmental damage liability. In the previous year, two non-binding texts were important milestones in the regulation of environmental liability: the "Green Paper", aimed at establishing the possibility of recourse in the field of environment and civil liability and the establishment of joint compensation and compensation funds, and the White Paper, which aimed to establish an environmental liability regime to improve the application of Community principles and ensure adequate remediation of the environment (Bergkamp 2000; Wilde 2001).

In 2002, the adoption of a Framework Directive was considered the most appropriate option. The Economic and Social Committee has explicitly advocated the establishment of an environmental liability system, which would seek to prevent damage and restore nature to its former state, in accordance with the "polluter pays" principle. The concept of environmental damage according to the above Proposal of the Directive should cover the damage in at least the following three elements of the natural environment: (a) biodiversity, (b) water and (c) soil.

However, the "polluter pays" principle was fully implemented on 21 April 2004, when the aforementioned Directive 35/2004 on environmental liability for the prevention and remediation of damage to the environment was signed (Ciochina 2012; Bergkamp and Goldsmith 2013). The noticeable difference between the regulations laid down in Directive 2004/35/EC in relation to other systems for the determination and imputation of environmental liability is the removal of the provisions of Directive 2004/35/EC from previous civil law schemes for environmental damage and sui generis framework of measures of administrative coercion with the aim of assuming the cost of prevention and remediation of damage to the environment per se by the respective operator.

"Environmental liability" means the system of provisions governing the obligation and the process of preventing and repairing damage which is threatened or has already been suffered by an environmental good, regarded either as an individual element of nature or as a set of natural elements; which are interdependent and interact with each other. The concept of "environmental liability" followed a long course until it crystallized in its current form. As a web of rules of law, regulations and procedures, environmental liability is universally regarded as a crucial institution of modern environmental law (Pozzo 2014). The scope of the regulations it introduces has made it a fundamental mechanism for assessing ecological damage, with the aim of preventively protecting the environment, restoring the spent environmental capital to its original state and sharing the cost of restoration with the imposition of sanctions, the amount of which often ranges in particularly high amounts of money.

According to Directive 2004/35/EC [2], environmental liability consists of a framework of rules based on the "polluter pays" principle, with the aim of preventing and remedying environmental damage. In accordance with the "polluter pays" principle, any operator who causes environmental damage or an immediate risk of such damage should, in principle, bear the cost of the necessary preventive or remedial measures (Munir 2013; Bugge 2009). This encourages operators to minimize the risk of environmental damage in order to avoid liability. It is noted that "operator" means any natural or legal person, whether under private or public law, who operates or controls a professional activity or, where this is provided for by national law, to whom decisive financial competence has been transferred in terms of technical operation of such activity, including the holder of a relevant license or authorization or any person registering or notifying such activity.

"Environmental damage" means any damage that has a materially adverse effect on the attainment or maintenance of a favorable conservation status of water, soil, protected species or natural habitats; "damage" means a measurable adverse change in natural resource or measurable degradation of a service linked to a natural resource that may occur directly or indirectly. "Immediate threat of damage" means the probable likelihood of environmental damage occurring in the near future.

Directive 2004/35/EC establishes a dual regime of liability: (a) objective liability for damage (or imminent threat of injury) from dangerous or potentially dangerous professional activities, as listed in Annex III to the Directive; and (b) subjective liability for cases of damage to species and natural habitats protected by EU legislation for all occupational activities except those listed in Annex III to the Directive.

The objectives of the Directive are to prevent environmental pollution and the loss of biodiversity in parts of the European Community, to minimize pollution of water, soil, protected species and habitats, to promote the principles of "polluter pays" and the assumption of financial responsibility by operators, ie natural or legal persons, whose activity causes or may cause damage to the environment (Anile 2019). For example, entities engaged in certain occupational activities, such as the transport of hazardous substances or activities involving discharges into water, must take precautionary measures in the event of an imminent threat to the environment. If damage has already been occurred, they must take the appropriate steps to repair it and pay the cost. It is noted that the application scope of the directive has been extended three times to include the management of extractive waste, the operation of geological storage sites and the safety of offshore oil and gas extraction activities.

An important advantage of the Directive is that it sets a wide minimum level of protection, allowing Member States to regulate the details of its implementation, and establishes a system for incurring environmental costs by the operator. The Directive innovates by stipulating, inter alia, that persons, including non-governmental organizations, who are affected or may be affected by environmental damage, have the right to require the competent public authority to take action (Ciochina 2012).

The main disadvantages of the Directive were the exclusion of "historical" forms of pollution and liability from diffuse pollution, as it avoided including liability for damage caused by nuclear energy, activities related to national defense and international security, ie pollution from indeterminate dumps, as is usually the case in agriculture, in the exploitation of energy and mineral resources, in atmospheric deposition, etc. (Pozzo 2014).

It is, of course, worth noting at this point that, while regulations are legislation that applies automatically and uniformly to all EU Member States as soon as they enter into force, without having to be transposed into national law, the directives require EU achieve a specific result, but leave them free to choose how to achieve this result. EU countries must adopt measures to incorporate them into national law (transpose) in order to achieve the objectives set by the directive. National authorities must communicate these measures to the European Commission. Transposition into national law must take place by the deadline set when the directive is adopted (generally within 2 years). When a Member State does not comply, the European Commission may initiate infringement proceedings.

Therefore, each EU Member State has, in the first instance, the responsibility for the correct, complete and timely transposition into national law, application and compliance, in this case, of companies operating in its territory, to the above—and in each case—EU environmental regulations, and consequently the implementation of the objectives of environmental CSR. In other words, the effectiveness of EU environmental policy, in general but also in the context of corporate environmental responsibility, is greatly influenced by its implementation at national, regional and local level. Monitoring both the state of the environment and the level of implementation of EU environmental legislation is vital.

2.5 Environmental Performance in EU Countries and Sectors

This section is devoted to providing a decomposition of main corporate environmental performance data of listed and unlisted firms domiciled in EU countries. The analysis is performed comparatively between countries and economic sectors, as well as during the research period 2003–2018. The main environmental performance indicators have been extracted from Datastream database in an annual basis. The relative scores include EN_INNOV which denotes each company's environmental innovation score, reflecting the firm's ability to mitigate the environmental costs towards its customers or even creating new market and product opportunities via enhanced environmental technologies and processes. EMM denotes companies' emission reduction score estimated via their efforts and effectiveness in reducing emissions from their production and operation. RES_USE denotes the firms' resource use score and indicates firms' ability to reduce the usage of hazardous materials, energy, water and other valuable natural resources. Finally, we have estimated CER which is the firms' annual firm-specific average of the aforementioned three scores. All environmental scores range between zero (0) and 100 so the higher the score the more environmentally responsible and efficient is the firm.

Table 2.1 and Figure 2.1 present the total averages (for the whole sample period 2003–2018) of the four environmental performance scores by country. As we can see, Malta, Romania and Cyprus are among the countries with the lowest environmental innovation scores. On the contrary, Poland has the lowest score among the EU sample countries in terms of the emission reduction score and again Malta presents the lowest score in resource usage. Overall, the lower ranked countries on the average CER score are again Malta, Poland and Romania, indicating that

Table 2.1 Average of environmental performance scores per country (2003–2018)

Countries	EN_INNOV	EMM	RES_USE	CER
Austria	59.64	54.75	52.68	55.69
Belgium	58.03	56.24	58.79	57.69
Cyprus	39.52	70.20	65.55	57.42
Czech Republic	46.02	58.64	52.83	52.49
Denmark	51.09	57.27	68.10	58.82
Finland	66.05	69.67	71.19	68.97
France	68.38	80.37	80.01	76.52
Germany	64.06	69.59	71.16	68.27
Greece	54.89	62.71	64.34	60.64
Hungary	70.99	73.61	62.44	69.01
Ireland	55.55	53.59	56.26	55.14
Italy	60.2	62.97	64.41	62.52
Luxemburg	60.67	73.33	66.93	66.98
Malta	36.05	55.25	30.48	40.59
Netherlands	61.51	68.59	74.68	68.26
Poland	46.22	44.38	44.36	44.99
Portugal	65.19	64.68	63.79	64.55
Romania	29.54	56.61	52.12	46.09
Slovakia	50.09	57.98	68.11	58.73
Slovenia	65.35	71.78	53.92	63.69
Spain	66.32	74.5	73.3	71.37
Sweden	63.85	65.24	66.28	65.12
UK	67.03	74.1	73.34	71.49

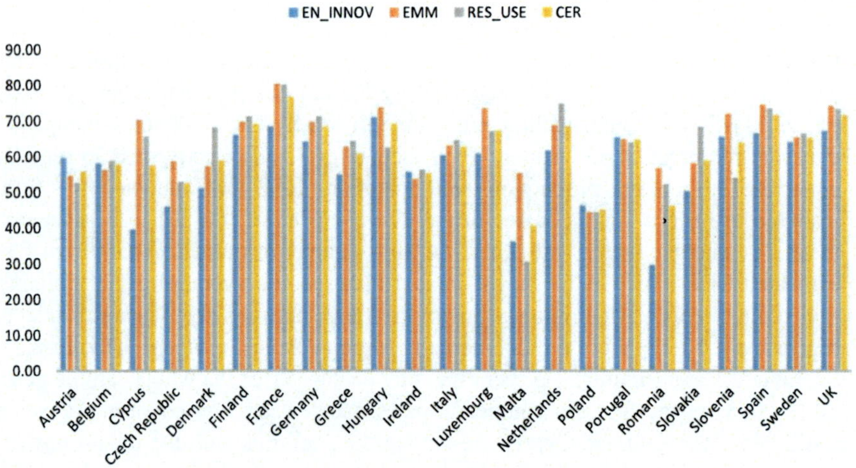

Fig. 2.1 Environmental performance averages per EU country (2003–2018)

corporations in these countries have not fully been investing on environmental activities and performance. The fact that are countries which have joined the EU fairly recently may be a valid reason for lagging behind the oldest EU member countries on those merits. On the contrary, corporations in France, UK, Netherlands and Luxemburg have the highest averages across all environmental performance indicators. Especially, French firms have the highest score (almost 80) followed by Spain and UK (both of them present an average CER score close to 70). The rest of the countries rank between those top and bottom ranked countries. Overall, there seems to be significant discrepancies between EU member countries where some take slower pace on incorporating relative regulations on environmental protection and activities, so firms on those countries take time to conform to the new norms.

Moreover, Table 2.2 and Fig. 2.2 indicate the evolution of the average CER score over the period 2003–2018. Malta and Poland (which score among the lower ranked countries in terms of CER) indicate significant fluctuations in CER over the sample period. Specifically, Polonaise firms reduced their environmental performance up until 2010 and then an upward trend is evidenced. Firms in Malta present the opposite picture with a positive upward trend (with fluctuations from 1 year to the other) up until 2013 and afterwards we observe a significant reduction, scoring below 40 in 2018. Irish firms present the most stable evolution of CER performance which averages around 50 for the whole period under investigation. Same picture is also evident for the highly-ranked countries such as Spain, UK, France and Netherlands where in all of those countries we observe a stable fixed time-series of average CER for the examination period, indicating that firms on those countries have well established corporate environmental strategies.Furthermore, Table 2.3 and Fig. 2.3 present the evolution of the four environmental performance scores from 2003 to 2018 for all countries. All scores present a stable evolution over the years (with a small reduction up until 2012 and a steady increase afterwards). RES_USE presents the highest score among the main three indicators followed by emission reduction score (EMM). It seems that EU firms invest more on reducing hazardous waste and emissions mainly because those scores have a direct impact on their daily operations, costs (both operating and litigation costs) and profitability potential. Environmental innovation (EN_INNOV) presents the smallest numbers yet with a stable trend over the sample years. This could be attributed to the fact that innovative activities require time to prosper, are highly volatile in nature and require significant funds, thus not all firms are able to adhere to those conditions.

Additionally, in order to extract more information on the decomposition of the four environmental performance indicators, we estimated the average of the four scores after distinguishing firms following the Statistical Classification of Economic Activities of the European Community (NACE Rev. 2) in order to assign each firms into a specific sector. Eleven sectors were determined as depicted on Table 2.4 and we estimated the average environmental performance scores for the whole period 2003–2018. As we can see in Fig. 2.4, the Consumer Discretionary sector presents the highest CER score along with Information Technology sector. On the contrary, sectors with a more significant environmental footprint (more polluting and hazardous sectors) such as Energy, Industrial and Material firms are lagging behind in

Table 2.2 Average CER score per country and sample year

Countries	2003	2004	2005	2006	2007	2008	2009	2010	2011	2012	2013	2014	2015	2016	2017	2018
Austria	55.13	55.81	56.37	56.41	57.23	55.69	56.18	56.12	55.58	55.43	56.09	55.58	54.83	54.70	54.36	55.35
Belgium	59.70	57.75	57.33	56.64	56.40	57.30	57.59	58.61	59.28	59.11	57.96	56.56	57.16	57.12	57.75	57.60
Cyprus	58.16	56.40	58.24	60.28	60.07	58.05	58.80	60.75	59.66	58.68	59.37	57.40	57.56	59.05	55.93	56.87
Czech Republic	52.32	52.60	52.83	52.70	52.53	51.69	52.47	52.92	51.65	52.29	52.92	53.43	52.69	52.09	51.76	52.57
Denmark	62.45	59.86	60.60	58.39	59.33	57.59	55.97	58.29	58.13	58.43	58.62	59.69	58.33	58.72	58.74	60.65
Finland	68.26	68.82	69.50	68.60	68.71	68.36	67.92	68.12	68.94	68.94	69.09	68.97	69.79	69.69	69.20	69.96
France	75.97	76.36	77.13	76.99	77.73	77.19	76.77	75.87	75.44	75.80	75.59	76.35	75.95	76.09	75.12	75.40
Germany	65.15	67.81	68.98	68.59	69.78	69.75	69.75	70.18	68.63	66.76	66.00	67.11	67.76	67.20	68.03	68.20
Greece	61.75	60.25	60.06	60.01	61.30	61.18	61.86	60.60	61.10	59.70	61.19	60.69	61.11	60.16	60.11	60.25
Hungary	70.65	69.69	69.23	69.36	68.64	68.15	68.10	68.96	70.04	70.46	69.95	68.21	68.25	68.20	67.70	69.20
Ireland	53.60	54.30	56.25	55.89	55.28	54.46	55.00	55.26	55.04	55.26	55.88	54.26	55.31	55.08	54.63	55.28
Italy	57.37	62.99	61.73	62.00	61.26	61.41	61.88	62.18	62.26	62.31	63.98	64.07	63.66	62.85	62.73	63.26
Luxemburg	68.53	66.77	66.81	66.07	66.00	66.59	65.84	66.47	67.67	67.98	67.14	67.29	66.81	67.32	67.66	67.75
Malta	38.28	36.67	35.61	39.26	37.44	39.94	37.96	40.16	41.59	46.26	47.02	45.56	44.97	43.47	39.37	37.44
Netherlands	66.32	67.60	68.22	68.45	68.53	68.76	68.28	68.58	68.60	68.40	68.45	68.01	68.16	68.33	68.74	68.62
Poland	48.33	46.02	44.62	44.16	42.92	42.07	42.46	40.86	44.57	45.81	47.70	47.43	46.65	45.89	46.03	50.18
Portugal	65.52	65.50	65.91	65.54	66.02	64.66	63.91	62.58	63.54	64.04	63.34	63.96	64.19	64.14	64.64	65.47
Romania	42.17	43.56	51.02	39.83	40.09	47.03	55.01	47.42	42.87	44.89	44.46	51.82	45.75	47.42	45.43	52.35
Slovakia	62.32	61.34	59.14	55.74	53.09	54.55	55.59	55.16	57.58	58.61	59.70	59.25	61.51	62.34	63.65	62.80
Slovenia	63.69	62.54	58.47	56.22	54.14	55.07	55.89	56.77	59.56	60.25	60.47	61.37	62.98	63.22	62.78	61.48
Spain	70.14	71.84	72.14	71.73	71.93	71.55	71.20	71.08	70.85	70.95	70.83	71.43	72.27	71.45	70.63	71.21
Sweden	66.24	66.65	65.70	65.89	66.12	64.58	62.57	62.65	64.01	63.27	65.11	65.89	66.01	66.23	66.00	65.49
UK	70.01	71.08	71.09	71.23	72.33	71.97	72.76	72.13	72.44	71.79	71.26	72.52	70.77	70.42	70.87	70.54

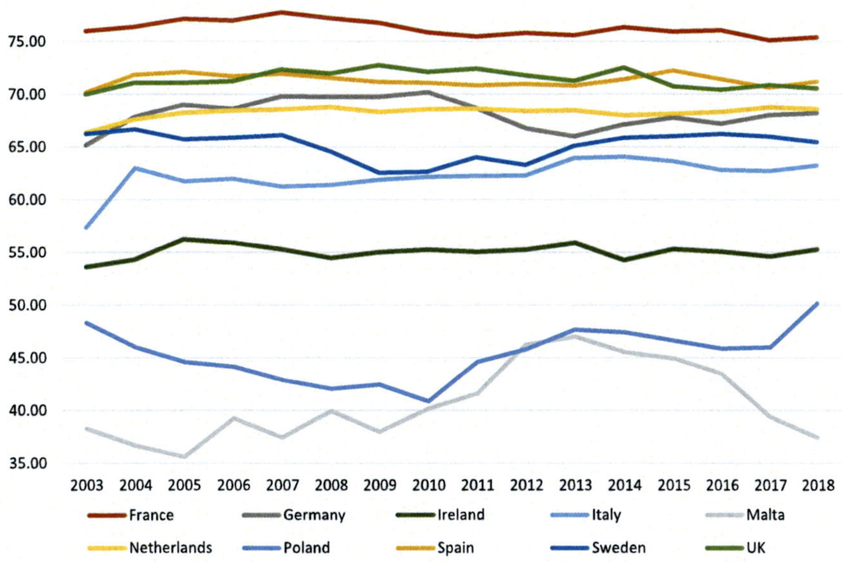

Fig. 2.2 Evolution of average CER score of some EU countries

Table 2.3 Average of environmental performance scores per year (2003–2018)

Year	EN_INNOV	EMM	RES_USE	CER
2003	60.58	65.13	65.36	63.69
2004	61.48	65.98	66.65	64.70
2005	61.67	66.30	67.01	65.00
2006	61.33	65.95	66.66	64.65
2007	61.76	65.93	66.73	64.81
2008	61.32	65.74	66.40	64.49
2009	61.20	65.51	66.31	64.34
2010	60.86	65.59	66.26	64.24
2011	61.25	65.86	66.30	64.47
2012	61.41	65.83	66.12	64.45
2013	61.36	65.94	66.76	64.68
2014	61.18	65.87	66.82	64.63
2015	61.27	66.34	66.85	64.82
2016	61.34	65.99	66.52	64.62
2017	61.2	65.89	66.23	64.44
2018	61.59	66.4	66.92	64.97

terms of average environmental performance. The lowest scores among all four indicators are evidenced in Health Care and Communication Services sectors.

A similar conclusion can be reached if we observe the annual decomposition of average environmental score (CER) per sector as depicted in Table 2.5 and Fig. 2.5. Health care, Industrial and Materials sectors all present a significant decrease on the average CER score over the sample period. On the contrary, Information

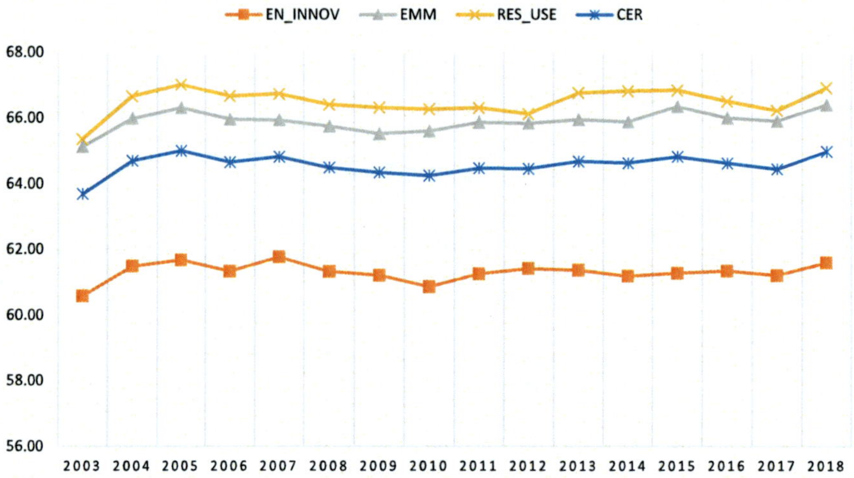

Fig. 2.3 Annual evolution of average environmental scores in the EU countries

Table 2.4 Average environmental scores per economic sectors (2003–2018)

Sector	EN_INNOV	EMM	RES_USE	CER
Communication Services	59.94	64.86	65.04	63.28
Consumer Discretionary	62.11	66.75	67.81	65.56
Consumer Staples	61.16	64.92	64.81	63.63
Energy	60.63	66.71	66.22	64.52
Financials	61.36	65.81	66.81	64.66
Health Care	59.81	64.44	65.07	63.11
Industrials	62.01	66.20	66.77	64.99
Information Technology	62.37	66.92	67.41	65.57
Materials	60.92	64.70	65.43	63.68
Real Estate	61.50	65.88	67.35	64.91
Utilities	60.12	65.87	65.58	63.86

Technology and Consumer Discretionary sectors present an upward trend for the period and both sectors score the highest numbers relative to the other sectors. Nevertheless, the most important increase on CER was reported in the Energy sector, especially from 2011 until 2017, but still lagging behind relative to less polluting sectors such as Consumer Discretionary and Information Technology.

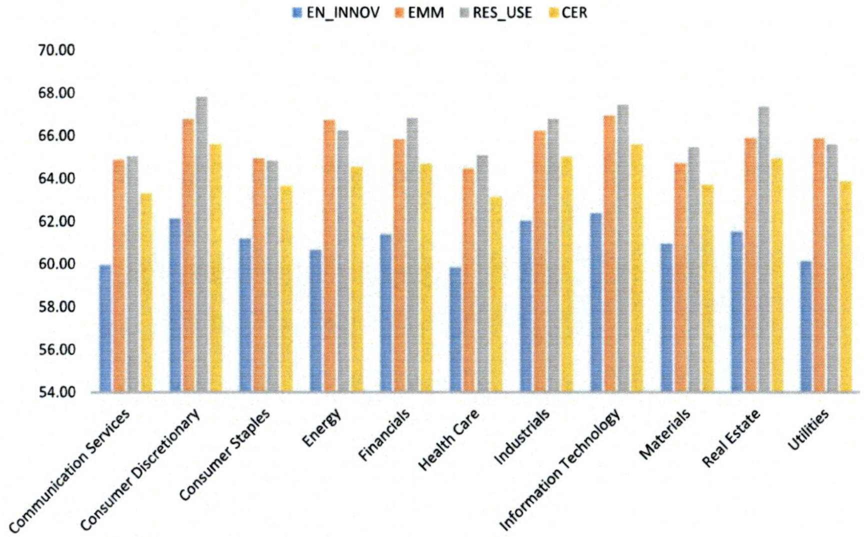

Fig. 2.4 Environmental performance averages per sector (2003–2018)

2.6 Conclusion

The goal of this chapter was to examine the main regulatory interventions made by the EU officials over the course of the previous decades which shaped the current framework of corporate environmental responsibility (CER). Being more specific we analyzed how from the Paris 1972 convention, where environmental protection was first raised, the issue evolved into specific principles set by the European community in terms of environmental protection. Later on, EU officials moved into more concrete regulations on environmental responsibility (almost a decade ago). These regulations were targeting key issues such as climate changes, water management, air pollution, resource allocation, chemical reduction and even sustainable production and consumption.

The results of these efforts (as the previous section suggests) was that European corporations have steadily abided to these regulations and the overall shift of behavior into a more responsible and liable corporate activity is evidence in the majority of EU countries. Surely there are observed differences between various economic sectors (with some of them leading while others lagging behind) nevertheless the overall trend is promising and seems that EU firms are starting to invest significant resources on CER. Whether these resources provide benefits to corporations remains an open empirical question, which we will try to answer in the following chapters of the book.

Table 2.5 Average CER score per economic sector and sample year

Sector	2003	2004	2005	2006	2007	2008	2009	2010	2011	2012	2013	2014	2015	2016	2017	2018
Communic. Services	63.85	63.99	64.49	62.85	63.04	63.17	64.11	63.80	64.20	62.75	62.57	62.84	63.41	62.94	62.39	62.49
Consumer Discret.	64.80	65.82	65.61	65.48	64.85	64.64	64.42	64.79	66.08	66.63	65.75	65.42	65.86	66.39	66.29	65.81
Consumer Staples	61.28	63.11	62.65	62.93	63.15	63.20	62.55	62.52	63.88	63.47	65.48	64.26	64.78	64.39	63.56	65.27
Energy	62.63	65.27	64.58	63.19	64.90	63.95	63.93	64.19	62.95	64.14	64.16	64.82	65.61	66.29	65.60	64.96
Financials	64.83	64.65	65.72	65.07	65.09	66.10	64.92	64.92	64.60	64.14	64.34	64.38	64.02	63.39	64.22	64.70
Health Care	60.96	63.27	64.89	62.98	62.94	63.17	63.51	63.59	63.35	63.14	63.15	62.80	61.88	62.26	61.89	63.80
Industrials	64.65	64.97	65.79	65.69	65.92	65.27	64.66	64.48	64.28	64.61	65.00	65.02	64.90	65.05	64.75	64.95
Information Technology	66.40	65.94	65.02	65.46	65.74	65.30	65.21	64.78	66.15	65.60	65.63	65.83	66.08	65.82	64.90	65.28
Materials	61.48	64.24	64.86	64.21	64.08	63.31	63.06	62.84	63.30	63.38	62.78	63.28	63.92	63.17	64.24	65.19
Real Estate	67.22	66.02	65.74	64.93	65.61	65.60	65.20	63.54	64.08	66.15	64.71	65.17	64.70	64.38	62.95	64.42
Utilities	60.82	63.51	63.74	64.21	64.43	64.59	64.00	64.53	62.90	63.65	64.27	64.05	64.11	63.78	63.68	64.83

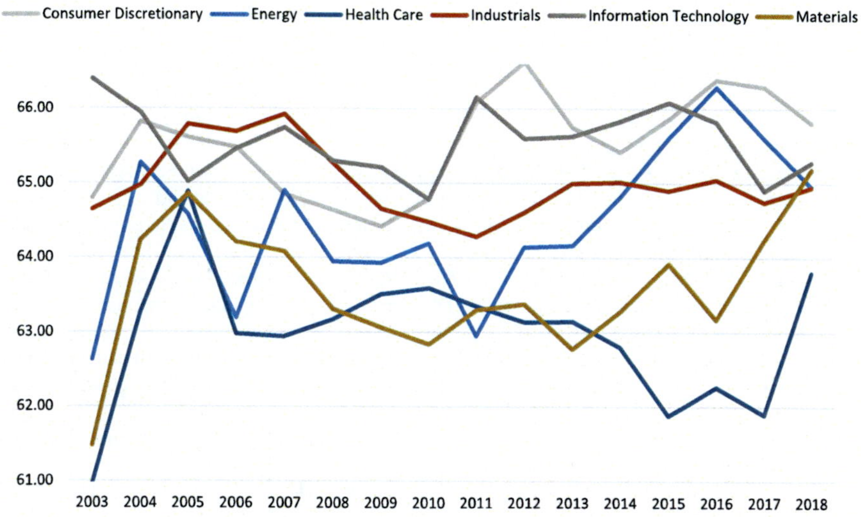

Fig. 2.5 Evolution of average CER score of some economic sectors

References

Anile, F. 2019. Pollution victim. In *Dictionary of statuses within EU law*, 427–432. Cham: Springer.

Aragão, A. 2018. Environmental principles in the EU. In *Elgar Encyclopedia of environmental law*, 449–462. Cheltenham, UK: Edward Elgar Publishing Limited.

Baker, S. 2012. The evolution of European Union environmental policy. In *Politics of sustainable development*, 101–116. New York: Routledge.

Barral, V., and P.M. Dupuy. 2015. *Principle 4: Sustainable development through integration*, 157–180. Oxford, UK: Oxford University Press.

Bergkamp, L. 2000. Commission's White Paper on Environmental liability: A weak case for an EC strict liability regime. *The European Environmental Law Review* 9: 105.

Bergkamp, L., and B. Goldsmith. 2013. *The EU environmental liability directive: A commentary*. Oxford, UK: Oxford University Press.

Bodansky, D. 1994. The precautionary principle in US environmental law. Interpreting the precautionary principle, 203.

Bosselmann, K. 2016. *The principle of sustainability: Transforming law and governance*. London: Taylor & Francis.

Bousta, R. 2013. Who said there is a right to good administration: A Critical Analysis of Article 41 of the Charter of Fundamental Rights of the European Union. European Public Law, 19: 481

Bugge, H.C. 2009. The polluter pays principle: Dilemmas of justice in national and international contexts. *Environmental Law and Justice in Context* 2009: 411–428.

Calleja, I., B. Rebollo, and J. Hemmelskamp. 2000. Urban wastewater treatment Directive 91/271/ EEC–Case study Spain. Technology and Environmental policy. UNU–Merit, Maastricht.

Cameron, J., and J. Abouchar. 1991. The precautionary principle: A fundamental principle of law and policy for the protection of the global environment. *BC International and Competitive Law Review* 14: 1.

Capitano, C., G. Peri, and G. Rizzo. 2014. Is the eo-label EU decision for hard coverings really capable of capturing the environmental performances of the marble productive chain? A field verification by means of a life cycle approach. *The International Journal of Life Cycle Assessment* 19 (5): 1022–1035.

Ciochina, D. 2012. Civil liability for environmental damages. *EIRP Proceedings* 7.

Colombo, L.A., M. Pansera, and R. Owen. 2019. The discourse of eco-innovation in the European Union: An analysis of the Eco-Innovation Action Plan and Horizon 2020. *Journal of Cleaner Production* 214: 653–665.

Crawley, D., and I. Aho. 1999. Building environmental assessment methods: Applications and development trends. *Building Research and Information* 27 (4–5): 300–308.

Csák, C. 2011. Thoughts about the problems of the enforcement of the "polluter pays" principle. *European Integration Studies* 1: 27–40.

Dagtoglou, P. 1985. *European Community Law*. Vol. I, 77ff. Athens: Ant. Sakkoulas Publishers.

De Chiara, A. 2016. Eco-labeled products: Trend or tools for sustainability strategies? *Journal of Business Ethics* 137 (1): 161–172.

De Sadeleer, N. 2009. The precautionary principle as a device for greater environmental protection: Lessons from EC courts. *Review of European Community and International Environmental Law* 18 (1): 3–10.

———. 2012. Principle of subsidiarity and the EU environmental policy. *Journal for European Environmental and Planning Law* 9 (1): 63–70.

———. 2020. *Environmental principles: From political slogans to legal rules*. Oxford, UK: Oxford University Press.

———. 2013. The principle of a high level of environmental protection in EU law: Policy principle or general principle of law? Mijörättsliga Perspektiv Och Tankevändor, Vänbok Till Jan Darpö and Gabriel Michanek, Uppsala, Iustus Förlag, 447–465

Directive, W. F. 2003. Common implementation strategy for the water framework directive (2000/60/EC). Guidance document, (7)

Douma, W.T. 2000. The precautionary principle in the European Union. *Review of European Competitive and International Environmental Law* 9: 132.

Doussis, A.D., V. Delitheou, and C.G. Athanassopoulos. 2019. The principle of subsidiarity. *International Relations* 7 (07): 317–324.

Ehlermann, C.D. 1987. Internal Market following the Single European Act. *The Common Market Law Review* 24: 361.

Elliott, J. 2012. *An introduction to sustainable development*. New York: Routledge.

Emas, R. 2015. The concept of sustainable development: Definition and defining principles. Brief for GSDR, 2015.

Eriksen, E.O., J.E. Fossum, and A. Menéndez. 2004. *Developing a constitution for Europe*. New York: Routledge.

Esty, D.C., and A. Winston. 2009. *Green to gold: How smart companies use environmental strategy to innovate, create value, and build competitive advantage*. Boca Raton, FL: Wiley.

Fischer, C. 2011. The development and achievements of EU waste policy. *Journal of Material Cycles and Waste Management* 13 (1): 2–9.

Fisher, E.C., J.S. Jones, and R. von Schomberg. 2006. *Implementing the precautionary principle: Perspectives and prospects*. Cheltenham, UK: Edward Elgar Publishing.

Gaines, S.E. 1991. The polluter-pays principle: From economic equity to environmental ethos. *Texas International Law Journal* 26: 463.

Gergen, M.J. 1994. The failed promise of the polluter pays principle: An economic analysis of landowner liability for hazardous waste. *NYUL Review* 69: 624.

Gharfalkar, M., R. Court, C. Campbell, Z. Ali, and G. Hillier. 2015. Analysis of waste hierarchy in the European waste directive 2008/98/EC. *Waste Management* 39: 305–313.

Giljum, S., T. Hak, F. Hinterberger, and J. Kovanda. 2005. Environmental governance in the European Union: Strategies and instruments for absolute decoupling. *International Journal of Sustainable Development* 8 (1–2): 31–46.

Glaesner, H.J. 1986. The Single European Act: Attempt at an appraisal. *Fordham International Law Journal* 10: 446.

Grossman, M.R. 2006. Agriculture and the polluter pays principle: An introduction. *Oklahoma Law Review* 59: 1.

Gupta, J., and M.J. Grubb. 2000. *Climate change and European leadership: A sustainable role for Europe?* Vol. 27. New York: Springer.

Handl, G. 2012. Declaration of the United Nations conference on the human environment (Stockholm Declaration), 1972 and the Rio Declaration on Environment and Development, 1992. United Nations Audiovisual Library of International Law, 11.

Holzleitner, M., S. Moser, and S. Puschnigg. 2020. Evaluation of the impact of the new Renewable Energy Directive 2018/2001 on third-party access to district heating networks to enforce the feed-in of industrial waste heat. *Utilities Policy* 66: 101088.

Hooghe, L., and G. Marks. 1999. The making of a polity: The struggle over European integration. In *Continuity and change in contemporary capitalism*, 35. Oxford: Oxford University Press.

Huulgaard, R.D., R. Dalgaard, and S. Merciai. 2013. Ecodesign requirements for televisions—Is energy consumption in the use phase the only relevant requirement? *The International Journal of Life Cycle Assessment* 18 (5): 1098–1105.

Jans, J.H. 1994. Self-sufficiency in European waste law. *European Environmental Law Review* 3: 223.

———. 1999. The status of the self-sufficiency and proximity principles with regard to the disposal and recovery of waste in the European Community. *Journal of Environmental law* 11 (1): 121–156.

Jordan, A. 1998. Step change or stasis? EC environmental policy after the Amsterdam Treaty.

Kanska, K. 2004. Towards administrative human rights in the EU. Impact of the charter of fundamental rights. *European Law Journal* 10 (3): 296–326.

Kaiser, Z. 2016. Article 37 Charter of Fundamental Rights of the European Union: On the way to a fundamental right upgrade?

Karabelas, A.J., K.V. Plakas, E.S. Solomou, V. Drossou, and D.A. Sarigiannis. 2009. Impact of European legislation on marketed pesticides—A view from the standpoint of health impact assessment studies. *Environment International* 35 (7): 1096–1107.

Katz, P. 2003. The Treaty of Nice and European Union enlargement: The political, economic, and social consequences of ratifying the Treaty of Nice. *University of Pennsylvania Journal of International Economic Law* 24: 225.

Kim, S., Y.K. Lee, and Y. Choi. 2015. How to Improve the Reliability of MSDS and Labels?(2): A Review on the EU CLP (REGULATION (EC) No 1272/2008). *Journal of Korean Society of Occupational and Environmental Hygiene* 25 (2): 115–125.

Koppen, I. J. 1988. The European Community's environment policy: From the summit in Paris, 1972 to the Single European Act, 1987.

Kozien, E., and A. Kozien. 2018. The sustainability development concept under the regulations in force of the treaty on European Union and the treaty on the functioning of the European Union–legal and economical view. *Economic and Social Development: Book of Proceedings*, 402–412.

Kramer, L. 1987. Single European Act and Environmental Protection: Reflections on several new provisions in Community Law. *The Common Market Law Review* 24: 659.

———. 1992. *Focus on European environmental law*, 76. London: Sweet & Maxwell.

———. 2005. Discussions on Directive 2004/35 Concerning Environmental Liability. *Journal for European Environmental and Planning Law* 2 (4): 250–256.

———. 2018. Environmental principles and the EU Court of Justice. In *Elgar Encyclopedia of Environmental Law*, 587–598. Cheltenham, UK: Edward Elgar Publishing Limited.

Kulovesi, K., and S. Oberthür. 2020. Assessing the EU's 2030 climate and energy policy framework: Incremental change toward radical transformation? *Review of European, Comparative and International Environmental Law* 29 (2): 151–166.

Liberatore, A. 1997. *The integration of sustainable development objectives into EU policy-making*, 107–126. London: Routledge.

Lindhout, P.E., and B. Van den Broek. 2014. The polluter pays principle: Guidelines for cost recovery and burden sharing in the case law of the European court of justice. *Utrecht Law Review* 10: 46.

Luppi, B., F. Parisi, and S. Rajagopalan. 2012. The rise and fall of the polluter-pays principle in developing countries. *International Review of Law and Economics* 32 (1): 135–144.

Luttenberger, A., and B. Rukavina. 2004. The right to access to information on environmental matters. In 17th Biennal International Congress "'Tourism and Hospitality Industry".

Makovetska, Y., and T. Omelyanenko. 2014. Waste Management in the Context of the Requirements of the Industrial Emissions Directive 2010/75/EU. Economics of Nature and the Environment, 10–12.

Marchi, B.D. 1991. The Seveso Directive: An Italian pilot study in enabling communication. *Risk Analysis* 11 (2): 207–215.

Moon, J. 2007. The contribution of corporate social responsibility to sustainable development. *Sustainable Development* 15 (5): 296–306.

Morgera, E., and G.M. Durán. 2014. Article 37–Environmental Protection. In *The EU Charter of Fundamental Rights*, 1026–1047. Baden-Baden: Nomos Verlagsgesellschaft mbH and Co. KG.

Munir, M. 2013. History and evolution of the polluter pays principle: How an economic idea became a legal principle? Available at SSRN 2322485.

Nash, H.A. 2009. The European Commission's sustainable consumption and production and sustainable industrial policy action plan. *Journal of Cleaner Production* 17 (4): 496–498.

O'Mahony, M.T., D. Doolan, A. O'Sullivan, and M. Hession. 2008. Emergency planning and the Control of Major Accident Hazards (COMAH/Seveso II) Directive: An approach to determine the public safety zone for toxic cloud releases. *Journal of Hazardous Materials* 154 (1–3): 355–365.

Oberthür, S. 2019. Hard or soft governance? The EU's climate and energy policy framework for 2030. *Politics and Governance* 7 (1): 17–27.

Pilone, E., M. Demichela, and G. Camuncoli. 2017. Seveso Directives and LUP: The mutual influence of natural and anthropic impacts. *Journal of Loss Prevention in the Process Industries* 49: 94–102.

Pogutz, S., and V. Micale. 2011. Sustainable consumption and production: An effort to reconcile the determinants of environmental impact. *Society and Economy* 33 (1): 29–50.

Portuese, A. 2010. The principle of subsidiarity as a principle of economic efficiency. *Columbia Journal of European Law* 17: 231.

Pozzo, B. 2014. The Environmental Liability Directive and the problem of terminological consistency. In *Criminal proceedings, languages and the European Union*, 123–129. Berlin, Heidelberg: Springer.

Proelss, A. 2016. Principles of EU Environmental Law: An appraisal. In *Contemporary issues in Environmental Law*, 29–45. Tokyo: Springer.

Reese, M. 2018. The proximity principle. In *Elgar Encyclopedia of Environmental Law*, 219–233. Cheltenham, UK: Edward Elgar Publishing Limited.

Renn, O., T. Webler, and P. Wiedemann. 2013. *Fairness and competence in citizen participation: Evaluating models for environmental discourse*. New York: Springer.

Rezza, G. 2006. The principle of precaution-based prevention: A Popperian paradox? *The European Journal of Public Health* 16 (6): 576–577.

Schaltegger, S., and R. Burritt. 2005. Corporate sustainability (Doctoral dissertation, Edward Elgar).

Scholl, G., F. Rubik, H. Kalimo, K. Biedenkopf, and Ó. Söebech. 2010. Policies to promote sustainable consumption: Innovative approaches in Europe. In Natural resources forum 34, 1, pp. 39-50). Oxford, UK: Blackwell Publishing Ltd.

Sheldon, J.C. 1998. Environmental effects of codecision under the Maastricht Treaty. *Boston College International and Comparative Law Review* 21: 247.

Storck, V., D.G. Karpouzas, and F. Martin-Laurent. 2017. Towards a better pesticide policy for the European Union. *Science of the Total Environment* 575: 1027–1033.

Swann, D., ed. 2017. *The single European market and beyond: A study of the wider implications of the Single European Act.* New York: Routledge.
Toth, A.G. 1992. Principle of subsidiarity in the Maastricht Treaty. *Common Market Law Review* 29: 1079.
UNION, P. 2008. Directive 2008/50/EC of the European Parliament and of the Council of 21 May 2008 on ambient air quality and cleaner air for Europe. *Official Journal of the European Union.*
Vasovic, D., J. Malenovic Nikolic, and G. Janackovic. 2016. Evaluation and assessment model for environmental management under the Seveso III, IPPC/IED and water framework directive. *Journal of Environmental Protection and Ecology* 17 (1): 356–365.
Vedder, H. 2010. The Treaty of Lisbon and European environmental law and policy. *Journal of Environmental Law* 22 (2): 285–299.
Verma, A.K. 2019. Sustainable development and environmental ethics. *International Journal on Environmental Sciences* 10 (1): 1–5.
Vícha, O. 2011. The polluter-pays principle in OECD recommendations and its application in international and EC/EU law. *Czech Yearbook of Public and Private International Law* 2: 57–67.
Von Bogdandy, A. 2000. European Union as a Human Rights Organization-Human Rights and the Core of the European Union. *The Common Market Law Review* 37: 1307.
Wasmeier, M. 2001. Integration of environmental protection as a general rule for interpreting Community Law. *The Common Market Law Review* 38: 159.
Weale, A., and A. Williams. 1992. Between economy and ecology? The single market and the integration of environmental policy. *Environmental Politics* 1 (4): 45–64.
Wibisana, A.G. 2006. *Three principles of environmental law: The polluter-pays principle, the principle of prevention, and the precautionary principle*, 24–76. Northampton, UK: Edward Elgar.
Wilde, M. 2001. The EC Commission's white paper on environmental liability. *Journal of Environmental Law* 13 (1): 21–37.
Wilkinson, D. 1992. Maastricht and the environment: The implications for the EC's environment policy of the treaty on European Union. *Journal of Environmental Law* 4 (2): 221–239.
Williams, E.S., J. Panko, and D.J. Paustenbach. 2009. The European Union's REACH regulation: A review of its history and requirements. *Critical Reviews in Toxicology* 39 (7): 553–575.

Chapter 3
Literature Review on Corporate Environmental Responsibility

Following the discussion on Chap. 2, we saw that the European Union (EU) pursued a decisive and organized plan on improving environmental regulations and rules with the member countries. Over the course of the previous decades, EU officials proceeded on several regulatory interventions which shaped the current framework of corporate environmental responsibility (CER). Starting from the Paris 1972 convention, the European community has moved into more concrete regulations on environmental responsibility. These regulations were targeting key issues such as climate change, water management, air pollution, resource allocation, chemical reduction and even sustainable production and consumption, and were incorporated within business strategies along the way. The scope of this chapter is to provide a contemporary literature review on corporate environmental responsibility (CER) by offering a comprehensive theoretical framework and a timeline on its development within the corporate world. Also, the chapter will present a brief discussion on the motivation for firms to adopt CER activities via environmental management tools (EMTs). The final section of the chapter is devoted on discussing the main empirical findings on the literature regarding the association of CER with accounting and financial issues relating to financial performance, innovation, capital structure decisions, firm risk, cash holdings, payout policies and the quality of published accounting information (value relevance, earnings management, and accounting conservatism).

3.1 Introduction

The neoclassical view on corporations regarding their role in society, considers their responsibility for achieving objectives including maximizing profit and their accountability to stakeholders (Werther and Chandler 2010). Until a decade ago, corporate managers were primarily oriented toward the maximization of

P. Dimitropoulos, K. Koronios, *Corporate Environmental Responsibility, Accounting and Corporate Finance in the EU*, CSR, Sustainability, Ethics & Governance, https://doi.org/10.1007/978-3-030-72773-4_3

shareholders' profits. They considered that expenditures or investments related to stakeholders' value maximization were not causally related to profit maximization, at least in the short run, thus they were considered as costs to be minimized. More recently, however, business leaders have encountered increasing public concern regarding their firms' social responsibility and the sustainability of their business methods. For example, managers have faced growing public demand for firms to reduce the pollutants they produce and increase the degree to which they engage in environmentally responsible activities. In this vein, Corporate Environmental Responsibility (CER) has been increasingly gaining recognition as one of the most important factors in firms' long-term value and sustainability.

Recent studies have shown that most managers believe that Corporate Environmental Responsibility is a critical factor for their companies' long-term survival and value. Furthermore, many corporate managers claim that a significant investment in new technologies is required in order to achieve environmental sustainability within 5 years. Contemporary managers also realize that by engaging in socially responsible activities, they can generate favorable perceptions about their companies, thereby cultivating consumer loyalty and increasing sales (Mohr et al. 2001). Corporate Environmental Responsibility (CER) is a new form of cooperation between business and environment. Companies promoting environmental objectives generate economic, political, and social advantages. Consumers and competitors perceive responsible organizations as friendly; consumers prefer their products and services and believe that they are the result of responsible and sustainable business conduct. Also, corporate environmental responsibility can contribute to the development of sustainable businesses, while social responsibility actions can be profitable business strategies. However, in contrast to the recognition and application of CER-related issues in industry, the concept of CER has received relatively little attention within academia (Kim et al. 2017).

3.2 Corporate Environmental Responsibility

The concept of Corporate Responsibility (CR) is well established in international management literature and has a social and an environmental aspect (Lo et al. 2008). According to Matten and Moon (2008), Corporate Responsibility means all clearly oriented and communicated policies and practices of firms that reflect business responsibility for some of the wider societal and ecological good.

Since the industrial society, the environmental problem has been increasingly highlighted: ecological environment deterioration, unusual global climate, resources conflict upgrade, international trade disputes, and energy crisis. Statistics indicates that more than one-third of the world annual industrial water and field irrigation water is polluted and more than 157 kinds of waste pollutant are found in industrial waste (Wang 2010). Such significant changes in global climate and environmental conditions result in a growing public awareness of the inherent element of Corporate Responsibility, making Corporate Environmental Responsibility (CER) an important topic for both the business world and academic literature (Gunningham 2009; Hart 1995; Starik and Rands 1995).

Corporate Environmental Responsibility deals with a company's relationship with the environment. It contains the obligations of decision makers to take responsible actions which aim to protect and improve the environment as a whole, and which are also in line with their own interests (Huckle 1995). Gunningham (2009) defines CER as "*practices that benefit the environment (or mitigate the adverse impact of business on the environment) that go beyond those that companies are legally obliged to carry out*". This notion is in line with the one's of World Business Council for Sustainable Development, which states that Corporate Environmental Responsibility encompasses "*environmentally friendly actions not required by law, also referred to as going beyond compliance, the private provision of public goods, or voluntarily internalizing externalities*" (World Business Council for Sustainable Development 2000).

The concept of Corporate Environmental Responsibility (CER) has been developed as one particular dimension of Corporate Social Responsibility (CSR). Consequently, CER is now recognized as an important subset of CSR (El Ghoul et al. 2011; Jo et al. 2015). Kitzmueller and Shimshack (2012) explain that CER is the outcome dimension of CSR, suggesting that the former is more specific and less abstract than the latter. Given its specificity, academics argue that one can measure CER more precisely than the vague concept of CSR. They also find that stakeholders' environmental preferences make CER an important part of corporate strategy and that CER investments are significantly related to corporate profits and sustainability. Research proves that enterprises can find the development opportunity in the corporate environmental responsibility (CER) practice (Schmidheiny and Timberlake 1992). Therefore, enterprises, the main force of industrial development and social progress, take an important role to protect the environment.

Despite the clear importance of CER (as a subdomain of the CSR index) academic research, however, has largely neglected its study. Few studies explore CER-related issues, and unlike the consensus that has emerged regarding the positive effect of CSR on firm performance or value, extant CER studies have thus far failed to provide consistent findings concerning CER's role and effectiveness. For example, Derwall et al. (2005), Konar and Cohen (2001), Miles and Covin (2000), and Russo and Fouts (1997) all argue that environmental management (e.g., reduction of pollution emissions) and environmental responsibility positively affect the future performance of firms. Other studies, however, claim that CER does not increase firms' values and is often abused by managers in order to build personal reputations (Brammer et al. 2006; Walley and Whitehead 1994). Further, prior studies on CER focus primarily on the roles and results of CER activities. Despite the fundamentality and scholastic immediacy of the determinants of CER, very few studies explore these issues. In short, the empirical study of CER studies is at an early stage.

Scholarly literature highlights some contradictions on the impact of corporate social responsibility. As Friedman (2007) contends "*an action counts as an act of CSR only if it is unprofitable. Socially beneficial actions that increase profits are merely hypocritical window dressing*". On the other hand, Baron (2009) differentiates between corporate social responsibility (unprofitable and driven by altruistic motives) and strategic corporate social responsibility (profitable). Reinhardt and Stavins (2010), emphasize the idea that corporate social responsibility requires sacrificing the profit for the social interest. A transition occurred from the idea that "*the only social responsibility of a company is to use its resources and engage in business designed to increase profits, while obeying the rules of the game*" (Friedman 2007) to the idea that caring for the environment and for the people should be a matter of concern. Corporate social responsibility is an umbrella for sustainable practices and the adoption of "green" management is often done due to strategic reasons (Babiak and Trendafilova 2011).

3.2.1 Corporate Environmental Responsibility (CER): Timeline Development

Starting with 1997, the year the Kyoto protocol was initially adopted until the end of 2010 when the Cancun Climate Change Conference took place, comprises the evolvement of the Corporate Environmental Responsibility (CER) notion coverage. The Kyoto protocol, aimed at fighting global warming, was a milestone event in environmental protection and triggered a growing perception of environmental issues along with increasing environmental responsibility (Brewer 2008; Kolk 2005). The Cancun Climate Change Conference encompassed a 14-year time span with the aim to provide a framework for climate change mitigation, causing an even faster growing recognition of environmental problems and the ongoing consequences (UNFFC Reports 2010). Wheeler and Elkington (2001), being known for his interest in corporate social responsibility and sustainable development, outlined three stages in the emergence and development of the concept of corporate social responsibility interacting with the environment. Each stage highlights the on-going construction of the concept of corporate environmental responsibility:

(a) The period between 1960 and 1978 saw numerous changes in the legislation of Western states, changes aimed at reducing the environmental consequences of economic activities and encouraging the responsible use of natural resources. International environmental standards were created, and all businesses were to comply with them.
(b) A shift towards sustainable consumption of natural resources and "green" production occurred between 1980 and 1999. This moment coincides with the emergence and international expansion of the concept of sustainable development.
(c) After 1999, new concerns have marked the business world: developing social responsibility practices, relations with the stakeholders, improving corporate governance practices and implementing social responsibility strategies as integral parts of the overall business strategy.

3.2.2 Why Do Companies Engage in Corporate Environmental Responsibility (CER)

We can explain the motivations of organizations to engage in CER related actions by focusing on frontier analysis within theoretical economic research. Specifically, reducing pollution and costs creates opportunities for organizations and the environment, the new generation of "green" consumers is willing to pay a higher price for clean products, businesses tend to become proactive and react before public pressure occurs. It is clear that corporate environmental responsibility can cover a wide range of ecological and environmental objectives that can affect business decisions and policies (DesJardins 1998).

Some organizations believe that social responsibility actions regarding the environment lead to lower profits. "*This simplistic perception implies that, because of competition, a good manager has no other option in the market economy than to buy as cheap and sell as expensive as possible. The existence of a legal framework to comply with is only acknowledged without enthusiasm: under the law, all is allowed to reach the sole purpose of any serious business—profit maximization*" (Ganescu and Dindire 2014).

Conversely, supporters of sustainable business practices argue that corporate environmental responsibility is inextricably linked to achieving important long-term profits. "*The relationship between socially responsible activities and profitability may be best characterized as some firms will generate long-term profits from some socially responsible activities some of the time*" (Reinhardt and Stavins 2010).

Achieving high environmental performance is possible when an environmental management system is in place and allows the development, implementation, coordination and monitoring of activities with environmental impact, in order to reach two objectives: compliance with standards and reduce waste. This idea is supported by other studies that show significant relationships between the presence of an environmental management system and increased environmental performance leading to reduced costs, increased product quality and reduced waste (Melnyk et al. 2003).

3.3 Corporate Environmental Responsibility (CER) and Environmental Management Tools (EMT)

Pollution is a global threat that has challenged policy makers and firms to reduce the environmental effects of economic activities (Sarkis 2001; Shrivastava 1995). Since the mid-1990s, numerous companies across the globe have adopted a wide range of environmental management tools (EMTs)—including environmental management systems (EMSs) based on ISO 14001 standards or certain European regulations (Eco Management and Audit Scheme [EMAS]), as well as other environmental tools based on numerous ISO standards ranging from eco-labels and declarations to carbon and water footprints—to implement green processes and to develop environmentally friendly products.

In recent decades, the European Union has confronted the challenge of integrating environmental sustainability with economic goals while encouraging European manufacturing to exploit opportunities to innovate. Firm awareness of the importance of this factor seems to be in a transition phase. As a result of perceiving environmental responsibility as an element that influences business performance, many firms are recognizing their environmental efforts as an important strategic factor. Consequently, a growing number of companies consider expenses to address environmental responsibility not only as a cost but also as an investment that can affect long-term corporate profit. At the same time environmental issues are a part of corporate strategy decisions. From this perspective, it seems that companies are developing a more proactive attitude towards environmental issues, perceiving them as business opportunities rather than as burdens.

One of the many challenges within the field of Corporate Environmental Responsibility (CER) is to select and adopt the most appropriate Environmental Management Tools (EMT) to improve environmental performance. The adoption of EMSs as frameworks for integrating corporate environmental protection policies, actions and practices is currently growing among both domestic and multinational companies around the world. Many studies focus on formalized EMSs and their ability to improve organizational environmental performance. In this regard, because some EMSs require strong employee participation and environmental training programs, many firms report increased employee awareness of the environmental aspects of their jobs and their responsibilities in reducing negative environmental impacts.

An EMS is a structured management tool enabling an organization to identify, control and reduce the negative environmental impact of its activities to achieve legal compliance and pollution/waste reduction (Sayre 1996). An EMS involves a sequence of steps ranging from devising an articulated environmental policy and commitments to implementing programs, plans and activities to improve the effectiveness and efficiency of environmental management. This approach requires the integration of environmental issues into every aspect of business management (Tinsley and Pillai 2012). When adopting an environmental management approach, organizations choose the approach best suited to their needs and goals by using a comprehensive toolkit of diversified voluntary instruments based mainly on international standards. The most significant standards in the management and assessment of environmental issues used individually or in an integrated mode can be classified either as "system standards" (EMSs based on ISO 14001:2015 or EMAS based on EU Regulation 1221/2009) or as "product standards" (ISO 14020 series for Environmental Product Label, ISO 14040 series for Life Cycle Assessment, ISO 14067 for the quantification and communication of the carbon footprint, and ISO 14046 related to water footprint assessment).

However, the synergistic integration of various EMTs at the product, process and organizational levels is now used more frequently than before to exploit similarities and potential synergies because the product life cycle perspective can enrich and broaden the scope of an EMS. This innovative approach—also known as the product-oriented environmental management system—can improve the environmental performance of products and is considered an integral part of operations and strategies. The successful implementation by means of systematic procedures requires developing and inserting the system into a routine organizational activity, such as an EMS (Salomone et al. 2013).

The current literature highlights the positive influence of well-designed EMSs on the environmental performance of organizations (Iraldo et al. 2009) and on a growing commitment towards environmental improvement (Darnall et al. 2010), as well as on additional benefits, including enhanced reputation and improved opportunities in international markets (Melnyk et al. 2003). Other studies have found that implementing comprehensive EMTs is particularly important in motivating organizations to adopt measures that improve environmental performance related to resource input efficiency and, thereby, in waste generation and in pollution reduction (Schucht 2000). In addition, these tools can assist managers to identify economical ways of meeting environmental goals, which can result in improved global performance (Johnstone and Labonne 2009). EMTs typically generate information about regulatory requirements and internal environmental practices, supporting internal agency control issues that may lead to negative environmental impacts (Potoski and Prakash 2005; Grolleau et al. 2012), and may help risk management regarding environmental accidents (Johnstone 2004). All these systems constitute CER a vital strategic feature of modern corporations, which has direct impact on firms' accounting and financial decisions. The following section is devoted on discussed this connection in greater detail.

3.4 Empirical Evidence on CER, Accounting and Financial Issues

The current section is devoted on the discussion of the main empirical findings evident on the literature regarding the association and impact of CER on the main issues that will be further examined on the following two parts of the book. Specifically, we will provide a brief discussion on CER determinants and its association with corporate financial performance, R&D activities, capital structure, systematic risk, cash holding decisions and dividend policy, CER value relevance and contribution to accounting quality (earnings management mitigation and accounting conservatism).

3.4.1 Corporate Determinants of CER Performance

Firm size has proved to have a positive impact on CER performance because larger companies attract more publicity and attention by various stakeholders and also are considered to be highly polluters relative to smaller firms thus requiring enhanced legitimacy for their operations, leading them to report and engage on CER activities. Also, larger firms have more resources on their disposal to engage and even sustain CER activities and thus establish strong ties with their stakeholders (Monteiro and Aibar-Guzmán 2010; Tang and Luo 2016, Giannarakis 2014; Dienes et al. 2016; Crisóstomo et al. 2019). In addition, firms' financial leverage is another corporate characteristic affecting firms' motive to engage in CER activities. Lenders tend to exert pressure to the firm to meet their information needs so as to guarantee the repayment of their funds. Thus, CER reporting could be perceived as value relevant for debt holders since it can mitigate their information asymmetries and the uncertainty of debt financing. Empirical evidence provide contradictory results with some studies suggesting a positive association while other a negative impact of leverage on CER. So this factor is still remaining an open empirical question in our case. Moreover, profitability is another determinants of CER performance. CER performance can improve corporate image, brand value and further advance firms' relations with stakeholders, thus favoring firms' market share and profitability. Hannifa and Cooke (2005) and Giannarakis (2014) provide evidence that profitable firms tend to report more CER related information to the market so as to legitimize their existence and even avoid the undervaluation of their market value. Another significant financial determinant of CER performance is growth opportunities of firms. High growth firms need to signal to external investors, stakeholders and financial markets their sustainability and strategic CER focus so as to gain easier access to financing. According to Padgett and Galan (2010) and Crisóstomo et al. (2019) this will facilitate their ongoing activities and allow them to sustain their future profitability and prospects.

Beyond the abovementioned financial variables, corporate governance characteristics have been proven in the literature as significant determinants of CER activities and performance. The existence of a dual role of the CEO acting also as a board chairman has been proved to impact negatively on CER performance. Giannarakis et al. (2014), Giannarakis (2014), Michelon and Parbonetti (2012) and Gul and Leung (2004) suggest that CEO duality is negatively associated with voluntary CER disclosures and firms' engagement on CER related activities, because such CEOs control the corporate agenda and diminish directors' monitoring role. This fact could lead to lower levels of CER related disclosures and performance. Board size is another corporate governance characteristic associated with CER disclosures and performance. Esa and Mohd Ghazali (2012), Rahman et al. (2011) and Wang (2017) indicate a positive impact of board size on CER disclosures, since larger boards are able to perform more productive and efficient discussion on CER issues and thus enhance related corporate activities. Finally, the existence of a CSR or sustainability committee within the board of directors has been proved as an important determinant of CER performance. The establishment of such committees are considered as a mechanism to adhere to stakeholders and even monitor managers towards this end. According to Jain and Zaman (2019), the presence of a CER (or sustainability) committee could provide valuable resources to the board members and managers for overseeing CER related investments and the overall CER strategic focus of the firm.

Thus, such firms are expected to perform better on environmental issues and comply effectively with respective rules and regulations.

3.4.2 CER and Financial Performance

The association of CER activities with corporate financial performance (CFP) has been examined through the frameworks of different theories. The resourced-based view of the firm is one of the most commonly used theories suggesting a positive association of CER with CFP. This association is justified by the fact that CER activities create valuable resources and provide significant risk mitigation abilities in cases of adverse shocks (Russo and Fouts 1997; Flammer 2013; Shiu and Yang 2017). Moreover, signaling theory is an additional theoretical framework positing that CER activities and related disclosures are disseminated to the markets so as to improve stakeholders' awareness and so firms can gain the necessary legitimacy on their operations (Wei et al. 2017). A large number of empirical studies on the issue point to a positive association between CER and CFP. Jiang et al. (2018), Chen et al. (2018) and Fiadrino et al. (2019) have documented a positive impact of CER on CFP in Chinese businesses and European firms, arguing that CER generates competitive resources leading to competitive advantages and enhanced financial performance.

3.4.3 CER and R&D Activities

Following the previous sub-section, the CSR-CFP relation has been a topic of extant examination in academic literature and so the association between R&D activities and CFP. However, the link between CSR-CFP based on the corporate decisions on R&D has not been well explored. Empirical evidence and theoretical assumptions so far have pointed towards two opposing frameworks suggesting that (1) R&D exercises a mediation effect on the CSR-CFP relation and produces a positive impact on CFP but only for firms with significant CER activities, while (2) the opposing view argues that R&D plays a substitution role on the CSR-CFP relation pointing towards a negative impact of R&D on the CSR-CFP association. Luo and Bhattacharya (2006) argue that large investments in R&D activities could create favorable stances regarding corporate motivation to engage in CER activities, so pointing towards a positive impact of R&D on the CER-CFP association. On the contrary, Hull and Rothenberg (2008) document that firms which gain competitive advantages and differentiate themselves from competition through R&D activities have less incentives to engage in CER undertakings in order to achieve such differentiation, thus pointing towards a negative impact of R&D on the CER-CFP association. Now which of the two opposing theoretical views is true for the EU corporations remains an open empirical question.

3.4.4 CER and Corporate Capital Structure

Existing empirical research on the impact of CER on capital structure decisions has identified two streams of theoretical backgrounds and evidence pointing to a positive and a negative association between those variables. Yang et al. (2018) document that CER activities mitigate firms' idiosyncratic risk, information asymmetries, while simultaneously improve corporate reputation and moral capital among stakeholders. Alternatively, Yang et al. (2018) suggest that CER activities contribute to the reduction of long-term risk so under these circumstances creditors are willing to finance firms on a long-term basis. Thus, high CER performing firms tend to raise more debt relative to equity capital so a positive association is evidenced between CER and long-term debt. On the contrary, agency cost theory suggests that CER activities can be utilized in order managers and firm insiders to identify themselves to the public as socially responsible and enhance their reputation at the expense of shareholders. Godfrey (2005) and Benlemlih (2017) suggest that CER investments reduce the cost of debt only when they are below an optimal level. So the use of less long-term debt can help shareholders to control potential overinvestment on CER activities by the managers. According to Benlemlih (2017), high CER firms have higher stability, reduced perception of risk, and resort to short-term debt and equity instead of long-term debt financing when they want to finance long term investments. This means the high CER performing firms may be associated with lower levels of leverage so as to reduce stakeholders' concerns.

3.4.5 CER and Systematic Risk

The main theoretical streams of literature regarding the association between CER and systematic risk are divided in half, were one stream considers CER as value enhancing for shareholders and stakeholders, while on the other hand CER is considered as a value destroying activity. Starting from the first stream of the literature, Godfrey (2005) argues that CER engagement creates a moral capital which provides shareholders a protection against bad reputation and adverse effects of economic activities. Moreover, Cai et al. (2016) argue that ethical managers use CER activities as a communication mechanism in order to improve information transparency regarding firm's strategic focus and responsible investments. This fact can further improve firm's transparency contributing to reduced information asymmetries between the firm and its main stakeholders, so mitigating the investor's perception of firm risk (Dhaliwal et al. 2011; Suto and Takehara 2018). However, the opposite scenario could also be true, since shareholders and creditors could perceive CER investments and activities as a waste of valuable firm resources, enhancing firm's uncertainty about future operations and potential cash flows, leading to increased systematic risk. As discussed by Barnea and Rubin (2010), channeling corporate resources towards CER activities diverts those resources from investment activities with positive NPVs (building new products or services, improving production lines, or even taking advantage of profitable investment opportunities). This can reduce firm's future competitiveness making it more vulnerable to financial frictions or takeover activities by the competitors. Consequently, CER engagement could be translated into higher risk (or uncertainty) by the stakeholders. Evidence by Cespa and Cestone (2007) suggest that firms which systematically engage in environmentally responsible investments could be associated with higher managerial entrenchment which in turn increases the perceived risk of the firm (Nguyen and Nguyen 2015). Whether CER performance enhanced or diminishes systematic risk in the EU corporate environment remains an open empirical question that will be addressed on Chap. 8 of the book.

3.4.6 CER, Cash and Dividends

Empirical evidence on the impact of CER on corporate cash holding and dividend policy decisions are contradictory for both issues. Specifically, on cash holding decisions the CER motive and the precautionary motive of cash holdings are the main theoretical frameworks proposed to explain how CER decisions impact on cash holding motivation. Under the CER motive of cash holdings, Arouri and Pijorlet (2017), Huang et al. (2019) and Cheung (2016) argue that environmentally responsible firms are devoted in adhering to stakeholders' needs and demands, and in order to achieve such strategy they are more in need to hold cash and cash equivalents in their balance sheets. On the contrary, the precautionary argument for cash holdings dictates that firms tend to hold more cash when they face higher environmental uncertainty and risks, in order to be able to face those difficulties without liquidating assets or raising external financing, thus pointing to a negative association between CER and cash on the balance sheet.

Regarding the association between CER and dividend payments, still two theoretical perspectives are offered to the literature, Firstly, the agency theory of the firm and the signaling theory both argue that dividends are useful in controlling agency costs associated with free cash flows (Jensen 1986) by reducing the cash available to managers' discretion. CER investments and activities moderate this agency problem by the fact that CER performing firms finance activities that interest different stakeholder groups. Also, signaling theory complements the positive association between CER and dividends (as argued by the agency theory) documenting that firms devoted to the protection of the environment are keen on creating their wealth within ethical and sustainable limits, and so they are focused in satisfying the interests of financial and non-financial stakeholders. A prosperous dividend payout policy can create reputational benefits by communicating the firm's fairness and ethical stance on wealth distribution. On the contrary, the substitution view on payout decisions argues for a negative association between CER and dividend payments, since it considers dividends as a tool for reputation building through signals to the market about the decision of the firm not to waste free cash flows

(Hasan and Habib 2020). So higher social capital (with improved social and environmental behavior) can mitigate the free cash flow problem and thus substitute the need to build corporate reputation through cash dividend payments. The analysis on Chap. 9 will try to disentangle those issues within the EU corporate setting.

3.4.7 Value Relevance of CER Performance

Empirical evidence on the contribution of CER on corporate market valuation still remains an open empirical question since there are conflicting evidence on the matter. Gregory et al. (2016), Bollen (2007) and El Ghoul et al. (2011), argue that CER activities could enhance the value of the firm because many investors have a multi-facet utility function which dictates that their utility is not determined by simple financial measures (profits or dividends) but also from environmental responsibility dimensions of corporate activity. CER engagement creates a reputational capital which helps the firm to gain competitive advantages, improve organizational capabilities while communicating their enhanced future prospects to the market. So high CER performing firms tend to be perceived as caring less future risk (litigation and environmentally related costs) and so CER disclosures are financially relevant for investors (Suto and Takehara 2018). However, the opposite impact of CER on market valuation is also evident in the literature. According to Alipour et al. (2019) and Williams (1999), managers may use CER activities as a way to pursue their personal goals and interests and even to legitimize their decisions as giving the picture of a responsible and environmentally aware firm. Also, CER could also be used to mask the true economic condition of the firm (via earnings manipulation) leading to lower quality of financial reporting. Salewski and Zülch (2014) provide evidence of an opportunistic managerial behavior for firms engaging in CSR activities. So the association between CER performance and earnings quality will be further examined on the third part of this book.

3.4.8 CER, Earnings Management and Accounting Conservatism

Several arguments to the accounting literature are supporting that high CER performing firms are more probable to report higher quality financial information. Empirical evidence by Yoon et al. (2019), Kim et al. (2019) and Palacios-Manzano et al. (2019) verify these arguments within different country settings and economic sectors. In addition, more environmentally responsible firms are less probable to experience self-interest and discretionary managerial behavior which may contradict shareholders' interests. Also highly CER performing firms tend to be long-term oriented, less focused on short-term financial goals and disclose more non-financial voluntary information (Litt et al. 2014). So under this framework, CER performing firms are expected to be associated with higher transparency and thus will have established a corporate culture that would not likely resort to earnings manipulation in order artificially enhance their financial performance.

Moving on to accounting conservatism, theoretical underpinnings on the literature propose two different potential associations between CER and conservatism. The first proposal asserts that CER and conservatism are negatively associated because CER performing firms can signal lower levels of managerial opportunism to the markets and so the demand for conditional conservatism from contracting parties (stakeholders) to mitigate managerial discretion may be lower. As documented by Burke et al. (2020), Cho et al. (2020) and Hong (2020), CER performance reduces the demand for conservative financial reporting by reducing stakeholders' concerns relative to managerial opportunism and by minimizing information asymmetries in the markets. The second theoretical underpinning posits a positive association between CER and accounting conservatism. According to Fombrun and Shanley (1990), Kim et al. (2012) and Cheng and Kung (2016), the provision of transparent and reliable financial information is perceived as a social responsibility of corporations towards their stakeholders. Thus, conservative firms could be motivated to invest more on CER related activities. Empirical evidence by Pyo and Lee (2013), Cheng and Kung (2016), and Francis et al. (2013) support the above-mentioned arguments within different markets. Whether earnings management and accounting conservatism are associated to CER performance in the EU market, is an empirical examination that will be addressed on Chaps. 11 and 12 of the book.

3.5 Conclusion

Concluding the chapter, we can argue that CER is a valuable strategic stool for company policies and activities, since it contributes decisively to the achievement of its goals and objectives. Its actions are multilevel, affecting all business operations, while the benefits of CER implementation in recent years have become increasingly apparent. Corporate environmental responsibility is not a static and unambiguous concept. Its actions can incorporate innovation and adaptability to meet modern requirements, and through them it contributes to both improving financial results and increasing the value of the business. As CER grows rapidly it is necessary to make clear its relationship to the financial results of the company and other fundamental measures of the firm. To this end, a number of tools and indicators are taken into account, which aim to facilitate the examination and study of CER and business performance, in order to demonstrate how these two issues, go hand in hand and complement each other.

The current chapter offered a comprehensive review on corporate environmental responsibility (CER) by offering a sound theoretical framework and the process of CER development through time, starting from a pure voluntary or philanthropic perception of CER, into a more concrete strategic focus. Also, the chapter discussed about the motives of firms to engage and adopt CER activities via environmental management tools (EMTs). These tools have been evolved into important and useful allies to the enhancement of CER performance and reporting. They are considered as an innovative approach which can improve the environmental performance of products and is considered an integral part of operations and strategies. Finally, at the last section of the chapter we discussed the main empirical findings on the literature regarding the association of CER with accounting and financial issues relating to financial performance, innovation, capital structure decisions, firm risk, cash holdings, payout policies and the quality of published accounting information, specifically value relevance, earnings management, and accounting conservatism. All these topics will be further analyzed on the next two sections of the book by providing a robust empirical analysis using a large sample of EU domiciled corporations over a period of 15 years, thus yielding up-to-date inferences on those issues.

References

Alipour, M., M. Ghanbari, B. Jamshidinavid, and A. Taherabari. 2019. The relationship between environmental disclosure quality and earnings quality: A panel study of an emerging market. *Journal of Asia Business Studies* 13 (2): 326–347.

Arouri, M., and G. Pijorlet. 2017. CSR performance and the value of cash holdings: International evidence. *Journal of Business Ethics* 140: 263–284.

Babiak, K., and S. Trendafilova. 2011. CSR and environmental responsibility: Motives and pressures to adopt green management practices. *Corporate Social Responsibility and Environmental Management* 18 (1): 11–24.

Barnea, A., and A. Rubin. 2010. Corporate social responsibility as a conflict between shareholders. *Journal of Business Ethics* 97: 71–86.

Baron, D.P. 2009. A positive theory of moral management, social pressure, and corporate social performance. *Journal of Economics & Management Strategy* 18 (1): 7–43.

Benlemlih, M. 2017. Corporate social responsibility and firm debt maturity. *Journal of Business Ethics* 144 (3): 491–517.

Bollen, N. 2007. Mutual fund attributes and investor behavior. *Journal of Financial and Quantitative Analysis* 42: 683–708.

Brammer, S., C. Brooks, and S. Pavelin. 2006. Corporate social performance and stock returns: UK evidence from disaggregate measures. *Financial Management* 35 (3): 97–116.

Brewer, T.L. 2008. Climate change technology transfer: A new paradigm and policy agenda. *Climate Policy* 8 (5): 516–526.

Burke, Q., P.-C. Chen, and G.L. Lobo. 2020. Is corporate social responsibility performance related to conditional accounting conservatism? *Accounting Horizons*. https://doi.org/10.2308/horizons-18-111.

Cespa, G., and G. Cestone. 2007. Corporate social responsibility and managerial entrenchment. *Journal of Economics and Management Strategy* 16 (3): 741–771.

Chen, F., T. Ngniatedema, and S. Li. 2018. A cross-country comparison of green initiatives, green performance and financial performance. *Management Decision* 56 (5): 1008–1032.

Cheng, C.-L., and F.-H. Kung. 2016. The effects of mandatory corporate social responsibility policy on accounting conservatism. *Review of Accounting and Finance* 15 (1): 2–20.

Cheung, A. 2016. Corporate social responsibility and corporate cash holdings. *Journal of Corporate Finance* 37: 412–430.

Cho, S.Y., P.K. Kang, C. Lee, and C.K. Park. 2020. Financial reporting conservatism and voluntary CSR disclosure. *Accounting Horizons*. https://doi.org/10.2308/hotizons-17-093.

Crisóstomo, V.L., F. de Souza Freire, and M.F. de Oliveira Freitas. 2019. Determinants of corporate sustainability performance: Evidence from Brazilian panel data. *Social Responsibility Journal*. https://doi.org/10.1108/SRJ-04-2018-0102.

Darnall, N., I. Henriques, and P. Sadorsky. 2010. Adopting proactive environmental strategy: The influence of stakeholders and firm size. *Journal of Management Studies* 47 (6): 1072–1094.

Derwall, J., N. Guenster, R. Bauer, and K. Koedijk. 2005. The eco-efficiency premium puzzle. *Financial Analysts Journal* 61 (2): 51–63.

DesJardins, J. 1998. Corporate environmental responsibility. *Journal of Business Ethics* 17 (8): 825–838.

Dhaliwal, D., O. Li, A. Zhang, and Y. Yang. 2011. Voluntary nonfinancial disclosure and the cost of equity capital: The initiations of corporate social responsibility reporting. *The Accounting Review* 86 (1): 59–100.

Dienes, D., R. Sassen, and J. Fisher. 2016. What are the drivers of sustainability reporting? A systematic disclosure and corporate governance characteristics in Malaysian public listed companies. Social disclosures. *Journal of Accounting and Public Policy* 23 (5): 351–379.

El Ghoul, S., O. Guedhami, C.C. Kwok, and D.R. Mishra. 2011. Does corporate social responsibility affect the cost of capital? *Journal of Banking & Finance* 35 (9): 2388–2406.

Esa, E., and N.A. Mohd Ghazali. 2012. Corporate social responsibility and corporate governance in Malaysian government-linked companies. *Corporate Governance* 12 (3): 202–305.

Fiadrino, S., A. Devalle, and V. Cantino. 2019. Corporate governance and financial performance for engaging socially and environmentally responsible practices. *Social Responsibility Journal* 15 (2): 171–185.

Flammer, C. 2013. Corporate social responsibility and shareholder reaction: The environmental awareness of investors. *Academy of Management Journal* 56 (3): 758–781.

Fombrun, C., and M. Shanley. 1990. What's in a name? Reputation building and corporate strategy. *Strategic Management Journal* 33 (2): 233–258.

Francis, R.N., S. Harrast, J. Mattingly, and L. Olsen. 2013. The relation between accounting conservatism and corporate social performance: An empirical investigation. *Business and Society Review* 118 (2): 193–222.

Friedman, M. 2007. The social responsibility of business is to increase its profits. In *Corporate ethics and corporate governance*. Berlin, Heidelberg: Springer.

Ganescu, C., and L. Dindire. 2014. Corporate environmental responsibility - A key determinant of corporate reputation. *Computational Methods in Social Sciences* 2 (1): 48.

Giannarakis, G. 2014. Corporate governance, and financial characteristic effects on the extent of corporate social responsibility disclosure. *Social Responsibility Journal* 10 (4): 569–590.

Giannarakis, G., G. Konteos, and N. Sariannidis. 2014. Financial, governance and environmental determinants of corporate social responsible disclosure. *Management Decision* 52 (10): 1928–1951.

Godfrey, P.C. 2005. The relationship between corporate philanthropy and shareholder wealth: A risk management perspective. *Academy of Management Review* 30: 777–798.

Gregory, A., J. Whittaker, and X. Yan. 2016. Corporate social performance, competitive advantage, earnings persistence and firm value. *Journal of Business Finance and Accounting* 43 (1–2): 3–30.

Grolleau, G., N. Mzoughi, and S. Pekovic. 2012. Green not (only) for profit: An empirical examination of the effect of environmental-related standards on employees' recruitment. *Resource and Energy Economics* 34 (1): 74–92.

Gul, F.A., and S. Leung. 2004. Board leadership, outside directors' expertise and voluntary corporate disclosures. *Journal of Accounting and Public Policy* 23 (5): 351–379.

Gunningham, N. 2009. The new collaborative environmental governance: The localization of regulation. *Journal of Law and Society* 36 (1): 145–166.

Hannifa, R.M., and T.E. Cooke. 2005. The impact of culture and corporate governance on corporate social reporting. *Journal of Accounting and Public Policy* 24 (5): 391–430.
Hart, O. 1995. Corporate governance: Some theory and implications. *The Economic Journal* 105 (430): 678–689.
Hasan, M.M., and A. Habib. 2020. Social capital and payout policies. *Journal of Contemporary Accounting and Economics* 16: 100183.
Hong, S. 2020. Corporate social responsibility and accounting conservatism. *International Journal of Economics and Business Research* 19 (1): 1–18.
Huang, H.H., C. Liu, and L. Sun. 2019. Chemical releases and corporate cash holdings. *International Review of Financial Analysis* 64: 159–173.
Huckle, J. 1995. Using television critically in environmental education. *Environmental Education Research* 1 (3): 291–304.
Hull, C.E., and S. Rothenberg. 2008. Firm performance: The interactions of corporate social performance with innovation and industry differentiation. *Strategic Management Journal* 29 (7): 781–789.
Iraldo, F., F. Testa, and M. Frey. 2009. Is an environmental management system able to influence environmental and competitive performance? The case of the eco-management and audit scheme (EMAS) in the European union. *Journal of Cleaner Production* 17 (16): 1444–1452.
Jain, T., and R. Zaman. 2019. When boards matter: The case of corporate social irresponsibility. *British Journal of Management*. https://doi.org/10.1111/1467-8551.12376.
Jensen, M.C. 1986. Agency costs of free cash flow, corporate finance and takeovers. *American Economic Review* 76 (2): 323–339.
Jiang, Y., X. Xue, and W. Xue. 2018. Proactive corporate environmental responsibility and financial performance: Evidence from Chinese energy enterprises. *Sustainability* 10: 964.
Jo, H., H. Kim, and K. Park. 2015. Corporate environmental responsibility and firm performance in the financial services sector. *Journal of Business Ethics* 131 (2): 257–284.
Johnstone, M.J. 2004. Patient safety, ethics and whistleblowing: A nursing response to the events at the Campbelltown and Camden Hospitals. *Australian Health Review* 28 (1): 13–19.
Johnstone, N., and J. Labonne. 2009. Why do manufacturing facilities introduce environmental management systems? Improving and/or signaling performance. *Ecological Economics* 68 (3): 719–730.
Kim, Y., M.S. Park, and B. Wier. 2012. Is earnings quality associated with corporate social responsibility? *The Accounting Review* 87 (3): 761–796.
Kim, H., K. Park, and D. Ryu. 2017. Corporate environmental responsibility: A legal origins perspective. *Journal of Business Ethics* 140 (3): 381–402.
Kim, S.H., P. Udawatte, and J. Yin. 2019. The effects of corporate social responsibility on real and accrual-based earnings management: Evidence from China. *Australian Accounting Review* 23 (3): 580–594.
Kitzmueller, M., and J. Shimshack. 2012. Economic perspectives on corporate social responsibility. *Journal of Economic Literature* 50 (1): 51–84.
Kolk, A. 2005. Environmental reporting by multinationals from the triad: Convergence or divergence? *Management International Review* 45: 145–166.
Konar, S., and M.A. Cohen. 2001. Does the market value environmental performance? *Review of Economics and Statistics* 83 (2): 281–289.
Litt, B., D. Sharma, and V. Sharma. 2014. Environmental initiatives and earnings management. *Managerial Auditing Journal* 29 (1): 76–106.
Lo, C.W., C.P. Egri, and D.A. Ralston. 2008. Commitment to corporate, social, and environmental responsibilities: An insight into contrasting perspectives in China and the US. *Organization Management Journal* 5 (2): 83–98.
Luo, X., and C.B. Bhattacharya. 2006. Corporate social responsibility, customer satisfaction and market value. *Journal of Marketing* 70 (1): 1–18.

Matten, D., and J. Moon. 2008. "Implicit" and "explicit" CSR: A conceptual framework for a comparative understanding of corporate social responsibility. *Academy of Management Review* 33 (2): 404–424.

Melnyk, S.A., R.P. Sroufe, and R. Calantone. 2003. Assessing the impact of environmental management systems on corporate and environmental performance. *Journal of Operations Management* 21 (3): 329–351.

Michelon, G., and A. Parbonetti. 2012. The effects of corporate governance on sustainability disclosure. *Journal of Management and Governance* 16 (3): 477–509.

Miles, M.P., and J.G. Covin. 2000. Environmental marketing: A source of reputational competitive, and financial advantage. *Journal of Business Ethics* 23 (3): 299–311.

Mohr, L.A., D.J. Webb, and K.E. Harris. 2001. Do consumers expect companies to be socially responsible? The impact of corporate social responsibility on buying behavior. *Journal of Consumer Affairs* 35 (1): 45–72.

Monteiro, S., and B. Aibar-Guzmán. 2010. Determinants of environmental disclosure in the annual reports of large companies operating in Portugal. *Corporate Social Responsibility and Environmental Management* 17: 185–204.

Nguyen, P., and A. Nguyen. 2015. The effect of corporate social responsibility on firm risk. *Social Responsibility Journal* 11 (2): 324–339.

Padgett, R.C., and J.J. Galan. 2010. The effect of R&D intensity on corporate social responsibility. *Journal of Business Ethics* 93: 407–418.

Palacios-Manzano, M., E. Gras-Gil, and J.M. Santos-Jaen. 2019. Corporate social responsibility and its effect on earnings management: An empirical research on Spanish firms. *Total Quality Management*. https://doi.org/10.1080/14783363.2019.1652586.

Potoski, M., and A. Prakash. 2005. Green clubs and voluntary governance: ISO 14001 and firms' regulatory compliance. *American Journal of Political Science* 49 (2): 235–248.

Pyo, G., and H.-Y. Lee. 2013. The association between corporate social responsibility activities and earnings quality: Evidence from donations and voluntary issuance of CSR reports. *The Journal of Applied Business Research* 29 (3): 945–962.

Rahman, N.H.W.A., M.M. Zain, and N.H.Y.Y. Al-Haj. 2011. CSR disclosures and its determinants: Evidence from Malaysian government link companies. *Social Responsibility Journal* 7 (2): 181–201.

Reinhardt, F.L., and R.N. Stavins. 2010. Corporate social responsibility, business strategy, and the environment. *Oxford Review of Economic Policy* 26 (2): 164–181.

Russo, M.V., and P.A. Fouts. 1997. A resource-based perspective on corporate environmental performance and profitability. *Academy of Management Journal* 40 (3): 534–559.

Salewski, M. and H. Zülch. 2014. The association between corporate social responsibility (CSR) and earnings quality – Evidence from European blue chips. Available at: https://papers.ssrn.com/sol3/papers.cfm?abstract_id=2141768

Salomone, R., M.T. Clasadonte, M. Proto, and A. Raggi. 2013. *Product-oriented environmental management systems (POEMS)*. Dordrecht: Springer.

Sarkis, J. 2001. Manufacturing's role in corporate environmental sustainability - Concerns for the new millennium. *International Journal of Operations & Production Management.* 21 (5/6): 666–686.

Sayre, D.A. 1996. *INSDE ISO 14000: The competitive advantage of environmental management.* Boca Raton, FL: CRC Press.

Schmidheiny, S., and L. Timberlake. 1992. *Changing course: A global business perspective on development and the environment*. Vol. 1. Cambridge: MIT Press.

Schucht, S. 2000. *The implementation of the environmental management and eco-audit scheme (EMAS) regulation in France*, 29–31. Kraków: IMPOL.

Shiu, Y.M., and S.L. Yang. 2017. Does engagement in corporate social responsibility provide strategic insurance-like effects? *Strategic Management Journal* 38: 455–470.

Shrivastava, P. 1995. Environmental technologies and competitive advantage. *Strategic Management Journal* 16 (S1): 183–200.

Starik, M., and G.P. Rands. 1995. Weaving an integrated web: Multilevel and multisystem perspectives of ecologically sustainable organizations. *Academy of Management Review* 20 (4): 908–935.

Suto, M., and H. Takehara. 2018. *Corporate social responsibility and corporate finance in Japan*. Singapore: Springer Nature.

Tang, Q., and L. Luo. 2016. Corporate ecological transparency: Theories and empirical evidence. *Asian Review of Accounting* 24 (4): 498–524.

Tinsley, S., and I. Pillai. 2012. *Environmental management systems: Understanding organizational drivers and barriers*. London: Taylor & Francis.

UNFFC Reports. 2010. Kyoto protocol and Copenhagen summit. Retrieved from https://unfccc.int/resource/docs/2009/cop15/eng/11a01.pdf

Walley, N., and B. Whitehead. 1994. It's not easy being green. *Reader in Business and the Environment* 36 (81): 4.

Wang, H. 2010. Factor analysis of corporate environmental responsibility. *Environment, Development and Sustainability* 12 (4): 481–490.

Wang, M.C. 2017. The relationship between firm characteristics and the disclosure of sustainability reporting. *Sustainability* 9 (4): 624.

Wei, J.C., Z. Ouyang, and H.P. Chen. 2017. Well known or well liked? The effects of corporate reputation on firm value at the onset of a corporate crisis. *Strategic Management Journal* 38: 2103–2120.

Werther, W.B., Jr., and D. Chandler. 2010. *Strategic corporate social responsibility: Stakeholders in a global environment*. Los Angeles: Sage.

Wheeler, D., and J. Elkington. 2001. The end of the corporate environmental report? Or the advent of cybernetic sustainability reporting and communication. *Business Strategy and the Environment* 10 (1): 1–14.

Williams, S.M. 1999. Voluntary environmental and social accounting disclosure practices in the Asia-Pacific region: An international empirical test of political economy theory. *The International Journal of Accounting* 34 (2): 209–238.

World Business Council for Sustainable Development. 2000. Corporate social responsibility: Making good business sense [Online]. Accessed December 12, 2013, from http://www.wbcsd.org/web/publications/csr2000.pdf/

Yang, S., F. He, Q. Zhu, and S. Li. 2018. How does corporate social responsibility change capital structure? *Asia-Pacific Journal of Accounting and Economics* 25 (3–4): 352–387.

Yoon, B., B. Kim, and J.H. Lee. 2019. Is earnings quality associated with corporate social responsibility? Evidence from the Korean market. *Sustainability* 11: 4116.

Chapter 4
Firm-Specific Determinants of Corporate Environmental Responsibility

Following the discussion on Chap. 2, corporate environmental responsibility (CER) has been widespread among European firms during the last 15 years. Listed corporations in some countries have been engaged themselves more strategically on environmental activities relating to emission reduction, environmental innovative activities and resource usage for the protection of the environment, while firms in other countries lack behind but with significant improvements over the last 10 years. The scope of this chapter is to examine the firm-specific determinants of CER which relate to financial and governance variables. Moreover, the chapter will examine whether the countries' legal environment (code law vs common law countries) impacts the firm-specific determinants of CER.

4.1 Introduction

The issue of corporate environmental responsibility (CER) has gained increased attention by academics, researchers and regulators during the past 25 years. The development of CER is mainly attributed to the increased pressure that firms face from various stakeholders in order to adhere to general societal concerns (Monteiro and Aibar-Guzmán 2010). Practically, firms are operating within the society and create value for the society and their operation is under scrutiny by various stakeholder groups. Corporations' success is dependent on how efficiently they address several stakeholders concerns including environmental issues. This has led several corporations to devote themselves into responsible corporate activities (investments, relationships with the communities and environment protection). They also devote significant efforts to communicate these activities to the public by publishing a large amount of voluntary CER reports and information (Dienes et al. 2016).

According to Suto and Takehara (2018), the main theoretical perspective that explains corporate stakeholder engagement and CER in specific is stakeholder

P. Dimitropoulos, K. Koronios, *Corporate Environmental Responsibility, Accounting and Corporate Finance in the EU*, CSR, Sustainability, Ethics & Governance, https://doi.org/10.1007/978-3-030-72773-4_4

theory. Under this theory several perspectives of CER have been developed in the literature. The resource-based view (Russo and Fouts 1997; Hall 1992; Roberts 1992) suggests that CER activities require specific corporate resources both tangible (financial and physical assets) and intangible (human resources, knowledge and experience). Practically, the creation of sustainable stakeholder relationships requires the creation of intangible assets which in turn could help the firm gain a competitive advantage through enhanced stakeholder relationships which can protect the firm by mitigating conflicts between their stakeholders (Suto and Takehara 2018). This outcome is substantiated in the legitimacy theory of the firm (Patten 1992, 2005; Tang and Luo 2016) which argues that CER is associated to the social pressure that firms face regarding their environmental performance. This fact urges firms to increase their CER related reports and information to the public in order to gain legitimacy for their operations and decisions and consequently to mitigate the risk which arises from conflicts of interests between stakeholders. This fact in turn can reduce the bias of information among various investors and thus environmentally responsible firms can gain significant advantages in the financial markets (either through the reduction of financial risks, access to financing and enhancement of corporate market value) (Huang and Watson 2015).

Under this framework, there is an extensive accounting and corporate finance literature which tries to examine the determinant factors of CER (see the study by Dienes et al. 2016 for further analysis). This literature includes studies focusing on several corporate attributes such as financial variables, corporate governance mechanisms and firms' ownership structure (Tang and Luo 2016; Dienes et al. 2016; Crisóstomo et al. 2019). According to Monteiro and Aibar-Guzmán (2010) these determinant factors are grouped in firm size variables (assets, number of employees, revenues etc.), company-specific variables (sector, ownership, governance etc.) financial status variables (leverage, profitability, growth opportunities etc.) and other variables (visibility, legal structure, audit quality etc.)

The scope of this chapter is to examine the financial, size and governance determinants of CER activities of 7313 listed corporations originating from 24 EU countries over the period 2003–2018. Empirical analysis suggested that profitability, leverage, size and growth opportunities impact positively on the CER and its components, while CEO duality had a negative impact on CER performance. Moreover, the level of CER performance (devotion and engagement) is associated with the existence of a CSR committee within the firm which provides strategic direction and focus on corporate activities. Finally, the separation of the sample firms between code law and common law firms indicated that the aforementioned associations are more significant for code law firms suggesting that countries' legal system has an impact on corporate decisions regarding CER. In the following section we provide the theoretical background on firm-specific determinants and the research hypotheses on those factors. The third section discusses the data selection procedure and the research design. The fourth section provides the discussion of empirical results and the last section concludes the chapter.

4.2 Firm-Specific Determinants and Testable Hypothesis

The present section is devoted to the discussion of the determinant factors that will be utilized in the research design of the chapter. Following the existing literature we have considered seven corporate determinants of CER which have been used by other studies in this context. Specifically we examine firms' profitability, leverage, growth opportunities, size, CEO duality, board size and the existence of a sustainability committee which captures firms' CER strategic focus. The following discussion presents the research hypothesis for each determinant and its impact on CER performance.

4.2.1 *Firm Size*

Firm size (measured via total assets, number of employees, revenues etc.) has been considered by several studies worldwide (Monteiro and Aibar-Guzmán 2010; Tang and Luo 2016; Giannarakis 2014; Dienes et al. 2016). The majority of those studies reported a positive association between size and CER. There are several reasons in the literature which point to that positive association. At first, larger companies attract more publicity and attention by various stakeholders and consider to be highly polluters relative to smaller firms thus requiring enhanced legitimacy for their operations, leading them to report and engage on CER activities (Monteiro and Aibar-Guzmán 2010). Crisóstomo et al. (2019) argue that larger firms have more resources on their disposal to engage and even sustain CER activities on the medium to long term periods, thus establishing strong ties with stakeholders. Practically, larger firms are more able to bear the costs of CER reporting relative to smaller corporations and this is the reason for the positive impact of size on CER performance. On the contrary, smaller firms tend to withheld CER related information from the public eye due to their sensitivity to competition and their inability to afford such costs. Empirical evidence in emerging markets as well provide support for the abovementioned arguments. Dyduch and Krasodomska (2017) in Poland, Kolsi and Attayah (2018) in UAE and Syed and Butt (2017) in Pakistan all point to a positive impact of firm size on environmental responsibility activities and disclosure. So based on this discussion the first research hypothesis is stated as follows:

H1: Firm size impacts positively on CER performance.

4.2.2 *Leverage*

Firms' financial leverage indicates the dependence of the firm to external creditors. Creditors are an effective mechanism which can exert pressure to the firm to meet their information needs so as to guarantee the repayment of their funds (Crisóstomo et al. 2019). Under this framework, corporate environmental reporting could be perceived value relevant for debt holders since it can mitigate their information asymmetry and the uncertainty of debt financing. Nevertheless, Tang and Luo (2016) argue that CER disclosure is dependent on the result of the cost-benefit analysis of such decisions leading to a differential impact of leverage on CER performance. Empirical evidence on the impact of leverage on CER are contradictory. For instance, Branco and Rodrigues (2008), Andrikopoulos and Kriklani (2013), and Giannarakis (2014) indicate that financial leverage is negatively associated with CER due to the fact that the firms' engagement to CER activities is a costly procedure and highly leveraged firms are not having the required cash flow to support such activities on the long term.

On the contrary, Esa and Mohd Ghazali (2012), Tang and Luo (2016) and Kolsi and Attayah (2018) document that financial leverage is positively associated with corporate environmental and responsibility activities. This result is attributed to the fact that highly leveraged firms are motivated to disclose more CER information in order to reduce the agency costs with their creditors and in turn their cost of capital and debt financing. Therefore, based on the aforementioned discussion and considering the contradictory evidence in the literature, we state the second research hypothesis as follows:

H2: Financial leverage is associated with CER performance.

4.2.3 Profitability

According to the stakeholder theory, firms' engagement on environmentally responsible activities creates potential for enhanced firm performance and market value. The reason for that is because when companies exert efforts to accommodate the interests and needs of various stakeholders (via the commitment to participate in social and environmental activities) this fact will generate a virtuous cycle which can enhance competitive advantages leading to financial performance (Harrison et al. 2010; Crisóstomo et al. 2019). Practically CER performance can improve corporate image, brand value and even improve firms' relations with key players in the market, thus favoring firms' market share and profitability. Moreover, Hannifa and Cooke (2005) and Giannarakis (2014) argue that profitable firms tend to report more CER related information to the market so as to legitimize their existence and even avoid the undervaluation of their market value. According to Dienes et al. (2016) highly profitable firms also publish more CER information in order to signal to the market their success and future prospects. Finally, profitable companies have more financial resources available to support and finance CER activities and even adhere to the costs of publishing CER reports.

Empirical evidence in the literature are somehow contradictory since some studies published by Giannarakis (2014) in the US, Syed and Butt (2017) in Pakistan indicated that financial performance has a positive impact on social and environmental performance and disclosures, while on the contrary Giannarakis et al. (2014), Crisóstomo et al. (2019), Monteiro and Aibar-Guzmán (2010) and Dienes et al. (2016) found an insignificant relationship between profitability and CER performance. Consequently, following the above-mentioned discussion and considering the theoretical arguments on the relation between profitability and CER, we consider appropriate to hypothesize a positive contribution of profitability on CER performance. Thus, the third research hypothesis will be stated as follows:

H3: Profitability has a positive impact on CER performance.

4.2.4 Growth Opportunities

Firms' growth opportunities is another determinant factor of CER engagement and performance. According to empirical evidence from Padgett and Galan (2010) and Crisóstomo et al. (2019), growth opportunities have a positive impact on CER engagement and disclosure. This result yields supports to previous arguments in the literature (Artiach et al. 2010) that innovative firms and those with significant growth potential tend to be related to higher levels of CER commitment. This outcome is justified by the fact that growth firms need to signal to external investors and financial markets their sustainability and strategic CER focus so as to gain easier access to financing, which in turn will facilitate their ongoing innovative activities. Consequently, growth firms have to show that they are a committed organization towards environmental issues, stakeholders' management, and business ethical behavior in order for them to maximize or at least enhance their growth options. Consequently, we assess that growth opportunities will have a positive effect on CER performance. The fourth research hypotheses is stated as follows:

H4: Growth opportunities have a positive impact on CER performance.

4.2.5 CEO Duality

CEO duality refers to the situation where the company's CEO serves also as the board chairman and has been considered by previous studies by Elsayed (2007) and Jizi et al. (2014) as a governance characteristic which can have a positive impact of corporate decisions (faster and more efficient decision making ability) and CSR related disclosures. However, the opposite opinion is also overwhelmingly manifested in the literature. Giannarakis et al. (2014), Giannarakis (2014), Michelon and Parbonetti (2012) and Gul and Leung (2004) all indicate that the existence of CEO duality is negatively associated with voluntary CER related disclosure and firms' engagement on CER related activities. The main reason for these findings is that when the CEO also occupies the chair at the board of directors, he/she could control the corporate agenda and thus diminishes the directors' monitoring role. This fact could lead to lower levels of CER related disclosures and performance.

Early studies on this topic (Chaganti et al. 1985) argued about the necessity of separating the CEOs and board chairman's roles in order to sustain board independence. Also a CEO who is not only concentrated on his/her main duties and has various others roles to fulfil within the corporate operations, tends not to have enough time to focus on CER strategic planning and coordinated related activities. Therefore, the presence of CEO duality within corporate governance structure is expected to have a negative impact on CER performance. So the fifth research hypothesis is stated as follows:

H5: CEO duality has a negative impact on CER performance.

4.2.6 Board Size

The size of the board of directors is another corporate governance determinant of CER disclosures and related performance. Even though there are contradictive opinions in the literature regarding the impact of board size on CER, empirical evidence tend to support a positive association. According to Giannarakis (2014) and Dyduch and Krasodomska (2017) larger boards are an indication of enhanced monitoring since board members bring in the decision process their human and social capital which can benefit corporate decisions and activities. This fact can constitute firms' more able to adhere to global challenges and coordinate their efforts to fulfil stakeholder's needs and concerns, including social and environmental issues. Empirical evidence by Esa and Mohd Ghazali (2012), Rahman et al. (2011) and Wang (2017) indicate a positive impact of board size on CSER disclosures, suggesting that larger boards are able to perform more productive and efficient discussion on CER issues and thus more investment could be financed towards those activities. Therefore the sixth research hypothesis is state as follows:

H6: Board size impact positively on CER performance.

4.2.7 CER Sustainability Committee

Finally, the last determinant factor under study is the firms' CER strategic focus. Firms tend to establish social or sustainability related board committees as a way to adhere to stakeholders expectations towards performance standards and even monitor managers towards this end. According to Jain and Zaman (2019), the presence of the CER committee could serve as an effective monitoring mechanism which can provide valuable resources to the board members and managers for overseeing CER related investments and the overall CER strategic focus of the firm. This focus is operationalized by incorporating CER related standards on their daily operations and even establishing a related CER or sustainability committee which is responsible for monitoring the implementation of the abovementioned sustainability decisions and strategies. The existence of such committees either as independent departments within the firm or under the control of the board of directors, provide evidence that the firm is willing to utilize human and physical resources so as to achieve specific goals regarding social and environmental strategies. Consequently, firms which established such committees within their governance structure are expected to be associated with enhanced levels of CER performance. Empirical evidence in the literature by Kent and Monem (2008) and Giannarakis (2014) support the abovementioned argument suggesting that sustainability committee's impact positively on companies' decision to publish their social and environmental behavior to the public and incorporate permanent CER strategies on their operations (Dienes et al. 2016). Thus the seventh research hypothesis is stated as follows:

H7: The existence of a sustainability committee will have a positive impact on CER performance.

4.3 Data Selection and Research Design

4.3.1 Data Selection Procedure

The current chapter (and the forthcoming chapters) utilize a sample of corporations (both listed and unlisted) from 23 EU-member countries over the period 2003–2018. All financial, governance and environmental responsibility data have been extracted from Datastream database. The sample collection procedure started by including all corporations that have been covered by the environmental, social and governance (ESG) scores provided by Datastream during the sample period. 2003 has been selected as the initial sample year since that year was the first year with significant coverage (taking into consideration that this database officially was initiated in 2002). From a total of 656,734 firm-year observations we excluded those firms without environmental score data for at least five consecutive years in order to avoid bias in the empirical results and also to facilitate the estimation of differenced variables. Furthermore, we excluded firms with incomplete financial and governance data and those that do not close their fiscal year on December. In addition, we winsorized the upper and down 1 per cent of the data distribution in order to mitigate the influence of significant outliers in the empirical analysis. After this procedure was finalized, we ended up with a final sample of 121,154 firm-year observations. Table 4.1 present the sample selection process in more detail.

Table 4.2 presents the sample distribution per year and country. As we can see the environmental and social disclosure of EU firms has increased over the years with 2004 been the year with the most significant increase and afterwards the coverage remains stable with a few improvements, however a very small decrease in coverage is observed from 2014 to 2018, yet the overall sample distribution over the years is

Table 4.1 Sample selection process

Description	Observations
For our sample selection we focused on EU firms covered by the ESG Datastream database over the period 2003–2018.	656,734
Firms with less than 5 years of consecutive ESG data (not continuous evaluation and coverage).	−305,447
Firms with incomplete financial and governance data.	−124,355
Available observations with complete ESG and financial data	***226,952***
Firms which do not close their fiscal year on December	−103,375
Winsorized the upper and down 1% of the data distribution	−2423
Final sample	***121,154***

fairly stable. The countries with the higher coverage and thus participation in the overall sample are Finland and Netherlands with something more than 12,000 observations while third follows Spain with 11,769 observations. The smallest contribution in the overall sample have the Balkan countries (Romania, Greece, and Hungary) and the Baltic counties (Slovenia, Slovakia) and Malta. This distribution is quit logical since most of those countries have entered the EU relatively recently and their market economy is not well developed relative to the other EU member countries, thus CER related activities have not been fully aware by companies in those countries and they have not incorporated CER within their corporate strategy. Data on Table 4.2 are presented graphically on Figs. 4.1 and 4.2.

4.3.2 Research Design

In order to examine the validity of the previously mentioned research hypotheses we will estimate the following panel regression model with Generalized Least Squares (GLS) random effects. The panel random effect estimation method was chosen due to the fact that panel analysis contains more information with more variability and less collinearity among the variables, providing more efficient estimates and precise parameters of model estimation, allowing us to detect many effects that are not detectable in the cross-sectional data analysis. In order to choose the random effect estimation we performed the Breusch-Pagan Lagrange multiplier test for random effect which tests the null hypothesis that the residuals variance is equal to zero. The test produced a highly significant statistic (Chi^2 37141.95, $p < 0.00001$) leading us to reject the null hypothesis stating that the variance of residuals is equal to zero, and thus the random effects estimation was considered as the most appropriate method for our case. The functional form of the model includes the average CER for each firm and year estimated as the annual average of the ESG resource use score, emissions reduction score and environmental innovation score, where i denotes the firm and t the year and e is the error term:

Table 4.2 Sample distribution per year and country

Firms	2003	2004	2005	2006	2007	2008	2009	2010	2011	2012	2013	2014	2015	2016	2017	2018	Total
Austria	227	454	455	457	457	446	445	445	446	435	435	451	454	469	462	460	6998
Belgium	155	492	488	495	499	499	494	504	507	497	488	493	488	482	484	487	7552
Cyprus	25	89	86	82	86	90	78	75	69	75	79	86	78	83	93	81	1255
Czech Republic	82	149	141	133	122	117	119	135	136	146	150	148	143	140	140	128	2129
Denmark	45	154	155	162	167	150	151	153	154	156	152	155	154	156	155	146	2365
Finland	138	786	797	776	768	792	801	788	799	799	811	823	811	805	806	787	12,087
France	90	502	516	495	490	498	499	485	489	500	489	496	492	486	490	492	7509
Germany	111	407	407	415	421	428	426	393	394	379	372	365	356	353	345	324	5896
Greece	94	346	342	337	341	344	341	338	336	350	352	352	347	346	341	328	5235
Hungary	46	152	166	157	166	170	164	163	163	175	176	160	150	147	137	134	2426
Ireland	229	543	558	571	574	556	593	603	600	599	565	559	562	558	581	552	8803
Italy	81	471	458	466	474	479	491	473	480	476	478	477	485	471	479	459	7198
Luxemburg	139	331	326	330	333	342	347	365	343	337	336	330	338	344	349	335	5225
Malta	7	31	32	30	28	28	18	23	25	22	20	23	31	33	35	25	411
Netherlands	230	814	804	805	811	807	815	835	824	830	826	833	809	816	787	744	12,390
Poland	12	161	181	186	187	190	184	200	215	194	190	155	142	142	145	148	2632
Portugal	188	480	474	460	452	459	454	455	453	452	451	458	451	478	471	449	7085
Romania	0	6	2	2	3	1	1	4	5	3	5	5	5	4	4	3	53
Slovakia	29	78	75	72	70	71	65	68	66	66	68	69	70	72	68	39	1046
Slovenia	0	2	2	1	4	1	0	1	2	0	1	3	1	2	0	4	24
Spain	373	789	792	780	770	756	726	730	746	744	752	751	745	762	774	779	11,769
Sweden	52	401	387	383	380	372	372	371	369	372	373	382	386	398	392	396	5786
UK	114	323	343	346	343	349	345	351	350	348	334	324	357	349	349	355	5280
Total	2467	7961	7987	7941	7946	7945	7929	7958	7971	7955	7903	7898	7855	7896	7887	7655	121,154

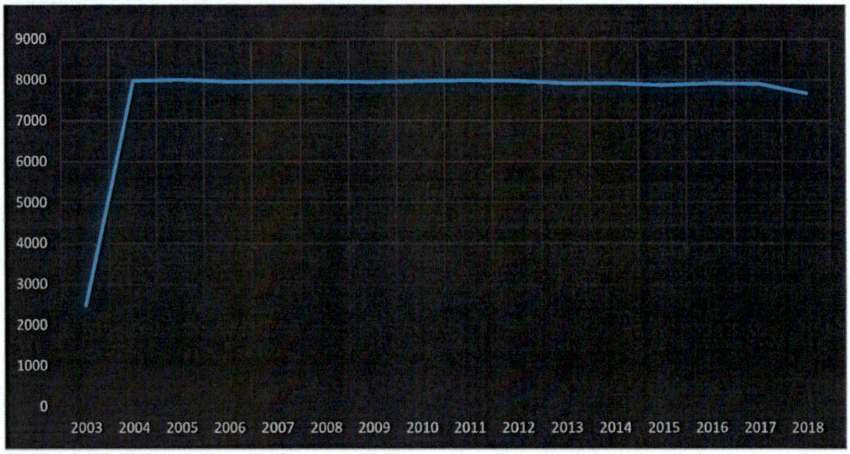

Fig. 4.1 Observations per year

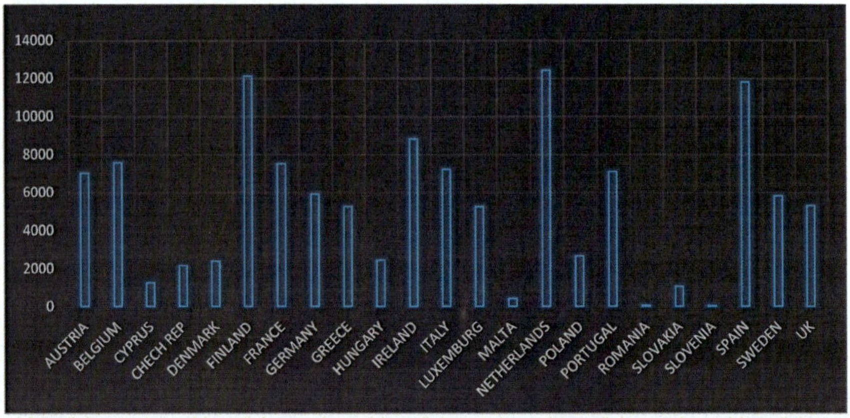

Fig. 4.2 Observations per country

$$\begin{aligned} CER_{it} = {} & a_0 + a_1 ROE_{it} + a_2 LEV_{it} + a_3 SIZE_{it} + a_4 GROWTH_{it} \\ & + a_5 DUAL_{it} + a_6 BDSIZE_{it} + a_7 CER\_COM_{it} + Year\ F.E \\ & + Industry\ F.E. + Country\ F.E. + e_{it} \end{aligned} \tag{4.1}$$

$$\begin{aligned} EN\_INNOV_{it} = {} & a_0 + a_1 ROE_{it} + a_2 LEV_{it} + a_3 SIZE_{it} + a_4 GROWTH_{it} \\ & + a_5 DUAL_{it} + a_6 BDSIZE_{it} + a_7 CER\_COM_{it} + Year\ F.E \\ & + Industry\ F.E. + Country\ F.E. + e_{it} \end{aligned} \tag{4.1a}$$

$$EMM_{it} = a_0 + a_1 ROE_{it} + a_2 LEV_{it} + a_3 SIZE_{it} + a_4 GROWTH_{it} + a_5 DUAL_{it} + a_6 BDSIZE_{it} + a_7 CER\_COM_{it} + Year\ F.E + Industry\ F.E. + Country\ F.E. + e_{it} \quad (4.1b)$$

$$RES\_USE_{it} = a_0 + a_1 ROE_{it} + a_2 LEV_{it} + a_3 SIZE_{it} + a_4 GROWTH_{it} + a_5 DUAL_{it} + a_6 BDSIZE_{it} + a_7 CER\_COM_{it} + Year\ F.E + Industry\ F.E. + Country\ F.E. + e_{it} \quad (4.1c)$$

EN_INNOV denotes each company's environmental innovation score, which reflects the firm's ability to mitigate the environmental costs towards its customers or even creating new market and product opportunities via enhanced environmental technologies and processes. EMM denotes companies' emission reduction score estimated via their efforts and effectiveness in reducing emissions from their production and operation. RES_USE denotes the firms' resource use score and indicates firms' ability to reduce the usage of hazardous materials, energy, water and other valuable natural resources. Finally, CER is the firms' annual average of the aforementioned three scores. All environmental scores range between zero (0) and 100 so the higher the score the more environmentally responsible and efficient is the firm.

ROE is return on equity and is estimated as net income before tax to common equity. If H3 is valid and profitability impacts positively on CER (and its components) we expect a positive coefficient in this variable. LEV is financial leverage capturing firm risk and is measured as the ratio of total debt to total assets. A significant coefficient on this variable will verify H2. SIZE is the natural logarithm of total assets and captures firm size. If coefficient a_3 is positive and statistically significant this will give support to H1. GROWTH captures firms' growth opportunities and is estimated as the annual percentage change of sales revenues. Firms with higher revenue growth are more able to finance CER related activities. So if H4 is true we expect a positive and significant coefficient on this variable. DUAL is a dummy variable taking unity (1) if the roles of CEO and board chairman are occupied by the same individual and zero (0) otherwise. If coefficient a_5 is negative and significant this will verify H5. Additionally, BDSIZE is the number of directors serving on the board and CER_COM is a dichotomous variable receiving unity (1) if the firm has established and official environmental responsibility (or sustainability) committee. If hypotheses 6 and 7 are true we expect positive and significant coefficients on these two variables. Finally the model includes year, country and industry specific effects in order to capture any variability in the data due to different country and sector settings not captured by the independent variables. The definitions of these determinants have been based on previous studies on the field such as Giannarakis (2014), Giannarakis et al. (2014), Tang and Luo (2016), Crisóstomo et al. (2019), Kolsi and Attayah (2018) and Dyduch and Krasodomska (2017).

4.4 Empirical Results

4.4.1 Descriptive Statistics and Correlations

Table 4.3 presents the descriptive statistics of the variables utilized in the regression models. First of all, the mean CER is 64.5 (with a standard deviation of 21.1) indicating that the sample firms have a satisfactory environmental performance score. The highest average score is that of resources usage (66.5) followed by emission reduction activities (65.9) while environmental innovation activities lag behind with an average of 61.3. The sample firms are highly profitable (mean ROE 79.5) and present significant growth opportunities (mean GROWTH at 58.9). However, our sample firms are highly leveraged since the average total debt covers more than four times the sample firms' total assets. Finally, in the majority of the observations (71.7 per cent) our sample firms have a governance structure were the CEO serves also as the board chairman, boards are comprised by 12 members but the most important is that more than in half of our sample firms there is an established environmental responsibility committee (average 0.55) suggesting that the majority of the sample firms have been considered CER strategically and invest resources towards that direction.

Table 4.4 presents the Pearson correlation coefficients between the sample variables. The majority of the correlation coefficients are significant with economic meaning, and are not very high indicating the absence of multicollinearity on the data. The VIF estimation of the sample variables provided values that were far below the threshold of 10 indicating no multicollinearity. As we can see, CER is negatively and significantly associated with SIZE and CEO duality, suggesting that larger firms and those with a CEO serving as board chairman are associated with lower CER performance. Nevertheless, firms with larger boards and especially those with a sustainability committee tend to perform better on the CER score. The same arguments stand for the rest three environmental performance scores. The only difference is observed in the environmental innovation score (EN_INNOV) which is positively

Table 4.3 Descriptive statistics of the sample variables

Variables	Mean	Standard deviation	Min	Max
CER	64.597	21.167	3.161	99.429
EN_INNOV	61.333	25.962	0.202	99.919
EMM	65.915	26.189	0.161	99.919
RES_USE	66.542	25.730	0.805	99.844
ROE	79.511	34.815	−33.315	118.315
LEV	4.625	4.319	−0.682	25.691
SIZE	20.496	2.848	2.302	30.311
GROWTH	58.915	141.720	−69.711	409.071
DUAL	0.717	0.450	0.000	1.000
BDSIZE	12.271	4.703	1	38
CER_COM	0.552	0.497	0.000	1.000

Table 4.4 Pearson correlation coefficients of sample variables

Variables	1	2	3	4	5	6	7	8	9	10	11
1. CER	1										
2. EN_INNOV	**0.728**	1									
3. EMM	**0.854**	**0.377**	1								
4. RES_USE	**0.863**	**0.405**	**0.710**	1							
5. ROE	0.002	0.054	0.001	−0.002	1						
6. LEV	0.059	**0.006**	0.001	**0.006**	−0.002	1					
7. SIZE	**−0.012**	**0.013**	−0.002	**−0.040**	**0.056**	**−0.046**	1				
8. GROWTH	−0.002	−0.005	−0.001	0.002	−0.001	−0.001	0.001	1			
9. DUAL	**−0.077**	**−0.055**	**−0.079**	**−0.053**	−0.004	0.004	**−0.081**	0.002	1		
10. BDSIZE	**0.059**	**0.060**	**0.055**	**0.029**	**0.016**	0.001	0.004	0.004	**−0.125**	1	
11. CER_COM	**0.508**	**0.317**	**0.443**	**0.481**	−0.005	0.001	**−0.038**	0.003	**−0.011**	**0.037**	1

Note: Correlation coefficients in bold indicate statistical significance at least at the 5% significance level

correlated with SIZE, a result which is economically logical since larger firms have more resources to finance such innovative activities. Finally, profitability and growth opportunities did not provide any significant correlation coefficients with the CER related scores. Of course correlations do not indicate causality or direction of association so we will proceed with the regression analysis results in order to extract more specific inferences.

4.4.2 Regression Results

Table 4.5 presents the regression results from the estimation of model (4.1) for the average CER as the dependent variable and the three other scores of environmental innovation, emissions and resource use. The Wald x^2 is highly significant suggesting that the random GLS estimation produced significant results and the explanatory power of the model is very satisfactory considering the size of the sample and the relative studies already published in the literature (Tang and Luo 2016; Syed and Butt 2017; Monteiro and Aibar-Guzmán 2010). Moving on to the firms-specific determinants of CER related performance, SIZE produced a positive and significant coefficients only in the environmental innovation model suggesting that larger firms invest more on innovative activities that aim to the protection of the environment. The relative coefficient in the other models produced insignificant results. Thus this

Table 4.5 Regression results on the determinants of CER and its components

Variables	CER	EN_INNOV	EMM	RES_USE
Constant	51.562*** (32.91)	46.503*** (23.02)	51.802*** (27.14)	56.882*** (30.58)
ROE	0.002*** (4.29)	0.007 (1.55)	0.023*** (7.12)	0.025*** (11.36)
LEV	0.015*** (6.41)	0.018*** (5.62)	0.009*** (2.59)	0.017*** (4.12)
SIZE	0.088 (1.21)	0.244*** (2.60)	0.114 (1.30)	−0.119 (−1.39)
GROWTH	0.010** (2.49)	0.004*** (4.24)	−0.003*** (−3.74)	0.018*** (3.08)
DUAL	−0.451 (−1.43)	−0.011 (−0.03)	−1.292*** (−3.20)	−0.387 (−0.98)
BDSIZE	0.022 (0.68)	0.047 (1.13)	0.044 (1.11)	−0.010 (−0.26)
CER_COM	18.339*** (67.60)	15.170*** (42.58)	19.954*** (56.81)	20.242*** (59.23)
R^2	0.2623	0.1031	0.2063	0.2319
Wald-x^2	4648.03***	1875.51***	3285.18***	3639.98***
Observations	50,757	50,757	50,757	50,757

Note: **, *** indicate statistical significance at the 5% and 1% significance level respectively, z-statistics in the parenthesis

finding partially supports H1 and corroborates arguments by Monteiro and Abar-Guzmán (2010) in Portugal and Syed and Butt (2017) in Pakistan. On the contrary, financial leverage (LEV) found to have a positive and highly significant impact on CER and its components in all model estimations. This result provides strong support to H2 and verifies arguments in the literature by Esa and Mohd Ghazali (2012), Tang and Luo (2016) and Kolsi and Attayah (2018), suggesting that highly leveraged firms are motivated to disclose more CER related information in order to reduce the agency costs with their creditors and in turn their cost of capital and debt financing.

Moreover, the coefficients on ROE were statistically significant (except the environmental innovation model) giving support to H3 and suggesting that more profitable firms are associated with more CER related performance. Our findings is consistent with previous evidence by Giannarakis (2014) in the US, Syed and Butt (2017) in Pakistan indicated that financial performance has a positive impact on CER performance and disclosures, suggesting that CER performance can improve corporate image, brand value and even improve firms' relations with key players in the market, thus favoring firms' market share and profitability. Regarding growth opportunities, GROWTH variable produced a positive and significant coefficient in all models except the emissions score model. These evidence corroborates H4 and previous evidence by Crisóstomo et al. (2019) arguing that the firms' need to take advantage of growth opportunities will force them to devote more resources and effort to environmental and sustainability concerns. This will be important for them in order to seek for financing to support their investment projects. So GROWTH presents a significantly positive impact on environmental innovation and resource use, but there is a negative impact of GROWTH in emission score. This results could be attributed to the fact that emissions are closely connected with the productive abilities of manufacturing corporations thus is may be even more difficult for them to devote resources towards reducing emissions without affecting their financial results. So it is easier for them to focus their efforts on other merits of CER performance.

Moving on to the governance variables, CEO duality produced a negative and significant coefficient only on the emissions model, providing partial support for H5. This finding is related to former studies by Giannarakis et al. (2014) and Said et al. (2009) who argued that CEO duality is associated with a weak governance structure and thus it affects CER related disclosures and performance negatively. BDSIZE did not provide any significant coefficient leading us to reject H6 and argue that the size of the board of directors does not impact on CER performance in our sample firms. Finally, the existence of a CER related committee has a significant impact on CER performance as evidenced by a highly significant and positive coefficient on the CER_COM variable in all models estimation, yielding strong support to H7. This finding corroborated evidence by Giannarakis (2014) and Kent and Monem (2008) suggesting that sustainability committee's impact positively on companies' decision to publish their social and environmental behavior to the public and incorporate permanent CER strategies on their operations (Dienes et al. 2016).

4.4.3 Sensitivity Analysis

In order to check the robustness of the main empirical results, we performed a number of sensitivity tests regarding the firms' legal environment, their listing status and various definitions of the determinant variables. At first we considered each EU member country's legal origin as potential confounding factor which can have an impact on the decision of firms to engage in CER related activities. La Porta et al. (1998, 1999) argued that firms in code law countries are oriented towards stakeholder engagement since they are characterized by more concentrated ownership, higher legal protection of outside investors or minority shareholders and even more managers are associated with the firms through managerial shareholdings. All these characteristics may motivate managers in code law firms to engage in CER related activities so as to improve firms' long term investments, via improving the interests' alignment between owners, managers and external stakeholders (Kim et al. 2017).

On the contrary, firms from common law countries are more shareholder-oriented organizations having as top priority the protection of investor rights. Under this framework, common law firms tend to focus on short-term profitability and performance and avoid to harm the wealth of outside investors. This fact creates less incentives for managers to engage in CER related activities since short-term investors may resist managerial decisions focusing on long-term perspectives and require the reduction of short-term financial performance so as to achieve higher future sustainability. Empirical evidence by Kim et al. (2017) indicate that code law firms invest more on environmental protection activities and projects compared to their common law counterparts. This result is attributed to the fact that managers of such firms tend to adhere to stakeholder-oriented ideologies and face less pressure when they decide to invest in CER projects. In order to examine this confounding factor we re-estimated all models by separating our sample firms based on their countries' legal origin. As common law firms are characterized those originating from UK, Cyprus, Ireland and Malta, while the rest of our sample firms are denoted as originating from code law countries.

Table 4.6 presents the descriptive statistics of the sample variables after separating the firms to code law and common law. As we can see the average CER performance and the separated environmental performance variables (EN_INNOV, EMM and RES_USE) are higher for code law firms relative to common law firms. Moreover, code law firms have higher profitability, size, and growth opportunities and are less leveraged compared to their common law counterparts. Moreover, code law firms incorporate CEO duality and establish CER committees in their governance structure in a higher extent relative to common law firms. Overall, descriptive evidence provide support to the evidence in the literature (Kim et al. 2017) suggesting that code law firms are associated with less environmental costs and they engage to CER related investments to a higher extent than common law firms.

Table 4.7 present the empirical results from the estimation of model (4.1) after separating the sample between common law and code law firms. Regarding the code law firms, empirical evidence are comparable to those on Table 4.5. Profitability

Table 4.6 Descriptive statistics of the sample variables between common law and code law EU country

	Code Law				Common Law			
Variables	Mean	St. dev.	Min	Max	Mean	St. dev.	Min	Max
CER	65.203	20.976	3.160	99.429	60.518	21.983	6.345	98.360
EN_INNOV	61.883	25.785	0.202	99.919	57.632	26.827	0.306	99.752
EMM	66.520	25.846	0.161	99.919	61.849	28.049	0.308	99.838
RES_USE	67.207	25.463	0.081	99.844	62.073	27.037	0.303	99.715
ROE	91.983	37.439	−33.315	118.315	−0.176	31.826	−80.394	81.958
LEV	4.355	27.359	−0.472	49.239	6.352	4.906	−0.682	25.690
SIZE	20.598	2.833	3.688	30.311	19.845	2.856	2.302	28.621
GROWTH	67.278	15.162	−37.836	409.071	11.064	76.152	−69.711	49.204
DUAL	0.734	0.441	0.000	1.000	0.602	0.489	0.000	1.000
BDSIZE	12.352	4.847	1	38	11.758	3.610	2	29
CER_COM	0.562	0.496	0.000	1.000	0.482	0.499	0.000	1.000

(ROE), financial leverage (LEV), firms size (SIZE) and growth opportunities (GROWTH) impact positively on almost all environmental performance variables. Board size (BDSIZE) produced insignificant coefficients on the code law firms but for common law firms, board size impact positively on CER related variables (except emissions). This result supports H6 for common law firms supporting previous evidence by Giannarakis (2014), Dyduch and Krasodomska (2017), Esa and Mohd Ghazali (2012), Rahman et al. (2011) and Wang (2017) who document that larger boards provide enhanced monitoring constituting firms' more able to adhere to global challenges and coordinate their efforts to fulfil stakeholder's needs and concerns, including social and environmental issues. Moreover, growth opportunities, size and profitability are not highly significant determinants for common law firms relative to code law firms, but the establishment of a CER or sustainability committee is also highly significant for common law firms. Overall, evidence on Table 4.7 provide supports to the arguments in the literature that the legal origin is a fundamental factor that builds the foundations of a business environment, along with managerial ideologies, corporate governance structures and investor protection regulations (La Porta et al. 2008). This framework creates different motives towards business risk and investments and different stances against various issues relating to internal and external stakeholders, such as environmental responsibility.

An additional sensitivity test included the control for firms listing status. Listed firms receive higher scrutiny from regulators and investors and usually are subjected to various reporting standards imposed by securities commissions. Listed firms are facing more pressures from internal and external stakeholders to engage in social and environmentally responsible investments in order to improve their image, gain their legitimacy on their operations and even enhance their stock market performance. Empirical evidence from Monteiro and Aibar-Guzmán (2010) indicated that listed firms are associated with more environmental disclosures on their annual reports

Table 4.7 Regression results on the determinants of CER and its components between common law and code law countries

	Code Law				Common Law			
Variables	CER	EN_INNOV	EMM	RES_USE	CER	EN_INNOV	EMM	RES_USE
Constant	51.38*** (30.62)	47.40*** (22.00)	49.48*** (24.36)	57.83*** (29.15)	50.93*** (11.79)	38.08*** (6.81)	65.49*** (11.99)	49.03*** (9.00)
ROE	0.002*** (4.53)	0.002 (1.61)	0.009*** (7.53)	0.003*** (11.07)	−0.007 (−0.27)	0.002 (0.03)	−0.024 (−0.68)	0.003 (0.10)
LEV	0.014*** (6.41)	0.017*** (4.70)	0.009*** (3.17)	0.015*** (3.56)	1.104*** (2.72)	1.341** (2.37)	1.019 (1.71)	0.934 (1.61)
SIZE	0.128 (1.64)	0.228** (2.28)	0.251*** (2.69)	−0.122 (−1.34)	−0.151 (−0.76)	0.381 (1.51)	−0.74*** (−2.91)	−0.124 (−0.51)
GROWTH	0.001** (2.27)	0.004*** (5.47)	−0.003*** (−5.16)	0.002*** (2.76)	0.003 (1.24)	−0.004 (−1.20)	0.009*** (2.97)	0.006 (1.16)
DUAL	−0.335 (−0.98)	0.281 (0.63)	−1.319*** (−3.01)	−0.288 (−0.67)	−1.656** (−2.16)	−2.566** (−2.37)	−1.283 (−1.29)	−1.629 (−1.72)
BDSIZE	−0.004 (−0.12)	0.011 (0.25)	0.031 (0.76)	−0.048 (−1.14)	0.338*** (3.19)	0.485*** (3.58)	0.147 (0.99)	0.461*** (3.36)
CER_COM	18.692*** (65.98)	15.81*** (42.34)	20.11*** (54.85)	20.461*** (56.94)	14.879*** (16.82)	8.819*** (8.11)	18.46*** (15.29)	18.004*** (16.71)
R^2	0.2632	0.1093	0.2035	0.2296	0.2709	0.083	0.2310	0.2645
Wald-x^2	4425.8***	1854.70***	3092.2***	3366.8***	308.61***	98.63***	266.7***	296.03***
Observations	45,517	45,517	45,517	45,517	5240	5240	5240	5240

Note: **, *** indicate statistical significance at the 5% and 1% significance level respectively, z-statistics in the parenthesis

Table 4.8 Sensitivity analysis on the determinants of CER and its components

Variables	CER	EN_INNOV	EMM	RES_USE
Constant	32.569*** (5.32)	36.122*** (4.48)	29.840*** (3.88)	33.629*** (4.35)
PROF_MARGIN	1.756*** (2.64)	2.055** (2.46)	1.145*** (2.66)	1.864** (2.32)
LT_DEBT	6.237*** (2.57)	5.092 (1.62)	4.185 (1.30)	8.702*** (2.58)
SIZE	0.927*** (3.42)	0.768** (2.09)	1.060*** (3.18)	0.859*** (2.58)
RnD_EXP	−2.792*** (−3.56)	−8.580*** (−10.90)	−1.478 (−0.81)	1.744*** (6.06)
LISTED	0.033 (0.04)	0.644 (0.50)	0.771 (0.71)	−1.208 (−1.20)
DUAL	−0.159 (−0.18)	−0.245 (−0.22)	−0.889 (−0.73)	0.435 (1.53)
BDSIZE	0.189 (1.50)	0.125 (0.79)	0.161 (0.73)	0.258 (0.33)
CER_COM	18.116*** (16.83)	14.610*** (10.01)	19.703*** (13.86)	20.995*** (13.73)
$\mathbf{R^2}$	0.3085	0.1425	0.2237	0.2661
Wald-$\mathbf{x^2}$	324.45***	247.88***	237.29***	228.05***
Observations	50,757	50,757	50,757	50,757

Note: **, *** indicate statistical significance at the 5% and 1% significance level respectively, z-statistics in the parenthesis

relative to unlisted corporations. Therefore, we re-estimated model (4.1) by including a dummy variable (LISTED) receiving unity (1) if the firm is listed on its national stock market in a given year and zero (0) otherwise. Moreover, growth opportunities were replaced by the ratio of research and development expenses to total revenues (RnD) since innovative firms tend to channel more funds to the development of new products and services and are less prone to finance CER related activities. Also ROE was replaced with net profit margin and LEV with the ratio of long-term debt to total assets. Empirical results from these modifications are presented on Table 4.8.

As we can see the main differences on the coefficients compared to those on Table 4.5 is on SIZE which has produced significantly positive coefficients in all model estimations. These results strongly verify H3. On the contrary, the listing status of the firm did not produce any significant results suggesting that listed firms do not have difference CER performance compared to non-listed firms. Moreover, firms with high research and development expenses tend to be associated with lower average CER performance and environmental innovation, but they have higher levels of efficient resource usage. These evidence are suggesting that innovative firms tend to invest less in environmental responsible activities since they channel more funds to the development of new products and services and are less prone to finance CER related investments. Finally, the existence of a sustainability committee impacts positively on CER performance. Overall, the sensitivity testing conducted produced qualitatively similar results compared with those on Table 4.5, thus our main conclusions hold after the aforementioned modifications in the model.

4.5 Conclusion

A vast amount of studies on the environmental responsibility research field have focused on examining the determinants and motivating factors that lead firms to engage in CER related activities and investments. An extensive number of papers within the domains of accounting and corporate finance have concluded that the determinant factors of CER are categorized in several groups such as financial variables, corporate governance mechanisms and firms' ownership structure (Tang and Luo 2016; Dienes et al. 2016; Crisóstomo et al. 2019; Monteiro and Aibar-Guzmán 2010).

The goal of this chapter was to examine the financial, size and governance determinants of CER activities of 7313 listed corporations originating from 24 EU countries over the period 2003–2018. Empirical analysis suggested that profitability, leverage, size and growth opportunities impact positively on the CER and its components, while CEO duality had a negative impact on CER performance. Moreover, the level of CER dedication and engagement is associated with the existence of a CSR committee within the firm which provides strategic direction and focus on corporate activities. Finally, the separation of the sample firms between code law and common law firms indicated that the aforementioned associations are more significant for code law firms suggesting that countries' legal system has an impact on corporate decisions regarding CER. The empirical results remain robust after several sensitivity tests regarding the definition of the independent variables and the firms' listing status.

References

Andrikopoulos, A., and N. Kriklani. 2013. Environmental disclosure and financial characteristics of the firm: The case of Denmark. *Corporate Social Responsibility and Environmental Management* 20 (1): 55–64.

Artiach, T., D. Lee, D. Nelson, and J. Walker. 2010. The determinants of corporate sustainability performance. *Accounting and Finance* 50 (1): 31–51.

Branco, M.C., and L.L. Rodrigues. 2008. Factors influencing social responsibility disclosure by Portuguese companies. *Journal of Business Ethics* 83 (4): 685–701.

Chaganti, R.S., V. Mahajan, and S. Sharman. 1985. Corporate board size, composition and corporate failures in retailing industry. *Journal of Management Studies* 22 (4): 400–417.

Crisóstomo, V.L., F. de Souza Freire, and M.F. de Oliveira Freitas. 2019. Determinants of corporate sustainability performance: Evidence from Brazilian panel data. *Social Responsibility Journal*. https://doi.org/10.1108/SRJ-04-2018-0102.

Dienes, D., R. Sassen, and J. Fisher. 2016. What are the drivers of sustainability reporting? A systematic review. *Sustainability Accounting, Management and Policy Journal* 7 (2): 154–189.

Dyduch, J., and J. Krasodomska. 2017. Determinants of corporate social responsibility disclosure: An empirical study of Polish listed companies. *Sustainability* 9: 1934.

Elsayed, K. 2007. Does CEO duality really affect corporate performance? *Corporate Governance: An International Review* 15 (6): 1203–1214.

Esa, E., and N.A. Mohd Ghazali. 2012. Corporate social responsibility and corporate governance in Malaysian government-linked companies. *Corporate Governance* 12 (3): 202–305.

Giannarakis, G. 2014. Corporate governance, and financial characteristic effects on the extent of corporate social responsibility disclosure. *Social Responsibility Journal* 10 (4): 569–590.

Giannarakis, G., G. Konteos, and N. Sariannidis. 2014. Financial, governance and environmental determinants of corporate social responsible disclosure. *Management Decision* 52 (10): 1928–1951.

Gul, F.A., and S. Leung. 2004. Board leadership, outside directors' expertise and voluntary corporate disclosures. *Journal of Accounting and Public Policy* 23 (5): 351–379.

Hall, R. 1992. The strategic analysis of intangible resources. *Strategic Management Journal* 13 (2): 135–144.

Hannifa, R.M., and T.E. Cooke. 2005. The impact of culture and corporate governance on corporate social reporting. *Journal of Accounting and Public Policy* 24 (5): 391–430.

Harrison, J.S., D.A. Bosse, and R.A. Phillips. 2010. Managing for stakeholders, stakeholder utility functions and competitive advantage. *Strategic Management Journal* 31 (1): 58–74.

Huang, X., and L. Watson. 2015. Corporate social responsibility research in accounting. *Journal of Accounting Literature* 34 (1): 1–16.

Jain, T., and R. Zaman. 2019. When boards matter: The case of corporate social irresponsibility. *British Journal of Management*. https://doi.org/10.1111/1467-8551.12376.

Jizi, M.I., A. Salama, R. Dixon, and S. Stratling. 2014. Corporate governance and corporate social responsibility disclosure: Evidence from the US banking sector. *Journal of Business Ethics* 125: 601–615.

Kent, P., and R. Monem. 2008. What drives TBL reporting: Good governance or threat to legitimacy? *Australian Accounting Review* 18 (4): 297–309.

Kim, H., K. Park, and D. Ryu. 2017. Corporate environmental responsibility: A legal origin perspective. *Journal of Business Ethics* 140: 381–402.

Kolsi, M.C., and O.F. Attayah. 2018. Environmental policy disclosures and sustainable development: Determinants, measure and impact on firm value for ADX listed companies. *Corporate Social Responsibility and Environmental Management* 25 (5): 807–818.

La Porta, R., F. Lopez-de-Silanes, A. Shleifer, and R. Vishny. 1998. Law and finance. *Journal of Political Economy* 106: 1113–1155.

———. 1999. The quality of government. *Journal of Law Economics and Organization* 15: 222–279.

La Porta, R., F. Lopez-de-Silanes, and A. Shleifer. 2008. Economic consequences of legal origins. *Journal of Economic Literature* 46: 285–332.

Michelon, G., and A. Parbonetti. 2012. The effects of corporate governance on sustainability disclosure. *Journal of Management and Governance* 16 (3): 477–509.

Monteiro, S., and B. Aibar-Guzmán. 2010. Determinants of environmental disclosure in the annual reports of large companies operating in Portugal. *Corporate Social Responsibility and Environmental Management* 17: 185–204.

Padgett, R.C., and J.J. Galan. 2010. The effect of R&D intensity on corporate social responsibility. *Journal of Business Ethics* 93: 407–418.

Patten, D. 1992. Intra-industry environmental disclosures in response to the Alaskan oil spill: A note on legitimacy theory. *Accounting, Organizations and Society* 17 (5): 471–475.

———. 2005. The accuracy of financial projections of future environmental capital expenditures: A research note. *Accounting, Organizations and Society* 30: 457–468.

Rahman, N.H.W.A., M.M. Zain, and N.H.Y.Y. Al-Haj. 2011. CSR disclosures and its determinants: Evidence from Malaysian government link companies. *Social Responsibility Journal* 7 (2): 181–201.

Roberts, R.W. 1992. Determinants of corporate social responsibility disclosure: An application of stakeholder theory. *Accounting, Organizations and Society* 17 (6): 595–612.

Russo, M., and P. Fouts. 1997. A resource-based perspective on corporate environmental performance. *Academy of Management Journal* 40 (3): 54–559.

Said, R., Y.H. Zainuddin, and H. Haron. 2009. The relationship between corporate social responsibility disclosure and corporate governance characteristics in Malaysian public listed companies. *Social Responsibility Journal* 5 (2): 212–226.

Suto, M., and H. Takehara. 2018. *Corporate social responsibility and corporate finance in Japan*. Singapore: Springer Nature.

Syed, M.A., and S.A. Butt. 2017. Financial and non-financial determinants of corporate social responsibility: Empirical evidence from Pakistan. *Social Responsibility Journal* 13 (4): 780–797.

Tang, Q., and L. Luo. 2016. Corporate ecological transparency: Theories and empirical evidence. *Asian Review of Accounting* 24 (4): 498–524.

Wang, M.C. 2017. The relationship between firm characteristics and the disclosure of sustainability reporting. *Sustainability* 9: 624.

Chapter 5
Corporate Environmental Responsibility and Financial Performance

Until recently, the main purpose of businesses was to maximize the wealth of its shareholders without much focus on other key stakeholders' needs. Over time, this approach proved detrimental for business performance and longevity because maximizing shareholder benefit can harm other stakeholders' (environment, society, employees, suppliers, customers, etc.) interests and may inflict significant impact on corporate reputation, image and performance. While initial research on the relationship between corporate environmental responsibility (CER) and corporate financial performance (CFP) has shown that there is a negative association, due to the fact that the costs of CER activities are not compensated by benefits, there is an increased stream of research during the last 10 years showing that there may be positive and pervasive business outcomes from CER engagement, which can improve the production process and make firms more efficient. The purpose of this chapter is to study the impact of CER related activities on their financial performance taking into consideration potential endogeneity between the examined factors and their bidirectional association, evidenced by previous studies on the field.

5.1 Introduction

CER refers to the voluntary actions of the company, in addition to those required by law (McWilliams and Siegel 2001). These actions include activities which reduce the firm's burden to the environment, producing products that contribute to society and the protection of the natural resources, reducing CO_2 emissions, etc. (Barnett 2007; McWilliams and Siegel 2001). Many researchers view CER as a source of comparative advantage (Porter and Kramer 2006) and have shown that positively influences various forms of business performance, such as its reputation (Brammer and Millington 2005; Turban and Greening 1996), customer satisfaction (Luo and Bhattacharya 2006), employee attractiveness (Backhaus et al. 2002; Turban and

P. Dimitropoulos, K. Koronios, *Corporate Environmental Responsibility, Accounting and Corporate Finance in the EU*, CSR, Sustainability, Ethics & Governance, https://doi.org/10.1007/978-3-030-72773-4_5

Greening 1996) and higher organizational commitment to its employees (Peterson 2004).

Researchers have put forward various views on how a company's social and environmental performance affects its financial profitability (Clarkson 1995; Jones and Wicks 1999; McWilliams et al. 2006; Wood 1991). The most widespread view is that CER activities helps firms to build a good relationship based on mutual trust and cooperation with stakeholders, which enhances company's profitability. Researchers interested in how companies interact with different stakeholders have sought to determine whether aligning the needs of an organization with shareholders, beyond stakeholders, contributes to the ability of the company to gain and maintain profitability. To this end, several studies have been conducted to examine the relationship between the level of CER performance and corporate profitability and the majority of them proposes a positive relationship (Laplume et al. 2008; Margolis and Walsh 2003; Orlitzsky et al. 2003; Roman et al. 1999).

Godfrey and Hatch (2007) recommend that CER and CFO may have different association between firms because each business has unique social interests, different internal and external environments, and financial assets, and thus each business may derive different advantages from engaging in CER activities (Godfrey and Hatch 2007; Griffin and Mahon 1997). More specifically, they examine CSR through five different aspects: (a) employee relations, (b) product quality, (c) relations with society, (d) environmental issues, and (e) diversity issues (minorities and women). In relation to the environment, CER activities are captured by the company's support for the protection of the natural environment, such as using clean energy, providing environmentally friendly products and services, implementing recycling programs, and reducing waste by more efficient resources use and allocation.

Moreover, Kacperczyk (2009) showed that the company's attention to diversity, environment and society positively affects stakeholder value in the long run. Berman et al. (1999), have argued that the concern for environmental issues allow businesses to benefit from lower costs and risks through their compliance with environmental rules and improved management (Russo and Fouts 1997). Therefore, they have shown that high environmental efficiency is positively related to profitability through asset efficiency.

Earlier empirical studies presented CER actions within a one-dimensional way, while many scholars insist on its multidimensional nature, where each one is represented by a set of activities (Clarkson 1995; Godfrey and Hatch 2007; Waddock and Graves 1997). According to Carroll (1979, 1999) the voluntary actions of a business can be divided into two aspects: ethical and charitable responsibility. Ethical responsibility refers to the actions of the business that are not required by law but society expects the business to engage them, while charitable actions are at its discretion. Alternatively, according to Clarkson (1995), the multidimensional form of CER can be better approached by stakeholder theory, which considers how companies handle their relationship with key stakeholders. Typical key stakeholders consist of owners, employees, suppliers, customers, society and the natural environment. Each of the above has different rights and interests within a company and the company is required to implement different actions and policies to satisfy each stakeholder and to achieve superiority in its financial performance (Peloza and Papania 2008).

In addition to the multidimensional nature of CER, CFP is also multidimensional (Griffin and Mahon 1997). Extensive studies have shown that there are two types of CFP measurement: short-term profitability and market estimation for future profitability (Cochran and Wood 1984; Luo and Bhattacharya 2006; McGuire et al. 1988). Return on assets represents the company's short-term performance or management capability and provides information on how the allocation of certain resources leads to current revenue (Hull and Rothenberg 2008). In contrast, market-based measurement, such as the Tobin's Q, reveals that investors value a company's ability to generate future revenue (Luo and Bhattacharya 2006; McGuire et al. 1988). Lee et al. (2013) in their research examined the relationship between CER and cost effectiveness by dividing CER actions into those related to the core business of the firm or and those which do not, providing an explanation of how some CER actions can be more beneficial than others to the value of the business during difficult economic times.

Tang et al. (2012) also followed the same direction and concluded that when a company takes CER actions slowly and steadily and those activities are tied to its purpose then the firm is benefited the most. However, the pace at which a CER strategy is undertaken does not alter the relationship between CER and CFP. However, numerous surveys have been conducted around the world on various sectors of the economy and even in times of crisis in order to study the possible relationship between CER activities undertaken by a business and its CFP and whether CER ultimately plays an important role especially during times of uncertainty. The results show that there is a positive relationship between the CER and the CFP at the beginning of the financial crisis (second half of 2007), but at the beginning of 2007 and after the first 6 months of the crisis the results show that there is no significant relationship between these two variables (Ducassy 2013).

The scope of this chapter is to examine the impact of CER related activities on their financial performance taking into consideration potential endogeneity between the examined factors and their bidirectional association, evidenced by previous studies on the field. We respond to the call for more research on the topic made by Chen et al. (2018), by covering a large multi-country sample of listed and unlisted European firms over a long period of time and also by considering the bidirectional association between CER and CFP. In the following section we provide the theoretical background on the impact of CER on CFP and the testable hypotheses. The third section discusses the data selection procedure and the research design. The fourth section provides the discussion of empirical results and the last section concludes the chapter.

5.2 CER-CFP Association and Testable Hypothesis

The impact of CER activities on financial performance has been examined by several researchers through the lenses of different theories and frameworks. The resourced-based view of the firm is the most commonly used theory which dictates that CER impacts positively on CFP arguing that the firms can gain competitive advantages by utilizing internal and external resources. CER activities create such resources and beyond that they provide risk mitigation abilities in cases of negative events (Russo and Fouts 1997; Flammer 2013; Shiu and Yang 2017). An alternative theory to the resource-based view is signaling theory which posits that CER activities and related information are disseminated to the market so as to improve stakeholders' awareness so as firms can improve their legitimacy on their operations (Wei et al. 2017). Moreover, the social identity theory refers to the firms' organizational identification or how people perceive themselves to fit within an organization. According to Farooq et al. (2017), CER has been evolved as an important factor of employee identification within firms, which in turn affects key strategic decisions and CER activities. Finally, the neo-institutional theory is the most recent theoretical framework which has been created in order to explain the nature and types of institutional processes (Oliver and Holzinger 2008; Marquis and Qian 2014). Under this framework CER is considered as a legitimacy mechanism through which firms can benefit from wider stakeholder acceptance, access to financial markets and financial performance (Jiang et al. 2018). All the above theories propose a positive association between CER activities and corporate financial performance.

Academic studies have shown that corporate strategies related to managing corporate relationships have evolved due to stakeholder pressure (Sarkar 2008) and environmental responsibility has been shown to have a positive impact on institutional investors (Wahba 2008). There are various forces that drive companies to adopt CER actions, such as the pressure they receive from environmental conscious customers and non-governmental organizations (Lyon and Mawell 2008). Suto and Takehara (2018) argue that corporate environmental initiatives prove that market forces are powerful indicators for improving corporate environmental responsibility. Other researchers have shown that the role and commitment of managers is an important factor in undertaking environmental practices (Kassinis and Panayiotou 2006; Lee and Ball 2003).

Companies have implemented various environmental actions such as: initiatives to reduce the impact on animals and natural resources, voluntary environmental remediation actions, investments and activities targeting waste reduction, the purchase of recyclable raw materials, the use of renewable energy, reducing energy consumption, reducing the risk of environmental accidents, etc. (Trendafilova and Babiak 2013). In recent years CER has been treated not only as a charitable effort but also as a strategy. Porter and Kramer (2006) support CER's strategic approach by defining the social impact of CER actions, whereby it creates opportunities for the business to reduce its negative characteristics and gain positive insights to achieve social and strategic recognition. There are many elements that influence a business to take environmental practices, such as: increasing profits, reducing operating costs, increasing operational efficiency, enhancing reputation, increasing sales, customer confidence, reducing administrative and financial risk, etc. (Monteil 2008).

CER should be considered as a strategic investment and CER actions should be expanded to achieve strategic advantages. Environmental protection activities have proven to impact positively on corporate value (Wahba 2008). Stakeholders around the world require a broader accountability framework within the traditional accounting process, thereby providing more accurate and reliable information about the business's financial position. This expanded accountability framework is based on the suggestion that corporate responsibility is not limited to maximizing profits (Waddock 2003). More recent studies (Berthelot et al. 2003) have found that voluntary publications on environmental data do not measure actual corporate environmental performance. Their credibility however is questionable. Previous studies have shown that companies do not always disclose environmental information, but do so only for strategic reasons. Companies are less likely to disclose environmental information if they are facing serious environmental problems (Li et al. 1997), because the increasing cost of producing such information may outweigh the potential gain from reduced investor uncertainty.

There is also evidence that voluntary environmental publications influence investor perceptions of a firm's future earnings and cash flow (Walden and Schwartz 1997). For example, a low investment in corporate environmental responsibility may indicate a weak governance and administrative structure of the company, which indicates that the company is not 'progressive' and responsible to its stakeholders (Orlitzsky and Benjamin 2001). This is related to management quality theory (Klassen and McLaughlin 1996) where it is stated that high level corporate environmental actions may indicate good management and better corporate financial performance.

Empirical evidence on the direct association between CER and CFP have been controversial with the majority of them pointing to a positive impact of CER on CFP. One stream of evidence points that CER activities consume valuable financial resources and their costs could not be compensated by the benefits leading to a negative impact of CER on CFP. Testa and D'Amato (2017), Xu et al. (2019) and Suto and Takehara (2018) provide empirical evidence that CER impacts negatively on CFP arguing that CER investments require massive expenditures which do not provide a positive return, leading to a negative impact on CFP. On the contrary, a large number of studies points to the opposite association. Jiang et al. (2018) have documented a positive impact of CER on CFP within the Chinese business environment, arguing that CER generates competitive resources (internal and external) leading to competitive advantages and enhanced financial performance (either accounting or market based). Corroborative evidence have been provided by Chen et al. (2018) using an international sample of firms and Fiadrino et al. (2019) using a large European sample, and all have found that green initiatives and green performance impact positively on financial performance. Moreover, several meta-analytical studies have verified the abovementioned arguments on the positive association between CER and CFP since social and environmental investments lead to improved financial returns enhancing profitability and future prospects (Friede et al. 2015; Revelli and Viviani 2015). So based on the above discussion and the overwhelming evidence on the positive association between CER and CFP we state the main research hypothesis as follows:

H1: CER performance will have a positive impact on CFP.

5.3 Data Selection and Research Design

5.3.1 Data Selection Procedure

The current chapter utilizes a sample of corporations (both listed and unlisted) from 23 EU-member countries over the period 2003–2018. All financial, governance and environmental responsibility data have been extracted from Datastream database. The sample collection procedure started by including all corporations that have been covered by the environmental, social and governance (ESG) scores provided by Datastream during the sample period. 2003 has been selected as the initial sample year since that year was the first year with significant coverage (taking into consideration that this database officially was initiated in 2002). From a total of 656,734 firm-year observations we excluded those firms without environmental score data for at least five consecutive years in order to avoid bias in the empirical results and also to facilitate the estimation of differenced variables. Furthermore, we excluded firms with incomplete financial and governance data and those that do not close their fiscal year on December. In addition, we winsorized the upper and down 1 per cent of the data distribution in order to mitigate the influence of significant outliers in the empirical analysis. After this procedure was finalized, we ended up with a final sample of 121,154 firm-year observations. The analysis and structure of the final sample is provided on Table 4.1 of the previous chapter.

5.3.2 Research Design

In order to examine the validity of the previously mentioned research hypothesis we will consider the case of a bidirectional association between CER and CFP. According to Testa and D'Amato (2017), highly profitable firms may have more resources available at their disposal to invest in CER related activities. This fact may increase their legitimacy to the market and improve their image towards their stakeholders leading to enhanced financial performance. So concurrently the improvement on CER performance can lead to firm performance and vice versa. This argument has been posited by Chen et al. (2018) after using a world sample of listed firms and they argue about the existence of a possible reverse relationship between CFP and CER. In order to control for this association we will estimate a system of seemingly unrelated regressions (SUR). This design allows us to impose or test cross-sectional restrictions, and most importantly to gain efficiency because we expect that the error terms across equations are contemporaneously correlated. In order to test whether the application of the SUR method yields a significant gain in efficiency (beyond the separate estimation of these equations) we performed the Breusch-Pagan test of independence, which LM statistic sums the squared correlations between residual vectors e and u with the null hypothesis of a zero contemporaneous covariance between the error terms of different equations (Baum 2006). The rejection of the null hypothesis indicates that the SUR estimation yields a significant gain compared to the separate estimation. The majority of the Breusch-Pagan tests provided on the result tables below are highly significant at least at the 5 per cent level indicating that the implementation of the SUR method is more efficient than the separate estimation as in previous studies on the field. The functional form of the system of equations includes the average CER for each firm and year estimated as the annual average of the ESG resource use score, emissions reduction score and environmental innovation score, where i denotes the firm and t the year and e is the error term:

$$\begin{aligned} CER_{it} = {} & a_0 + a_1 ROE_{it}(PROF_{it}) + a_2 LEV_{it} + a_3 SIZE_{it} + a_4 GROWTH_{it} \\ & + a_5 DUAL_{it} + a_6 BDSIZE_{it} + a_7 CER\_COM_{it} + Year\ F.E \\ & + Industry\ F.E. + Country\ F.E. + e_{it} \end{aligned} \tag{5.1}$$

$$\begin{aligned} ROE_{it}(PROF_{it})_{it} = {} & a_0 + a_1 CER_{it} + a_2 LEV_{it} + a_3 SIZE_{it} + a_4 GROWTH_{it} \\ & + a_5 DUAL_{it} + a_6 BDSIZE_{it} + a_7 CER\_COM_{it} \\ & + Year\ F.E + Industry\ F.E. + Country\ F.E. + u_{it} \end{aligned} \tag{5.2}$$

$$\begin{aligned} EN\_INNOV_{it} = {} & a_0 + a_1 ROE_{it}(PROF_{it}) + a_2 LEV_{it} + a_3 SIZE_{it} \\ & + a_4 GROWTH_{it} + a_5 DUAL_{it} + a_6 BDSIZE_{it} \\ & + a_7 CER\_COM_{it} + Year\ F.E + Industry\ F.E. \\ & + Country\ F.E. + e_{it} \end{aligned} \tag{5.1a}$$

$$\begin{aligned} ROE_{it}(PROF_{it}) = {} & a_0 + a_1 EN\_INNOV_{it} + a_2 LEV_{it} + a_3 SIZE_{it} \\ & + a_4 GROWTH_{it} + a_5 DUAL_{it} + a_6 BDSIZE_{it} \\ & + a_7 CER\_COM_{it} + Year\ F.E + Industry\ F.E. \\ & + Country\ F.E. + u_{it} \end{aligned} \tag{5.2a}$$

$$\begin{aligned} EMM_{it} = {} & a_0 + a_1 ROE_{it}(PROF_{it}) + a_2 LEV_{it} + a_3 SIZE_{it} + a_4 GROWTH_{it} \\ & + a_5 DUAL_{it} + a_6 BDSIZE_{it} + a_7 CER\_COM_{it} + Year\ F.E \\ & + Industry\ F.E. + Country\ F.E. + e_{it} \end{aligned} \tag{5.1b}$$

$$\begin{aligned} ROE_{it}(PROF_{it}) = {} & a_0 + a_1 EMM_{it} + a_2 LEV_{it} + a_3 SIZE_{it} + a_4 GROWTH_{it} \\ & + a_5 DUAL_{it} + a_6 BDSIZE_{it} + a_7 CER\_COM_{it} \\ & + Year\ F.E + Industry\ F.E. + Country\ F.E. + u_{it} \end{aligned} \tag{5.2b}$$

$$\begin{aligned} RES\_USE_{it} = {} & a_0 + a_1 ROE_{it}(PROF_{it}) + a_2 LEV_{it} + a_3 SIZE_{it} \\ & + a_4 GROWTH_{it} + a_5 DUAL_{it} + a_6 BDSIZE_{it} \\ & + a_7 CER\_COM_{it} + Year\ F.E + Industry\ F.E. \\ & + Country\ F.E. + e_{it} \end{aligned} \tag{5.1c}$$

$$\begin{aligned} ROE_{it}(PROF_{it}) = {} & a_0 + a_1 RES\_USE_{it} + a_2 LEV_{it} + a_3 SIZE_{it} \\ & + a_4 GROWTH_{it} + a_5 DUAL_{it} + a_6 BDSIZE_{it} \\ & + a_7 CER\_COM_{it} + Year\ F.E + Industry\ F.E. \\ & + Country\ F.E. + u_{it} \end{aligned} \tag{5.2c}$$

EN_INNOV denotes each company's environmental innovation score, which reflects the firm's ability to mitigate the environmental costs towards its customers or even creating new market and product opportunities via enhanced environmental technologies and processes. EMM denotes companies' emission reduction score estimated via their efforts and effectiveness in reducing emissions from their production and operation. RES_USE denotes the firms' resource use score and indicates firms' ability to reduce the usage of hazardous materials, energy, water and other valuable natural resources. Finally, CER is the firms' annual average of the aforementioned three scores. All environmental scores range between zero (0) and 100 so the higher the score the more environmentally responsible and efficient is the firm.

ROE is return on equity and is estimated as net income before tax to common equity and PROF is the net profit margin measured as net income before tax to sales revenue. If H1 is valid and profitability impacts positively on CER (and its components) we expect a positive coefficient in this variable. If the CER and its dimensions produce positive and significant coefficients, this will provide support to the bidirectional association between CER and CFP. LEV is financial leverage capturing firm risk and is measured as the ratio of total debt to total assets. Singh and Faircloth (2005) and Shin-Ping and Tsung-Hsien (2009) ague that highly leveraged firms face reduced investment opportunities which may have a negative impact on financial performance. So we expect a negative impact of LEV on CFP. SIZE is the natural logarithm of total assets and captures firm size. GROWTH captures firms' growth opportunities and is estimated as the annual percentage change of sales revenues. Orlitzky (2001) and Dimitropoulos and Tsagkanos (2012) document that firm size and growth opportunities are positively related to firm performance because larger and more growing firms can take advantage of economies of scale in their operations, can exert greater control on their external stakeholders and resources leading to enhanced financial performance.

DUAL is a dummy variable taking unity (1) if the roles of CEO and board chairman are occupied by the same individual and zero (0) otherwise. As Mitchell (2005) argues, when the CEO serves as the chairman of the board he/she may hide critical information from the directors (or even falsified) and so the CEO may influence and manipulate the decisions of the board for private benefit. This may have a detrimental impact on CFP. BDSIZE is the number of directors serving on the board. Larger boards are considered as a mechanism which can provide an increased pool of expertise, greater management monitoring and access to a wider range of contracts and resources (Goodstein et al. 1994; Reddy et al. 2010; Cheng 2008), thus making more efficient business decisions leading to enhance financial performance.

Finally, CER_COM is a dichotomous variable receiving unity (1) if the firm has established and official environmental responsibility (or sustainability) committee. Previous evidence provided by Winston (2009), Ambec and Lanoie (2008) and Chen et al. (2018) indicate that the existence of a CER or sustainability committee within the firm is associated with economic advantages and cost reductions (from preventing negative events which can be financially detrimental to the firm) leading enhanced financial performance. Finally the model includes year, country and industry specific effects in order to capture any variability in the data due to different country and sector settings not captured by the independent variables and in order to control for heteroscedasticity and serial correlation we calculated the robust standard errors for clustered data (Testa and D'Amato 2017).

5.4 Empirical Results

5.4.1 Descriptive Statistics and Correlations

Table 5.1 presents the descriptive statistics of the variables utilized in the regression models. First of all, the mean CER is 64.5 (with a standard deviation of 21.1) indicating that the sample firms have a satisfactory environmental performance score. The highest average score is that of resources usage (66.5) followed by emission reduction activities (65.9) while environmental innovation activities lag behind with an average of 61.3. The sample firms are highly profitable (mean ROE 79.5 and mean PROF 26.2) and present significant growth opportunities (mean GROWTH at 58.9). However, our sample firms are highly leveraged since the average total debt covers more than four times the sample firms' total assets. Finally,

Table 5.1 Descriptive statistics of the sample variables

Variables	Mean	Standard deviation	Min	Max
CER	64.597	21.167	3.161	99.429
EN_INNOV	61.333	25.962	0.202	99.919
EMM	65.915	26.189	0.161	99.919
RES_USE	66.542	25.730	0.805	99.844
ROE	79.511	34.815	−33.315	118.315
PROF	26.239	57.632	−16.753	189.149
LEV	4.625	4.319	−0.682	25.691
SIZE	20.496	2.848	2.302	30.311
GROWTH	58.915	141.720	−69.711	409.071
DUAL	0.717	0.450	0.000	1.000
BDSIZE	12.271	4.703	1	38
CER_COM	0.552	0.497	0.000	1.000

Table 5.2 CFP based on CER ranked portfolios

Variables	P1—High CER	P2—Medium CER	P3—Low CER	Difference (P1 − P3)	p-value
ROE	71.460	163.203	62.085	9.402	0.322
PROF	26.122	35.023	10.741	15.381***	0.001
LEV	5.143	1.376	1.514	3.629***	0.001
SIZE	20.440	20.856	20.807	−0.367***	0.001
GROWTH	60.036	77.723	2.685	57.351***	0.001

in the majority of the observations (71.7 per cent) our sample firms have a governance structure were the CEO serves also as the board chairman, boards are comprised by 12 members but the most important is that more than in half of our sample firms there is an established environmental responsibility committee (average 0.55) suggesting that the majority of the sample firms have been considered CER strategically and invest resources towards that direction.

Table 5.2 presents the differences between the main financial variables after separating firms within low, medium and high CER performance groups in order to examine the extent to which CER impacts on CFP. Following Suto and Takehara (2018) we followed the portfolio formation approach and estimated the average of the CER variable per year and country and assigned each firm as low CER firm if its annual score is below the 1st quartile, high CER firms are those that their CER score is above the 3rd quartile while all other firm observations lying between the 1st and 3rd quartiles are assigned as medium CER firms. The column under the title "difference" reports the average CER difference between the high CER portfolio firms and the low CER portfolio. The "p-value" column presents the corresponding probability values from estimating the Welch two-sample t-test and asterisks indicate statistical significance at the 1 per cent significance level (Suto and Takehara 2018). As we can see from the last column, high CER performance firms are characterized by statistically higher financial performance (profit margin), more leverage and growth opportunities, and smaller size. These findings corroborate arguments by Suto and Takehara (2018) that more socially responsible firms are those with higher financial performance. However, the analysis so far does not allow us to infer if the association runs from CFP to CER or the opposite. For this reason we will proceed with the SUR analysis in the following sub-section.

Table 5.3 presents the Pearson correlation coefficients between the sample variables. The majority of the correlation coefficients are significant with economic meaning, and are not very high indicating the absence of multicollinearity on the data. The VIF estimation of the sample variables provided values that were far below the threshold of 10 indicating no multicollinearity. As we can see, CER is positively associated with PROF, while is negatively and significantly associated with SIZE and CEO duality, suggesting that larger firms and those with a CEO serving as board chairman are associated with lower CER performance. Nevertheless, firms with larger boards and especially those with a sustainability committee tend to perform better on the CER score. Also, more profitable firms are positively associated with

Table 5.3 Pearson correlation coefficients of sample variables

Variables	1	2	3	4	5	6	7	8	9	10	11	12
1. ROE	1											
2. PROF	**0.302**	1										
3. CER	0.002	**0.008**	1									
4. EN_INNOV	0.005	**0.019**	**0.728**	1								
5. EMM	0.001	**0.011**	**0.854**	**0.377**	1							
6. RES_USE	−0.002	**−0.010**	**0.863**	**0.405**	**0.710**	1						
7. LEV	−0.002	**−0.039**	0.005	**0.006**	0.001	**0.006**	1					
8. SIZE	**0.056**	**0.144**	**−0.012**	**0.013**	−0.002	**−0.040**	**−0.040**	1				
9. GROWTH	−0.001	−0.002	−0.002	−0.005	−0.001	0.002	0.001	0.005	1			
10. DUAL	−0.004	**−0.040**	**−0.077**	**−0.055**	**−0.079**	−0.053	0.004	**−0.080**	0.002	1		
11. BDSIZE	**0.017**	**0.061**	**0.059**	**0.060**	**0.055**	**0.030**	0.001	**0.072**	0.004	**−0.125**	1	
12. CER_COM	−0.005	−0.002	**0.507**	**0.317**	**0.443**	**0.481**	0.001	**−0.038**	0.003	**−0.011**	**0.037**	1

Note: Correlation coefficients in bold indicate statistical significance at least at the 5% significance level

SIZE and board size (BDSIZE). The same arguments stand for the rest three environmental performance scores. The only difference is observed in the resource use score (RES_USE) which is negatively associated with PROF and with SIZE. Finally, growth opportunities did not provide any significant correlation coefficients with the CER related and financial performance. Of course correlations do not indicate causality or direction of association so we will proceed with the SUR regression analysis results in order to extract more specific inferences.

Table 5.4 Regression results on the bidirectional association between CER and CFP

	ROE			PROF		
Variables	Coef.	z-stat	P > | z|	Coef.	z-stat	P > | z|
Equation (5.1)—Dep. Variable: CER						
ROE/PROF	0.005***	4.59	0.001	0.003**	2.51	0.012
LEV	0.012	0.89	0.375	0.012	0.89	0.376
SIZE	0.067**	2.04	0.041	0.054	1.63	0.103
GROWTH	−0.001	−0.16	0.873	−0.001	−0.15	0.879
DUAL	−2.862***	−15.63	0.001	−2.838***	−15.50	0.001
BDSIZE	0.077***	4.52	0.001	0.076***	4.46	0.001
CER_COM	21.745***	133.63	0.001	21.752***	133.70	0.001
Constant	52.146***	70.61	0.001	52.408***	70.54	0.001
R-sq	0.265			0.265		
Chi^2 (p-value)	183.17*** (0.001)			181.19*** (0.001)		
Equation (5.2)—Dep. Variable: ROE/PROF						
CER	7.435***	4.59	0.001	0.312**	2.51	0.012
LEV	0.630	0.12	0.904	0.083	0.21	0.836
SIZE	25.583**	2.16	0.031	27.069***	29.36	0.001
GROWTH	−0.006	−0.08	0.937	−0.001	−0.020	0.842
DUAL	11.815	0.18	0.858	−33.07***	−6.42	0.001
BDSIZE	15.796***	2.56	0.010	5.193***	10.82	0.001
CER_COM	−23.98***	−3.52	0.001	−17.07***	−3.21	0.001
Constant	−97.61***	−3.50	0.001	−58.41***	−26.91	0.001
R-sq	0.1588			0.1218		
Chi^2 (p-value)	35.85*** (0.001)			114.03*** (0.001)		
Breusch-Pagan test of independence	5.270**			5.577**		
Chi^2 (p-value)	(0.021)			(0.017)		

Note: **, *** indicate statistical significance at the 5% and 1% significance level respectively, z-statistics in the parenthesis

5.4.2 Regression Results

Table 5.4 presents the regression results from the SUR regression estimation of models (5.1) and (5.2) using the CER variable as the environmental performance measure. The first column uses ROE as the CFP measure and the second column net profit margin (PROF). The Chi-square values are highly significant and the R^2s are very satisfactory considering the size of the sample and other relative studies published in the field. Also the Breusch-Pagan test of independence is statistically significant at the 5 per cent level suggesting that the SUR estimation provides informational value beyond that produced by estimating the relative models separately. The estimation of Eq. (5.1) provided positive and significant coefficients on both CFP variables on CER verifying evidence provided on the previous Chap. 4, suggesting that more profitable firms are associated with more CER related performance (Giannarakis 2014; Syed and Butt 2017). Moreover, the estimation of Eq. (5.2) provided positive and significant coefficients on the CER variable suggesting that CER impacts positively on CFP. This result verifies our main research hypothesis indicating that CER generates competitive resources leading to competitive advantages and enhanced financial performance, corroborating previous empirical evidence by Jiang et al. (2018), Chen et al. (2018) and Fiadrino et al. (2019). Also, the specification of our empirical design verifies arguments by Testa and D'Amato (2017) regarding the bidirectional association between CER and CFP. Our evidence confirm that highly profitable firms are associated with higher CER performance due to their enhanced ability to finance such activities. In turn, higher CER performance contributes to enhanced CFP which feeds back to higher social and environmental investments and CER performance.

Regarding the rest of the control variables, larger firms proved to be more profitable corroborating evidence by Orlitzky (2001) and Dimitropoulos and Tsagkanos (2012) documenting that larger firms can take advantage of economies of scale in their operations, can exert greater control on their external stakeholders and resources leading to enhanced financial performance. BDSIZE produced a positive and significant coefficient for both CFP variables and DUAL produced a negative and significant coefficient on the "PROF" column indicating that firms with larger boards and separate roles between the CEO and board chairman tend to be more profitable relative to firms with smaller boards and CEO duality structures. These findings verify analogous evidence by Mitchell (2005) on CEO duality and Goodstein et al. (1994), Reddy et al. (2010), and Cheng (2008) on board size. Finally, the existence of a CER or sustainability committee (CER_COM) produced highly significant but negative coefficients for both CFP measures. This results contradicts empirical evidence by Winston (2009), Ambec and Lanoie (2008) and Chen et al. (2018), suggesting that CER committees impose significant costs to firm operations deteriorating financial performance. This result is similar to the arguments by Chen et al. (2018) regarding the impact of green (CER related) activities and initiatives on CFP and specifically, CER initiatives are detrimental for CFP per

Table 5.5 Regression results on the bidirectional association between environmental innovation score and CFP

Variables	ROE Coef.	ROE z-stat	ROE P > |z|	PROF Coef.	PROF z-stat	PROF P > |z|
Equation (5.1)—Dep. Variable: EN_INNOV						
ROE/PROF	0.001	1.41	0.158	0.009***	4.66	0.001
LEV	0.018	0.94	0.346	0.018	0.94	0.348
SIZE	0.302***	6.81	0.001	0.272***	6.10	0.001
GROWTH	0.004	1.40	0.161	0.004	1.41	0.157
DUAL	−2.282***	−9.21	0.001	−2.240***	−9.05	0.001
BDSIZE	0.176***	7.62	0.001	0.172***	7.43	0.001
CER_COM	16.616***	75.53	0.001	16.639***	75.56	0.001
Constant	45.085***	45.15	0.001	45.694***	45.50	0.001
R-sq	0.1054			0.1056		
Chi^2 (p-value)	598.39*** (0.001)			601.71*** (0.001)		
Equation (5.2)—Dep. Variable: ROE/PROF						
EN_INNOV	1.671	1.41	0.158	0.429***	4.66	0.001
LEV	0.694	0.13	0.894	0.079	0.20	0.844
SIZE	25.579**	2.16	0.031	26.960***	29.23	0.001
GROWTH	−0.006	−0.09	0.928	−0.001	−0.23	0.818
DUAL	−5.406	−0.08	0.935	−32.99***	−6.41	0.001
BDSIZE	16.074***	2.69	0.009	5.141***	10.71	0.001
CER_COM	−10.786	−1.74	0.081	−17.41***	−3.61	0.001
Constant	−66.85**	−2.46	0.014	−58.71***	−27.80	0.001
R-sq	0.133			0.121		
Chi^2 (p-value)	16.76** (0.019)			115.58*** (0.001)		
Breusch-Pagan test of independence	0.498			5.432**		
Chi^2 (p-value)	(0.480)			(0.019)		

Note: **, *** indicate statistical significance at the 5% and 1% significance level respectively, z-statistics in the parenthesis

se but they can have a positive impact on CFP only if those initiatives lead to enhanced CER performance.

Tables 5.5, 5.6 and 5.7 presents the results from the estimation of the SUR regression models using the CER determinants and specifically the environmental innovation score (EN_INNOV), the emission score (EMM) and resource use score (RES_USE). All CER performance variables produced positive and significant coefficients in almost all tables (except Table 5.5 on the ROE model). These evidence indicate that higher environmental innovation score, emissions score and resource use score lead to higher profitability, corroborating existing evidence in the literature, thus accepting the main research hypothesis. Again, the bidirectional relationship between CER and CFP is corroborated, since financial performance

Table 5.6 Regression results on the bidirectional association between emission score and CFP

	ROE			PROF		
Variables	Coef.	z-stat	P > |z|	Coef.	z-stat	P > |z|
Equation (5.1)—Dep. Variable: EMM						
ROE/PROF	0.008***	4.98	0.001	0.007***	3.57	0.001
LEV	0.005	0.28	0.782	0.005	0.28	0.783
SIZE	0.139***	3.31	0.001	0.118***	2.80	0.005
GROWTH	−0.002	−1.01	0.313	−0.002	−1.00	0.316
DUAL	−3.762***	−16.09	0.001	−3.752***	−15.93	0.001
BDSIZE	0.101***	4.62	0.001	0.098***	4.49	0.001
CER_COM	23.756***	113.50	0.001	23.583***	113.56	0.001
Constant	51.189***	54.30	0.001	51.606***	54.41	0.001
R-sq	0.208			0.208		
Chi^2 (p-value)	133.43*** (0.001)			133.40*** (0.001)		
Equation (5.2)—Dep. Variable: ROE/PROF						
EMM	6.240***	4.98	0.001	0.348***	3.57	0.001
LEV	0.692	0.13	0.894	0.085	0.21	0.932
SIZE	25.206**	2.13	0.034	27.041***	29.33	0.001
GROWTH	−0.004	−0.06	0.952	−0.001	−0.18	0.853
DUAL	14.265	0.22	0.829	−32.66***	−6.34	0.001
BDSIZE	15.732**	2.55	0.011	5.182***	10.80	0.001
CER_COM	−22.71***	−3.46	0.001	−18.47***	−3.61	0.001
Constant	−91.24***	−3.33	0.001	−58.56***	−27.49	0.001
R-sq	0.103			0.122		
Chi^2 (p-value)	39.55*** (0.001)			114.67*** (0.001)		
Breusch-Pagan test of independence	6.196**			5.181**		
Chi^2 (p-value)	(0.013)			(0.033)		

Note: **, *** indicate statistical significance at the 5% and 1% significance level respectively, z-statistics in the parenthesis

impacts positively on CER and vice versa. The rest of the control variables produced similar coefficients in sign and significance compared with those on Table 5.4.

5.4.3 Sensitivity Analysis

In order to check the robustness of the main empirical results, we performed a number of sensitivity tests regarding the research design, and various definitions of the determinant variables. An important issue that has been raised by several studies (Testa and D'Amato 2017; Chen et al. 2018; Suto and Takehara 2018) is the existence of endogeneity meaning that CER performance might not only determine the firm's financial performance, but the causality mighty also go the other way

Table 5.7 Regression results on the bidirectional association between resource use score and CFP

	ROE			PROF		
Variables	Coef.	z-stat	P > |z|	Coef.	z-stat	P > |z|
Equation (5.1)—Dep. Variable: RES_USE						
ROE/PROF	0.006***	4.41	0.001	0.005***	2.66	0.008
LEV	0.015	0.83	0.408	0.015	0.83	0.406
SIZE	−0.240***	−5.83	0.001	−0.229***	−5.55	0.001
GROWTH	−0.002	−0.87	0.384	−0.002	−0.87	0.384
DUAL	−2.542***	−11.11	0.001	−2.551***	−11.15	0.001
BDSIZE	−0.045**	−2.10	0.035	−0.041**	−1.93	0.050
CER_COM	25.044***	123.19	0.001	25.036***	123.16	0.001
Constant	60.164***	65.20	0.001	59.924***	64.55	0.001
R-sq	0.233			0.234		
Chi^2 (p-value)	154.50*** (0.001)			154.45*** (0.001)		
Equation (5.2)—Dep. Variable: ROE/PROF						
RES_USE	5.649***	4.41	0.001	0.265***	2.66	0.008
LEV	0.640	0.12	0.902	0.082	0.23	0.821
SIZE	27.435**	2.31	0.021	27.024***	29.30	0.001
GROWTH	−0.005	−0.06	0.948	−0.001	−0.21	0.833
DUAL	5.149	0.08	0.938	−34.64***	−6.73	0.001
BDSIZE	16.617***	2.70	0.007	5.206***	10.85	0.001
CER_COM	−22.15***	−3.31	0.001	−3.630	−0.70	0.486
Constant	−93.29***	−3.37	0.001	−55.18***	−25.59	0.001
R-sq	0.093			0.121		
Chi^2 (p-value)	34.22*** (0.001)			114.10*** (0.001)		
Breusch-Pagan test of independence	4.864**			1.765		
Chi^2 (p-value)	(0.027)			(0.184)		

Note: **, *** indicate statistical significance at the 5% and 1% significance level respectively, z-statistics in the parenthesis

round. Under this circumstance, the SUR estimation of models (5.1) and (5.2) may partially control for that by producing consistent estimators, yet not efficiently as the Generalized Method of Moments (GMM). For this reason, all regression models controlling for the impact of CER on CFP (model 5.2) and by including all CER components in the model concurrently, were estimated using the GMM method. The GMM method is selected over the Two-Stage Least Squares (2SLS) because in order to determine the 2SLS estimator we assume that the errors are i.i.d., meaning that they are identically and independently distributed over the observations. Having cross-sectional data the departure from independence may reflect neighboring effects in the sense that observations are similar in value (as in the case of board size, CEO duality, CER_COM existence etc.) and this way they share a correlation with the disturbances (Baum 2006). Also, observations that are close in time may be correlated with the strength of the correlation increasing with proximity. Thus, in the case that the errors are not i.i.d., 2SLS estimators can be consistent but inefficient. In other words, the 2SLS estimator is a special case of the GMM estimator. The GMM method allows us to estimate efficient and consistent estimators in the case of non-i.i.d. errors.

The next step to the estimation of the models is to select proper instruments, meaning variables that are correlated with the CER and its components, but uncorrelated with CFP. In this case, we selected two instrumental variables that have the abovementioned characteristic. The first is a dummy variable COMMON receiving (1) if the firm is based on a common law country and (0) otherwise. Evidence on Chap. 4 indicate that firms' legal origin is a significant determinant of CER performance. The second instrument is the firm listing status dummy receiving (1) if the firms is listed in the stock market and (0) otherwise. Listed firms receive higher scrutiny from regulators and investors and usually are subjected to various reporting standards imposed by securities commissions. Listed firms are facing more pressures from internal and external stakeholders to engage in social and environmentally responsible investments in order to improve their image, gain their legitimacy on their operations and even enhance their stock market performance (Monteiro and Aibar-Guzmán 2010). All instrumental variables are measured on an annual basis and serve the scope of this study since they are a determinant of the CER performance and simultaneously these characteristics do not have an impact on the dependent variables (CFP). In order to examine the relevance of the instruments we estimated the Anderson-Rubin likelihood ratio test, which has the null hypothesis that the smallest canonical correlation is zero and assumes that the regressors are distributed multivariate normal. A failure to reject the null hypothesis indicates that the status of the estimated equation is in question (Baum 2006). In our case the Anderson-Rubin test produced a chi-square value of 65.12 ($p > \chi^2 = 0.0001$) indicating that the null hypothesis is rejected at the 1 percent significance level at least. In addition the Wald χ^2 statistic of weak instrument produced a value of 63.87 ($p > \chi^2 = 0.0001$) thus our instruments are relevant and valid for our research setting. Also, the test of over-identifying restrictions (Hansen J-test) was performed, which examines the joint null hypothesis that the excluded instruments are uncorrelated with the error term and is correctly excluded from the second-stage equation. The results of this tests suggested that the used instruments were statistically meaningful, as the Hansen statistic was insignificant in conventional levels.

Table 5.8 presents the GMM empirical results from the estimation of model (5.2) and by including the CER components simultaneously in the same model. The results corroborate the main findings on the previous tables and verify the main research hypothesis. CER produced positive and highly significant coefficients for both ROE and PROF dependent variables. The inclusion of the three CER components when are included concurrently in the model yield results that again corroborate previous findings. Only the EN_INNOV variable produced positive yet insignificant coefficients. Emissions score and resource use score impact positively on CFP supporting H1. The rest of the control variables have the expected sign and significance and are comparable to the results presented in the main Tables. Another sensitivity test regarding the selection of the instruments was to re-estimate models

Table 5.8 GMM Regression results on the association between CER, EN_INNOV, EMM and RES_USE on CFP

Variables	ROE			PROF		
	Coef.	z-stat	P > |z|	Coef.	z-stat	P > |z|
CER	3.863**	2.04	0.041	0.510***	3.22	0.001
LEV	0.300	0.75	0.455	−0.046**	−2.18	0.030
SIZE	76.56***	7.78	0.001	9.049***	5.97	0.001
GROWTH	0.001	0.35	0.729	−0.007***	−3.77	0.001
DUAL	−89.26***	−4.53	0.001	2.799	0.72	0.469
BDSIZE	2.181	0.38	0.703	2.323***	4.95	0.001
CER_COM	15.36***	87.54	0.001	16.25***	93.64	0.001
Constant	−17.70***	−7.18	0.001	−24.19***	−5.34	0.001
R-sq	0.236			0.105		
Chi^2 (p-value)	72.34*** (0.001)			95.54*** (0.001)		
EN_INNOV	12.39	0.67	0.502	4.321	0.97	0.331
EMM	13.25**	2.27	0.023	28.79***	3.23	0.001
RES_USE	11.32**	2.07	0.038	23.76**	2.54	0.011
LEV	−0.515	−0.66	0.511	−0.198	−1.54	0.123
SIZE	68.06***	3.62	0.001	35.54***	6.37	0.001
GROWTH	−0.021	−1.01	0.314	−0.005	−1.16	0.244
DUAL	−19.15***	−2.91	0.004	−71.68***	−4.45	0.001
BDSIZE	32.73**	2.56	0.010	8.436***	4.25	0.001
CER_COM	−23.98***	2.28	0.023	−19.51***	−3.45	0.001
Constant	−84.77***	−4.05	0.001	−72.30***	−7.81	0.001
R-sq	0.203			0.115		
Chi^2 (p-value)	34.83*** (0.001)			87.82*** (0.001)		

Note: **, *** indicate statistical significance at the 5% and 1% significance level respectively, z-statistics in the parenthesis

of Table 5.8 by using the 1 and 2 year lags of the CER and its components as proposed by Suto and Takehara (2018) and Testa and D'Amato (2017). Despite the fact that Larcker and Rusticus (2010) argue that the use of lagged endogenous regressors as the instrument assumes that the exogenous part of the regressor persists over time, but on the contrary, the endogenous part does not persist over time, suggesting that the use of lagged endogenous variables as instruments does not solve the problem of endogeneity. Empirical evidence provided qualitatively similar arguments based on those presented on Table 5.8, so the control for endogeneity and simultaneity should be controlled when examining the association between CER and CFP. Finally, additional sensitivity testing was performed following different variables definitions such as LEV was re-estimated as the ratio of long term debt to total assets, growth was estimated as the logarithm of total revenue, profitability was estimated as return on assets and also Tobin's Q was utilized as an additional performance variable. Results remained qualitatively unchanged compared with those on Tables 5.4, 5.5, 5.6, and 5.7.

5.5 Conclusion

Several studies on the environmental responsibility research field have focused on examining the impact of CER activities and performance on financial performance (CFP). Researchers have expressed different views on how a company's social and environmental performance affects its financial profitability (Clarkson 1995; Jones and Wicks 1999; McWilliams et al. 2006; Wood 1991). The most widely accepted view is that CER activities and performance assist firms to build strong stakeholder relationships based on mutual trust and cooperation, which in turn enhances company profitability through the creation of internal and external resources and gaining competitive advantages (Jiang et al. 2018). The goal of this chapter was to examine the impact of CER activities on the financial performance of 7313 corporations originating from 24 EU countries over the period 2003–2018.

Empirical analysis suggested that CER generates competitive resources leading to competitive advantages and enhanced financial performance, corroborating previous empirical evidence by Jiang et al. (2018), Chen et al. (2018) and Fiadrino et al. (2019). Also, the specification of our empirical design verifies arguments by Testa and D'Amato (2017) regarding the bidirectional association between CER and CFP. Our evidence confirm that highly profitable firms are associated with higher CER performance due to their enhanced ability to finance such activities. In turn higher CER performance contributes to enhance CFP which feeds back to higher social and environmental investments and CER performance. The empirical results remain robust after several sensitivity tests regarding the definition of the independent variables and after controlling for the impact of endogeneity.

References

Ambec, S., and P. Lanoie. 2008. Does it pay to be green? A systematic overview. *Academy of Management Perspectives* 22 (4): 45–62.

Backhaus, K.B., B.A. Stone, and K. Heiner. 2002. Exploring the relationship between corporate social performance and employer attractiveness. *Business & Society* 41: 292–328.

Barnett, M.L. 2007. Shareholder influence capacity and the variability of financial returns to corporate social responsibility. *Academy of Management Review* 32 (3): 794–816.

Baum, C.F. 2006. *An introduction to econometrics using Stata*. College Station, TX: Stata Press Books, StataCorp LP.

Berman, S.L., A.C. Wicks, and T.M. Jones. 1999. Does stakeholder orientation matter? The relationship between stakeholder management models and firm financial performance. *Academy of Management Journal* 42 (2): 488–506.

Berthelot, S., D. Cormier, and M. Magnan. 2003. Environmental disclosure research: Review and synthesis. *Journal of Accounting Literature* 22 (1): 1–44.

Brammer, S., and A. Millington. 2005. Corporate reputation and philanthropy: An empirical analysis. *Journal of Business Ethics* 61 (1): 29–44.

Carroll, A.B. 1979. A three-dimensional conceptual model of corporate performance. *Academy of Management Review* 4 (4): 495–505.

———. 1999. Corporate social responsibility evolution of a definitional construct. *Business and Society* 38 (3): 268–295.

Chen, F., T. Ngniatedema, and S. Li. 2018. A cross-country comparison of green initiatives, green performance and financial performance. *Management Decision* 56 (5): 1008–1032.

Cheng, S. 2008. Board size and the variability of corporate performance. *Journal of Financial Economics* 87 (1): 157–176.

Clarkson, M.B.E. 1995. A stakeholder framework for analyzing and evaluating corporate social performance. *Academy of Management Review* 20 (1): 92–117.

Cochran, P., and R. Wood. 1984. Corporate social responsibility and financial performance. *The Academy of Management Journal* 27 (1): 42–56.

Dimitropoulos, P.E., and A. Tsagkanos. 2012. Financial performance and corporate governance in the European football industry. *International Journal of Sport Finance* 7 (4): 280–308.

Ducassy, I. 2013. Does corporate social responsibility pay off in times of crisis? An alternate perspective on the relationship between financial and corporate social performance. *Corporate Social Responsibility and Environmental Management* 20: 157–167.

Farooq, O., D.E. Rupp, and M. Farooq. 2017. The multiple pathways which internal and external corporate social responsibility influence organizational identification and multifoci outcomes: The moderating role of cultural and social orientations. *Academy of Management Journal* 60: 954–985.

Fiadrino, S., A. Devalle, and V. Cantino. 2019. Corporate governance and financial performance for engaging socially and environmentally responsible practices. *Social Responsibility Journal* 15 (2): 171–185.

Flammer, C. 2013. Corporate social responsibility and shareholder reaction: The environmental awareness of investors. *Academy of Management Journal* 56 (3): 758–781.

Friede, G., T. Busch, and A. Bassen. 2015. ESG and financial performance: Aggregated evidence from more than 2000 empirical studies. *Journal of Sustainable Finance and Investment* 5 (4): 210–233.

Giannarakis, G. 2014. Corporate governance, and financial characteristic effects on the extent of corporate social responsibility disclosure. *Social Responsibility Journal* 10 (4): 569–590.

Godfrey, P.C., and N.W. Hatch. 2007. Researching corporate social responsibility: An agenda for the 21st century. *Journal of Business Ethics* 70 (1): 87–98.

Goodstein, L., K. Gautam, and W. Boeker. 1994. The effect of board size and diversity on strategic change. *Strategic Management Journal* 15: 241–250.

Griffin, J., and J. Mahon. 1997. The corporate social performance and corporate financial performance debate: 25 years of incomparable research. *Business & Society* 36 (1): 5–31.

Hull, C.E., and S. Rothenberg. 2008. Firm performance: The interactions of corporate social performance with innovation and industry differentiation. *Strategic Management Journal* 29: 781–789.

Jiang, Y., X. Xue, and W. Xue. 2018. Proactive corporate environmental responsibility and financial performance: Evidence from Chinese energy enterprises. *Sustainability* 10: 964.

Jones, T., and A. Wicks. 1999. Convergent stakeholder theory. *Academy of Management Review* 24: 208–221.

Kacperczyk, A. 2009. With greater power comes greater responsibility? Takeover protection and corporate attention to stakeholders. *Strategic Management Journal* 30: 261–285.

Kassinis, G.I., and A. Panayiotou. 2006. Perceptions matter: CEO perceptions and firm environmental performance. *Journal of Corporate Citizenship* 23 (23): 67–80.

Klassen, R.D., and C.P. McLaughlin. 1996. The impact of the environmental management on firm performance. *Management Science* 42 (8): 1199–1214.

Laplume, A.O., K. Sonpar, and R.A. Litz. 2008. Stakeholder theory: Reviewing a theory that moves us. *Journal of Management* 34: 1152–1189.

Larcker, D.F., and T.O. Rusticus. 2010. On the use of instrumental variables in accounting research. *Journal of Accounting and Economics* 49 (3): 186–205.

Lee, K., and R. Ball. 2003. Achieving sustainable corporate competitiveness: Strategic link between top management's (green) commitment and corporate environmental strategy. *GMI* 44: 89–104.

Lee, E.M., S.-Y. Park, and H.J. Lee. 2013. Employee perception of CSR activities: Its antecedents and consequences. *Journal of Business Research* 66 (10): 1716–1724.

Li, Y., G.D. Richardson, and D. Thornton. 1997. Corporate disclosure of environmental liability information: Theory and evidence. *Contemporary Accounting Research* 4 (3): 435–474.

Luo, X., and C.B. Bhattacharya. 2006. Corporate social responsibility, customer satisfaction and market value. *Journal of Marketing* 70 (1): 1–18.

Lyon, T.P., and J.W. Mawell. 2008. Corporate social responsibility and the environment: A theoretical perspective. *Review of Environmental Economics and Policy* 2 (2): 240–260.

Margolis, J.D., and J.P. Walsh. 2003. Misery loves companies: Rethinking social initiatives by business. *Administrative Science Quarterly* 48: 268–305.

Marquis, C., and C.L. Qian. 2014. Corporate social responsibility reporting in China: Symbol or substance? *Organizational Science* 25 (1): 127–148.

McGuire, J.B., A. Sundgren, and T. Schneeweis. 1988. Corporate social responsibility and firm financial performance. *Academy of Management Journal* 31: 854–872.

McWilliams, A., and D. Siegel. 2001. Corporate social responsibility: A theory of the firm perspective. *Academy of Management Review* 26 (1): 117–127.

McWilliams, A., D. Siegel, and P.M. Wright. 2006. Guest editor's introduction. Corporate social responsibility: Strategic implications. *Journal of Management Studies* 43 (1): 1–18.

Mitchell, L. 2005. Structural roles, CEOs and the missing link in corporate governance. *Brooklyn Law Review* 70: 1313–1368.

Monteil, I. 2008. Corporate social responsibility and corporate sustainability: Separate pasts, common futures. *Organization and Environment* 21 (3): 245–269.

Monteiro, S., and B. Aibar-Guzmán. 2010. Determinants of environmental disclosure in the annual reports of large companies operating in Portugal. *Corporate Social Responsibility and Environmental Management* 17: 185–204.

Oliver, C., and I. Holzinger. 2008. The effectiveness of strategic political management: A dynamic capabilities framework. *Academy of Management Review* 33: 496–520.

Orlitzky, M. 2001. Does firm size confound the relationship between corporate social performance and firm financial performance? *Journal of Business Ethics* 33: 167–180.

Orlitzsky, M., and J.D. Benjamin. 2001. Corporate social performance and firm risk: A meta analytic review. *Business and Society* 40 (4): 369–396.

Orlitzsky, M., F.L. Schmidt, and S.L. Rynes. 2003. Corporate social and financial performance: A meta analysis. *Organization Studies* 24: 403–441.

Peloza, J., and L. Papania. 2008. The missing link between corporate social responsibility and financial performance: Stakeholder salience and identification. *Corporate Reputation Review* 11: 169–181.

Peterson, D. 2004. The relationship between perceptions of corporate citizenship and organizational commitment. *Business and Society* 43 (3): 296–319.

Porter, M.E., and M.R. Kramer. 2006. Strategy and society: The link between competitive advantage and corporate social responsibility. *Harvard Business Review* 84 (12): 78–92.

Reddy, K., S. Locke, and F. Srcimgeour. 2010. The efficacy of principle-based corporate governance practices and firm financial performance: An empirical investigation. *International Journal of Managerial Finance* 6 (3): 190–219.

Revelli, C., and J.-L. Viviani. 2015. Financial performance of socially responsible investing (SRI): What have we learned? A meta analysis. *Business Ethics: A European Review* 24 (2): 158–185.

Roman, R.M., S. Hayibor, and B.R. Agle. 1999. The relationship between social and financial performance: Repainting a portrait. *Business and Society* 17: 595–612.

Russo, M., and P. Fouts. 1997. A resource-based perspective on corporate environmental performance. *Academy of Management Journal* 40 (3): 54–559.

Sarkar, R. 2008. Public policy and corporate environmental behaviour: A broader view. *Corporate Social Responsibility and Environmental Management* 15 (5): 281–297.

Shin-Ping, L., and C. Tsung-Hsien. 2009. The determinants of corporate performance: A viewpoint from insider ownership and institutional ownership. *Managerial Auditing Journal* 24 (3): 233–247.

Shiu, Y.M., and S.L. Yang. 2017. Does engagement in corporate social responsibility provide strategic insurance-like effects? *Strategic Management Journal* 38: 455–470.

Singh, M., and S. Faircloth. 2005. The impact of corporate debt on long term investment and firm performance. *Applied Economics* 37: 875–883.

Suto, M., and H. Takehara. 2018. *Corporate social responsibility and corporate finance in Japan*. Singapore: Springer Nature.

Syed, M.A., and S.A. Butt. 2017. Financial and non-financial determinants of corporate social responsibility: Empirical evidence from Pakistan. *Social Responsibility Journal* 13 (4): 780–797.

Tang, Z., C.E. Hull, and S. Rothenberg. 2012. How corporate social responsibility engagement strategy moderates the CSR – Financial performance relationship. *Journal of Management Studies* 49: 1274–1303.

Testa, M., and A. D'Amato. 2017. Corporate environmental responsibility and financial performance: Does bidirectional causality work? Empirical evidence from the manufacturing industry. *Social Responsibility Journal* 13 (2): 221–234.

Trendafilova, S., and K. Babiak. 2013. Understanding strategic corporate environmental responsibility in professional sport. *International Journal of Sport Management and Marketing* 13 (1/2): 1–24.

Turban, D., and D.W. Greening. 1996. Corporate social performance and organizational attractiveness to prospective employees. *Academy of Management Journal* 40 (3): 658–672.

Waddock, S.A. 2003. Myths and realities of social investing. *Organization and Environment* 16: 369–380.

Waddock, S.A., and S.B. Graves. 1997. The corporate social performance-financial performance link. *Strategic Management Journal* 18 (4): 303–310.

Wahba, H. 2008. Does the market value corporate environmental responsibility? An empirical examination. *Corporate Social Responsibility and Environmental Management* 15 (1): 89–99.

Walden, W.D., and B.N. Schwartz. 1997. Environmental disclosures and public policy pressure. *Journal of Accounting and Public Policy* 16: 125–154.

Wei, J.C., Z. Ouyang, and H.P. Chen. 2017. Well known or well liked? The effects of corporate reputation on firm value at the onset of a corporate crisis. *Strategic Management Journal* 38: 2103–2120.

Winston, A. 2009. Energize employees with green strategy. *Harvard Business Review* 87 (9): 24.

Wood, D.J. 1991. Corporate social performance revisited. *Academy of Management Review* 16 (4): 691–718.

Xu, J., J. Wei, and L. Lu. 2019. Strategic stakeholder management, environmental corporate social responsibility engagement, and financial performance of stigmatized firms derived from the Chinese special environmental policy. *Business Strategy and Environment* 28: 1027–1044.

Chapter 6
Corporate Environmental Responsibility and Innovative Activities

Innovation is a key to success in today's highly fickle and vague economic environment. Accordingly, the solution to a problem of how innovation arises and is brought has been addressed by academics and practitioners from various perspectives. Many academics in accounting, finance and management have, however, been focusing their attention chiefly on correlation, rather than causality, between certain financial or non-financial information (CER related disclosures) and innovation. On the other hand, researchers in business administration have looked into causality and attempted to reveal the impact of innovation (R&D investments) on voluntary social activities and disclosures. This chapter aims to bridge the gap between theories of accounting and innovation in order to better understand the interconnections between CER and R&D on the firm's financial performance.

6.1 Introduction

Innovation is regarded as what makes living standards, individuals, institutions, economy, and countries more progressive than in the past (OECD/Eurostat 2018). Since innovation has pervasive influences on society and, inter alia, on business environment, innovation in the context of business has been researched in depth by academics and practitioners. As the word "innovation" does not have a strict definition, it has been construed in many ways. For example, OECD/Eurostat (2018, p. 20) defines innovation as "*a new or improved product or process (or combination thereof) that differs significantly from the unit's previous products or processes and that has been made available to potential users (product) or brought into use by the unit (process)*". Whilst innovation is generally associated with something new and better, it is not adequate to analyze the mechanism of innovation without further classification.

P. Dimitropoulos, K. Koronios, *Corporate Environmental Responsibility, Accounting and Corporate Finance in the EU*, CSR, Sustainability, Ethics & Governance, https://doi.org/10.1007/978-3-030-72773-4_6

Innovation in business is frequently classified into two types: technology (product) and business models (process). According to OECD/Eurostat (2018, p. 21), a product innovation is "*a new or improved good or service that differs significantly from the firm's previous goods or services and that has been introduced on the market*", and a business process innovation is "*a new or improved business process for one or more business functions that differs significantly from the firm's previous business processes and that has been brought into use by the firm*". These kinds of taxonomy are instrumental in understanding how innovative technology should be deployed within a competitive environment. Innovation is not merely an invention in that it is necessary for the technology to be permeated widely into society in order to provide value added. Therefore, there are a variety of arguments over business models. Kim and Mauborgne (2005) determine how to differentiate a company's business model from those of competitors and to create uncontested market space called a "blue ocean". McGrath (2010) reveals that successful new business design concepts can be discovered only through a massive amount of experimentation. Rayna and Striukova (2016) integrate various academic researches on business models into a single framework to facilitate analysis. However, it remains unclear that in what situation what kind of business model works effectively. Radical innovation can transform a field through technology or enhanced and more efficient processes, and thus improve firm position in the market and even increase customer commitment towards the firms and its products (Lin 2017).

Innovation has also been researched from the perspective of accounting and finance. Researchers on management accounting endeavor to untangle the mechanism of innovation as a driver of competitiveness. In general, budgetary planning and control is accepted as one of the most useful accounting systems to manage and control a company's financial performance (Emmanuel et al. 2012). The budgetary system is, however, unsuitable for many businesses because the current business environment is too volatile for a manager to forecast based only on financial results that reflect past performance (Otley 1999). In spite of that, financial information referring to historical financial performance is often utilized as managers are likely to be obsessed with inertia (i.e. routines), past experiences, and prior knowledge (Hall 2010; Scapens 2006). A whole different framework is in need to conquer bad habits originated from the traditional managerial methodology.

Corporate social and environmental responsibility activities (CER) have been devised by several managers as a powerful tool for pursuing and achieving innovation while concurrently meeting the demands and needs of various stakeholders. In their seminal paper, McWilliams and Siegel (2000) argue that consumers in general want to purchase products and services that have socially responsible attributes and produced by socially responsible processes. Both these cases are closely connected with processes and product innovation. Also, many firms channel funds in CER activities in order to enhance product reputation and so customers could perceive firm's products as of higher quality. According to Lin et al. (2009), CSR activities create competitive advantages for firms and yield significant economic benefits through consumer positive product evaluations, awareness, recommendations and seems that many customers add value to CSR attributes and actions. This so called "halo effect" originates from the fact that many innovative activities (R&D investments) can produce CER-related processes and product innovations where many consumers' value and so CER activities can foster product differentiation and novelty. Lin et al. (2009) provide evidence of a significant positive impact of CSR activities on long-term financial performance for highly innovative firms in Taiwan. Also, Lin (2017) provide evidence for the opposite association of CSR and R&D in the Taiwanese market by documenting that R&D investments is positively associated with financial performance only for firms with engaged in CSR activities.

Moreover, Erhemjamts et al. (2013) examined the association between R&D investments and CSR activities and document that R&D intensive firms are more likely to engage in socially responsible activities and vice versa, while CSR performing firms tend to invest more on capital expenditures. In addition, a recent study by Zou et al. (2019) document that socially responsible investments of firms' within developed markets contribute positively to abnormal stock returns, arguing that investors perceive socially and environmentally responsible investments as less risky and with more prospects. Also, this positive impact is more strengthened by R&D expenditures. Consequently, the association between CER and CFP (as evidenced in the previous chapter) could be moderated by R&D investments since R&D could be considered an investment to firm's technical capital leading to knowledge advancement and this in turn contributes to product and process innovation (McWilliams and Siegel 2000). Thus, not including R&D investments as a potential factor affecting financial and environmental performance could lead to inefficient inferences.

The scope of this chapter is to examine the impact of R&D investments (capitalized R&D) on the CER-CFP association as argued by previous studies on the field (McWilliams and Siegel 2000; Lin 2017) while taking into consideration potential endogeneity between the examined factors. We respond to the call for more research on the topic made by Zou et al. (2019), by covering a large multi-country sample of developed economies, and of listed and unlisted European firms over a long period of time and also by considering the bidirectional association between CER and CFP while controlling for R&D capitalization instead of R&D expenses utilized in previous studies. In the following section we provide the theoretical background on the impact of R&D on the CER-CFP relation and the testable hypotheses. The third section discusses the data selection procedure and the research design. The fourth section provides the discussion of empirical results and the last section concludes the chapter.

6.2 R&D Impact on CER-CFP Relation and Testable Hypothesis

The innovation theory originates in the seminal work by Schumpeter and Nichol (1934), who argued that the driving force of economic growth are the innovation activities of firms. The probability of improved profitability and enhanced earnings prospects are the basic yet not the only sources of innovation. Large firms are considered by many researchers as the actual pioneers of innovative activities because they have sufficient resources to create a state of the art infrastructure and can quickly apply R&D output into practice (Mayhew 1980). These primary studies on innovation gave birth to the renowned resource based view of the firm (RBV), which argues that corporate financial performance is dependent on the firm's absorptive capacity to generate and utilize the necessary types of resources for achieving its goals and gaining competitive advantages. R&D investments could be considered as a crucial activity able to create innovation and further develop knowledge capabilities (Lome et al. 2016). According to Parcharidis and Varsakelis (2010), R&D investments can improve firms' sustainability and performance through the outcome of R&D activities (products or services) and via enhancing firm's absorptive capacity and assimilation of new knowledge within corporate processes. Thus, through R&D activities firms can exploit profitable investments opportunities leading to enhanced corporate performance, sustainability and reduced corporate risk (Suto and Takehara 2018; Del Monte and Papagni 2003; Wang and Tsai 2004; Wang et al. 2017; Lome et al. 2016; Nemlioglu and Mallick 2017). However, several researchers have made the exact opposite arguments regarding the impact of R&D on firms performance since, R&D investments are intangible assets which could not be collateralized as the rest of tangible assets in order to assist the firm raise external financing (debt), leading to increased cost of capital and less profitable investment opportunities. Also, R&D investments are medium to long-term investment projects requiring several years to prosper, and so R&D intensive firm must have the ability to cover the costs of such projects in order to achieve future financial value.

Unlike the CSR-CFP relation which has been examined extensively in the literature (as documented in the previous chapter) and the R&D-CFP relation has also been thoroughly considered by previous researchers, the link between CSR-CFP based on the corporate decisions on R&D has not been well explored. Evidence and theoretical assumptions so far have pointed towards two opposing frameworks suggesting that (1) R&D exerts a mediation effect on CSR-CFP relation and yields a positive impact on CFP but only for firms with significant CSR activities, thus pointing towards a positive association between innovation and CSR while (2) the opposing view argues that R&D plays a substitution role on CSR-CFP relation pointing towards a negative impact of R&D on the CSR-CFP association.

The first strand of literature originates in the theory of firm perspective which dictates that the goal of the firm is to maximize profits and ultimately shareholder value. Under this framework McWilliams and Siegel (2001) suggested that social activities could be seen as an investment or differentiation strategy abiding to both shareholders and stakeholders goals. Specifically, if firms wish to accommodate stakeholder demands they have to utilize CER related resources which could include capital investments in let's say new environmental friendly equipment and production facilities that reduce emissions and lead to efficient resource utilization and even further creating new products and services that are environmentally responsible (Erhemjamts et al. 2013). Thus, investing in CER firms can enjoy lower cost of capital allowing them to accept more investment projects pleasing investors and shareholders (El Ghoul et al. 2011). So under this theoretical strand of the literature, profit maximizing firms would engage in such CER activities and investments which the potential future revenues (or profits) could cover the increased costs of such activities. Moreover, Luo and Bhattacharya (2006) argue that large investments in R&D could generate favorable attributions regarding corporate motivation to engage in CER activities. Improved quality products (as the output of R&D activities) could generate customer satisfaction and loyalty leading to positive perceptions of stakeholders regarding firm's social-environmental responsibility. In other words, firm's commitment to R&D is positively perceived as a socially responsible activity (Zou et al. 2019). Thus, this strand of the literature argues for a mediating impact of R&D on the CER-CFP relation leading to a positive impact of R&D on CER and financial performance.

On the contrary, the second strand of the literature argues about a substitution effect of R&D on the CER-CFP relation which is based on the attribution theory of the firm. This theory postulates that a firm is perceived as legitimate when it satisfies both the social and pragmatic sides of legitimacy (Zou et al. 2019). This means that a firm which does not innovate and improves process and products via R&D investments, is less probable to generate a positive impact only through CER activities. This arguments originates in Hull and Rothenberg (2008) who document that firms which gain competitive advantages and differentiate themselves from competition through R&D have less need to achieve such differentiation via CER. So less innovative firms can resort to increased CER activities in order to gain such differentiation (or competitive advantages) and enhance their financial performance. Consequently, R&D could play a substitution role on the CER-CFP relation and so it is expected to have a negative impact on CSR and corporate financial performance. Due to the opposing views in the literature our main research hypotheses are formulated as follows:

H1a: R&D investments have a mediation effect on the CER-CFP relation.
H1b: R&D investments have a substitution effect on the CER-CFP relation.

6.3 Data Selection and Research Design

6.3.1 Data Selection Procedure

The current chapter utilizes a sample of corporations (both listed and unlisted) from 23 EU-member countries over the period 2003–2018. All financial, governance and environmental responsibility data have been extracted from Datastream database. The sample collection procedure started by including all corporations that have been covered by the environmental, social and governance (ESG) scores provided by Datastream during the sample period. 2003 has been selected as the initial sample year since that year was the first year with significant coverage (taking into consideration that this database officially was initiated in 2002). From a total of 656,734 firm-year observations we excluded those firms without environmental score data for at least five consecutive years in order to avoid bias in the empirical results and also to facilitate the estimation of differenced variables. Furthermore, we excluded firms with incomplete financial data namely net accounting income, intangible assets and R&D investments and also we excluded firms with zero total assets and sales and those firms that did not have at least four consecutive years of financial information. Related to short term debt lending, long-term lending and those that do not close their fiscal year on December. In addition, we winsorized the upper and down 1 per cent of the data distribution in order to mitigate the influence of significant outliers in the empirical analysis. After this procedure was finalized, we ended up with a final sample of 121,154 firm-year observations. The analysis and structure of the final sample is provided on Table 4.1 of Chap. 4.

6.3.2 Research Design

In order to examine the validity of the previously mentioned research hypotheses we will consider the case of a bidirectional association between CER and CFP as in the previous chapter and concurrently considering the capitalization of R&D investments. The reason for selecting the capitalized R&D instead of the expenses is because International Financial Reporting Standards (IFRS) dictate that R&D costs can be capitalized (reported in the balance sheet as intangible assets and depreciated over their useful lives) only if the firm is able to prove the (a) feasibility of the projects' completion and selling potential, (b) the future potential economic benefits that they will be generated by the firm and (c) to realistically measure the expenses attributed to the projects' development phases (IAS 38.57). This means that the capitalized R&D costs are denoted as those that are expected by the firm to create future economic benefits. On the contrary, expensed R&D (reported to the profit and loss account) do not fulfil the abovementioned IFRS criteria for capitalization, thus is considered as payments with less probable future economic benefits. So capitalized R&D investments are considered by the market as a signal for those projects that are more probable to be successful. Previous studies by Dinh et al. (2016) and Kreß et al. (2019) indicate that capitalized R&D is associated with higher market values, future profitability and lower cost of debt financing, suggesting that capitalized R&D indicate future benefits to market participants.

In order to control for this association we will estimate a system of seemingly unrelated regressions (SUR) as in Chap. 5. This design allows us to impose or test cross-sectional restrictions, and most importantly to gain efficiency because we expect that the error terms across equations are contemporaneously correlated. In order to test whether the application of the SUR method yields a significant gain in efficiency (beyond the separate estimation of these equations) we performed the Breusch-Pagan test of independence, which LM statistic sums the squared correlations between residual vectors e and u with the null hypothesis of a zero contemporaneous covariance between the error terms of different equations (Baum 2006). The rejection of the null hypothesis indicates that the SUR estimation yields a significant gain compared to the separate estimation. The majority of the Breusch-Pagan tests provided on the result tables below are highly significant at least at the 5 per cent level indicating that the implementation of the SUR method is more efficient than the separate estimation as in previous studies on the field. The functional form of the system of equations includes the average CER for each firm and year estimated as the annual average of the ESG resource use score, emissions reduction score and environmental innovation score, where i denotes the firm and t the year and e is the error term:

$$\begin{aligned} CER_{it} = {} & a_0 + a_1 ROE_{it} + a_2 R\&D\_D_{it} + a_3 R\&D\_D_{it} * ROE_{it} + a_4 SIZE_{it} \\ & + a_5 LEV_{it} + a_6 GROWTH_{it} + a_7 DUAL_{it} + a_8 BDSIZE_{it} \\ & + a_9 CER\_COM_{it} + Year\ F.E + Industry\ F.E. + Country\ F.E. \\ & + e_{it} \end{aligned} \tag{6.1}$$

$$\begin{aligned} ROE_{it} = {} & a_0 + a_1 CER_{it} + a_2 R\&D\_D_{it} + a_3 R\&D\_D_{it} * CER_{it} + a_4 SIZE_{it} \\ & + a_5 LEV_{it} + a_6 GROWTH_{it} + a_7 DUAL_{it} + a_8 BDSIZE_{it} \\ & + a_9 CER\_COM_{it} + Year\ F.E + Industry\ F.E. + Country\ F.E. \\ & + u_{it} \end{aligned} \tag{6.2}$$

$$\begin{aligned} EN\_INNOV_{it} = {} & a_0 + a_1 ROE_{it} + a_2 R\&D\_D_{it} + a_3 R\&D\_D_{it} * ROE_{it} \\ & + a_4 SIZE_{it} + a_5 LEV_{it} + a_6 GROWTH_{it} + a_7 DUAL_{it} \\ & + a_8 BDSIZE_{it} + a_9 CER\_COM_{it} + Year\ F.E \\ & + Industry\ F.E. + Country\ F.E. + e_{it} \end{aligned} \tag{6.1a}$$

$$\begin{aligned} ROE_{it} = {} & a_0 + a_1 EN\_INNOV_{it} + a_2 R\&D\_D_{it} + a_3 R\&D\_D_{it} \\ & * EN\_INNOV_{it} + a_4 SIZE_{it} + a_5 LEV_{it} + a_6 GROWTH_{it} \\ & + a_7 DUAL_{it} + a_8 BDSIZE_{it} + a_9 CER\_COM_{it} + Year\ F.E \\ & + Industry\ F.E. + Country\ F.E. + u_{it} \end{aligned} \tag{6.2a}$$

$$EMM_{it} = a_0 + a_1ROE_{it} + a_2R\&D\_D_{it} + a_3R\&D\_D_{it} * ROE_{it} + a_4SIZE_{it} + a_5LEV_{it} + a_6GROWTH_{it} + a_7DUAL_{it} + a_8BDSIZE_{it} + a_9CER\_COM_{it} + Year\ F.E + Industry\ F.E. + Country\ F.E. + e_{it} \quad (6.1b)$$

$$ROE_{it} = a_0 + a_1EMM_{it} + a_2R\&D\_D_{it} + a_3R\&D\_D_{it} * EMM_{it} + a_4SIZE_{it} + a_5LEV_{it} + a_6GROWTH_{it} + a_7DUAL_{it} + a_8BDSIZE_{it} + a_9CER\_COM_{it} + Year\ F.E + Industry\ F.E. + Country\ F.E. + u_{it} \quad (6.2b)$$

$$RES\_USE_{it} = a_0 + a_1ROE_{it} + a_2R\&D\_D_{it} + a_3R\&D\_D_{it} * ROE_{it} + a_4SIZE_{it} + a_5LEV_{it} + a_6GROWTH_{it} + a_7DUAL_{it} + a_8BDSIZE_{it} + a_9CER\_COM_{it} + Year\ F.E + Industry\ F.E. + Country\ F.E. + e_{it} \quad (6.1c)$$

$$ROE_{it} = a_0 + a_1RES\_USE_{it} + a_2R\&D\_D_{it} + a_3R\&D\_D_{it} * RES\_USE_{it} + a_4SIZE_{it} + a_5LEV_{it} + a_6GROWTH_{it} + a_7DUAL_{it} + a_8BDSIZE_{it} + a_9CER\_COM_{it} + Year\ F.E + Industry\ F.E. + Country\ F.E. + u_{it} \quad (6.2c)$$

R&D_D is a dummy variable receiving unity (1) if a firm capitalizes R&D investments within a given fiscal year and zero (0) otherwise. The rest of the independent variables are estimated as mentioned in Chap. 5. Specifically, EN_INNOV denotes each company's environmental innovation score, EMM denotes companies' emission reduction score and RES_USE denotes the firms' resource use score and indicates firms' ability to reduce the usage of hazardous materials, energy, water and other valuable natural resources. Finally, CER is the firms' annual average of the aforementioned three scores. All environmental scores range between zero (0) and 100 so the higher the score the more environmentally responsible and efficient is the firm. ROE is return on equity and is estimated as net income before tax to common equity. If H1a is valid and R&D investments impact positively on CER (given firm profitability) we expect a positive coefficient in the interaction variable R&D_D*ROE. If coefficient a_3 is negative and significant then the substitution hypothesis (H1b) will be accepted, suggesting that innovative firms resort less to CER activities for gaining competitive advantages, something which can be achieved through their R&D activities (Hull and Rothenberg 2008). As for the CER and its dimensions, if they produce positive and significant coefficients, this will provide support to the bidirectional association between CER and CFP.

The rest of the control variables have been utilized in the previous chapter as well. LEV is financial leverage capturing firm risk and is measured as the ratio of total debt to total assets. SIZE is the natural logarithm of total assets and captures firm size and according to our previous findings on Chap. 5, firm size is expected to have a positive impact on the examined association. GROWTH captures firms' growth opportunities and is estimated as the annual percentage change of sales revenues. DUAL is a dummy variable taking unity (1) if the roles of CEO and board chairman are occupied by the same individual and zero (0) otherwise and according to our previous findings is expected to be negatively associated to financial performance This may have a detrimental impact on CFP. BDSIZE is the number of directors serving on the board. Larger boards are considered as a mechanism which can provide an increased pool of expertise, greater management monitoring and access to a wider range of contracts and resources (Reddy et al. 2010; Cheng 2008), thus making more efficient business decisions leading to enhance financial performance as evidence on the previous chapter. Finally, CER_COM is a dichotomous variable receiving unity (1) if the firm has established and official environmental responsibility (or sustainability) committee. Empirical evidence in Chap. 5 suggest that CER committees are detrimental to financial performance but strongly associated to firm's CER focus and orientation. Finally the model includes year, country and industry specific effects in order to capture any variability in the data due to different country and sector settings not captured by the independent variables and in order to control for heteroscedasticity and serial correlation we calculated the robust standard errors for clustered data at the firm level, so after following these steps we can ensure that the results are based on statistical testing that effectively address deviations from standard regression assumptions (Testa and D'Amato 2017; Erhemjamts et al. 2013).

6.4 Empirical Results

Table 6.1 presents the descriptive statistics of the variables utilized in the regression models. First of all, the mean CER is 64.5 (with a standard deviation of 21.1) indicating that the sample firms have a satisfactory environmental performance score. The highest average score is that of resources usage (66.5) followed by emission reduction activities (65.9) while environmental innovation activities lag behind with an average of 61.3. The sample firms are highly profitable (mean ROE

Table 6.1 Descriptive statistics of the sample variables

Variables	Mean	Standard deviation	Min	Max
CER	64.597	21.167	3.161	99.429
EN_INNOV	61.333	25.962	0.202	99.919
EMM	65.915	26.189	0.161	99.919
RES_USE	66.542	25.730	0.805	99.844
ROE	79.511	34.815	−33.315	118.315
R&D_D	0.114	0.318	0	1
R&D_CAP	0.066	0.445	0.000	99.733
LEV	4.625	4.319	−0.682	25.691
SIZE	20.496	2.848	2.302	30.311
GROWTH	58.915	141.720	−69.711	409.071
DUAL	0.717	0.450	0.000	1.000
BDSIZE	12.271	4.703	1	38
CER_COM	0.552	0.497	0.000	1.000

Table 6.2 R&D_CAP based on CER ranked portfolios

Variables	P1—High CER	P2—Medium CER	P3—Low CER	Difference (P1 − P3)	p-value
R&D_CAP	0.061	0.084	0.095	−0.034***	0.001

79.5) and present significant growth opportunities (mean GROWTH at 58.9). However, our sample firms are highly leveraged since the average total debt covers more than four times the sample firms' total assets. Also, in the majority of the observations (71.7 per cent) our sample firms have a governance structure were the CEO serves also as the board chairman, boards are comprised by 12 members but the most important is that more than in half of our sample firms there is an established environmental responsibility committee (average 0.55) suggesting that the majority of the sample firms have been considered CER strategically and invest resources towards that direction. Finally, 11 per cent of our sample firms capital their R&D investments which cover up to 6.6 per cent of annual total assets (R&D_CAP) with a standard deviation of 0.44 indicating significant variability in the R&D capitalization rates in the sample firms.

Table 6.2 presents the differences between the R&D capitalization ratio (R&D investments to total assets) after separating firms within low, medium and high CER performance groups in order to examine the extent to which CER impacts on R&D investments. Following Suto and Takehara (2018) we followed the portfolio formation approach and estimated the average of the CER variable per year and country and assigned each firm as low CER firm if its annual score is below the 1st quartile, high CER firms are those that their CER score is above the 3rd quartile while all other firm observations lying between the 1st and 3rd quartiles are assigned as medium CER firms. The column under the title "difference" reports the average CER difference between the high CER portfolio firms and the low CER portfolio. The "p-value" column presents the corresponding probability values from estimating the Welch two-sample t-test and asterisks indicate statistical significance at the 1 per cent significance level (Suto and Takehara 2018). As we can see from the last column, high CER performance firms are characterized by statistically less R&D investments (0.061) compared to their low CER counterparts (0.095). These findings provide an initial indication supporting arguments by Hull and Rothenberg (2008) who document that firms which gain competitive advantages and differentiate themselves from competition through R&D have less need to achieve such differentiation via CER, so they invest less resources on such activities.

Table 6.3 presents the Pearson correlation coefficients between the main variables of interest. The majority of the correlation coefficients are significant with economic meaning, and are not very high indicating the absence of multicollinearity on the data. The VIF estimation of the sample variables provided values that were far below the threshold of 10 indicating no multicollinearity. As we can see, CER is negatively associated with R&D verifying the results of the previous table that less R&D intensive firms resort to CER activities in order to differentiate themselves and gain market advantages. Also, R&D is negatively and significantly correlated with

Table 6.3 Pearson correlation coefficients of sample variables

Variables	R&D_D	CER	EN_INNOV	EMM	RES_USE
1. R&D_D	1				
2. CER	**−0.006**	1			
3. EN_INNOV	0.001	**0.728**	1		
4. EMM	**−0.012**	**0.854**	**0.377**	1	
5. RES_USE	−0.003	**0.863**	**0.405**	**0.710**	1

Note: Correlation coefficients in bold indicate statistical significance at least at the 5% significance level

EMM (emission score) (−0.012) while the coefficients with the other two scores were not significant within conventional levels. The correlation coefficients of the rest independent variables included in the regression models (not tabulated in this chapter) are provided on Table 5.2 of Chap. 5 so they are not reported here for brevity. Of course correlations do not indicate causality or direction of association so we will proceed with the SUR regression analysis results in order to extract more specific inferences.

6.4.1 Regression Results

Table 6.4 presents the regression results from the SUR regression estimation of models (6.1) and (6.2) using the CER variable as the environmental performance measure. The Chi-square values are highly significant and the R^2s are very satisfactory considering the size of the sample and other relative studies published in the field. Also the Breusch-Pagan test of independence is statistically significant at the 5 per cent level suggesting that the SUR estimation provides informational value beyond that produced by estimating the relative models separately. The estimation of Eq. (6.1) provided positive and significant coefficient of ROE on CER verifying evidence provided on the previous Chap. 5, suggesting that more profitable firms are associated with more CER related performance (Giannarakis 2014; Syed and Butt 2017). Moreover, the estimation of Eq. (6.2) provided positive and significant coefficients on the CER variable suggesting that CER impacts positively on ROE. However, the R&D_D variable and its interaction with ROE yield insignificant coefficients in both model estimations leading us to accept H1b. This result corroborates arguments in the literature that less innovative firms can resort to increased CER activities in order to gain market differentiation (or competitive advantages) and enhance their financial performance. Consequently, R&D investments in our EU sample plays a substitution role on the CER-CFP relation and so it has a non-significant impact on CSR and corporate financial performance. This result verifies findings on Chap. 5 indicating that CER generates competitive resources leading to competitive advantages and enhanced financial performance for less R&D intensive firms. Also, the specification of our empirical design verifies arguments by

Table 6.4 SUR results on the bidirectional association between CER and CFP considering R&D investments

Variables	Coef.	z-stat	P > lzl
Equation (6.1)—Dep. Variable: CER			
ROE	0.005***	4.60	0.001
R&D_D	0.088	0.43	0.670
R&D_D*ROE	−0.020	−0.62	0.533
SIZE	0.066**	2.01	0.044
LEV	0.012	0.89	0.375
GROWTH	−0.001	−0.16	0.875
DUAL	−2.857***	−15.59	0.001
BDSIZE	0.077***	4.52	0.001
CER_COM	21.740***	133.56	0.001
Constant	52.140***	70.57	0.001
R-sq	0.264		
Chi^2 (p-value)	183.98*** (0.001)		
Equation (6.2)—Dep. Variable: ROE			
CER	7.824***	4.53	0.001
R&D_D	55.162	0.23	0.817
R&D_D*CER	−2.537	−0.72	0.472
SIZE	26.864**	2.26	0.024
LEV	0.605	0.12	0.907
GROWTH	−0.006	−0.09	0.932
DUAL	12.584	0.19	0.849
BDSIZE	15.059***	2.43	0.015
CER_COM	−23.92***	−3.51	0.001
Constant	−100.4***	−3.54	0.001
R-sq	0.103		
Chi^2 (p-value)	38.47*** (0.001)		
Breusch-Pagan test of independence	5.283**		
Chi^2 (p-value)	(0.021)		

Note: **, *** indicate statistical significance at the 5% and 1% significance level respectively, z-statistics in the parenthesis

Testa and D'Amato (2017) regarding the bidirectional association between CER and CFP and specifically, higher CER performance contributes to enhanced CFP which feeds back to higher social and environmental investments and CER performance, but for non-R&D intensive firms.

Regarding the rest of the control variables, larger firms proved to be more profitable corroborating evidence by Orlitzky (2001) and Dimitropoulos and Tsagkanos (2012) documenting that larger firms can take advantage of economies of scale in their operations, can exert greater control on their external stakeholders and resources leading to enhanced financial performance. BDSIZE produced a positive and significant coefficient and DUAL produced a negative and significant coefficient on the model (6.1) indicating that firms with larger boards and separate

Table 6.5 SUR results on the bidirectional association between EN_INNOV and CFP considering R&D investments

Variables	Coef.	z-stat	P > lzl
Equation (6.1a)—Dep. Variable: EN_INNOV			
ROE	0.006***	4.62	0.001
R&D_D	0.441	1.57	0.116
R&D_D*ROE	−0.021	−0.48	0.634
SIZE	0.298***	6.70	0.001
LEV	0.018	0.95	0.343
GROWTH	0.001	1.41	0.159
DUAL	−2.281***	−9.21	0.001
BDSIZE	0.178***	7.69	0.001
CER_COM	16.605***	75.47	0.001
Constant	45.065***	45.12	0.001
R-sq	0.105		
Chi^2 (p-value)	597.71*** (0.001)		
Equation (6.2a)—Dep. Variable: ROE			
EN_INNOV	1.824***	4.40	0.001
R&D_D	−66.38	−0.35	0.728
R&D_D*EN_INNOV	−0.684	−0.24	0.811
SIZE	26.656**	2.24	0.025
LEV	0.663	0.13	0.898
GROWTH	−0.007	−0.10	0.922
DUAL	−4.574	−0.07	0.945
BDSIZE	15.319**	2.47	0.013
CER_COM	−10.79*	−1.74	0.081
Constant	−67.078**	−2.44	0.015
R-sq	0.093		
Chi^2 (p-value)	18.93** (0.025)		
Breusch-Pagan test of independence	0.513		
Chi^2 (p-value)	(0.473)		

Note: **, *** indicate statistical significance at the 5% and 1% significance level respectively, z-statistics in the parenthesis

roles between the CEO and board chairman tend to be more profitable relative to firms with smaller boards and CEO duality structures. These findings verify analogous evidence by Mitchell (2005) on CEO duality and Goodstein et al. (1994), Reddy et al. (2010), and Cheng 2008 on board size. Finally, the existence of a CER or sustainability committee (CER_COM) produced highly significant but negative coefficients on ROE. This results contradicts empirical evidence by Winston (2009), Ambec and Lanoie (2008) and Chen et al. (2018), suggesting that CER committees impose significant costs to firm operations deteriorating financial performance. This result is similar to findings on Chap. 5 and the arguments by Chen et al. (2018).

Tables 6.5, 6.6 and 6.7 presents the results from the estimation of the SUR regression models using the CER determinants and specifically the environmental

Table 6.6 SUR results on the bidirectional association between EMM and CFP considering R&D investments

Variables	Coef.	z-stat	P > lzl
Equation (6.1b)—Dep. Variable: EMM			
ROE	0.007***	4.96	0.001
R&D_D	−0.442*	−1.67	0.095
R&D_D*ROE	0.022	0.55	0.584
SIZE	0.143***	3.40	0.001
LEV	0.005	0.27	0.788
GROWTH	−0.002	−1.01	0.310
DUAL	−3.753***	−16.05	0.001
BDSIZE	0.097***	4.46	0.001
CER_COM	23.573***	113.46	0.001
Constant	51.219***	54.31	0.001
R-sq	0.208		
Chi^2 (p-value)	133.33*** (0.001)		
Equation (6.2b)—Dep. Variable: ROE			
EMM	6.684***	4.91	0.001
R&D_D	61.036	0.30	0.762
R&D_D*EMM	−2.544	−0.89	0.376
SIZE	26.492**	2.23	0.026
LEV	0.673	0.13	0.897
GROWTH	−0.004	−0.06	0.948
DUAL	14.907	0.23	0.822
BDSIZE	15.036**	2.43	0.015
CER_COM	−22.594***	−3.44	0.001
Constant	−94.07***	−3.39	0.001
R-sq	0.105		
Chi^2 (p-value)	42.18*** (0.001)		
Breusch-Pagan test of independence	6.142**		
Chi^2 (p-value)	(0.013)		

Note: **, *** indicate statistical significance at the 5% and 1% significance level respectively, z-statistics in the parenthesis

innovation score (EN_INNOV), the emission score (EMM) and resource use score (RES_USE). All CER components produced positive and significant coefficients in almost all tables. These evidence indicate that higher environmental innovation score, emissions score and resource use score lead to higher profitability, corroborating existing evidence in the literature and findings on Chap. 5. The R&D_D variable and its interaction term with ROE produced insignificant coefficients. The only exception was on the estimation of model (6.1b) were the R&D_D variable produced a negative coefficient (−0.442) but slightly significant at the 10 per cent level corroborating our previous arguments regarding the substituting impact of R&D on environmental responsibility. Specifically, R&D intensive firms receive lower emissions score or invest less in emission reduction activities relative to their

Table 6.7 SUR results on the bidirectional association between RES_USE and CFP considering R&D investments

Variables	Coef.	z-stat	P > lzl
Equation (6.1c)—Dep. Variable: RES_USE			
ROE	0.006***	4.42	0.001
R&D_D	0.266	1.03	0.304
R&D_D*ROE	−0.062	−1.54	0.123
SIZE	−0.242***	−5.90	0.001
LEV	0.015	0.83	0.406
GROWTH	−0.002	−0.87	0.387
DUAL	−2.536***	−11.08	0.001
BDSIZE	−0.043**	−2.02	0.044
CER_COM	25.043***	123.16	0.001
Constant	60.136***	65.16	0.001
R-sq	0.233		
Chi^2 (p-value)	154.47*** (0.001)		
Equation (6.2c)—Dep. Variable: ROE			
RES_USE	6.002***	4.33	0.001
R&D_D	13.677	0.07	0.947
R&D_D*RES_USE	−1.837	−0.64	0.525
SIZE	28.639**	2.41	0.016
LEV	0.614	0.12	0.906
GROWTH	−0.005	−0.07	0.944
DUAL	5.686	0.09	0.932
BDSIZE	15.898***	2.5	0.010
CER_COM	−22.110***	−3.31	0.001
Constant	−95.230***	−3.40	0.001
R-sq	0.113		
Chi^2 (p-value)	36.81*** (0.001)		
Breusch-Pagan test of independence	4.894**		
Chi^2 (p-value)	(0.027)		

Note: **, *** indicate statistical significance at the 5% and 1% significance level respectively, z-statistics in the parenthesis

less R&D intensive counterparts. So H1b is accepted. Again, the bidirectional relationship between CER and CFP is corroborated, since financial performance impacts positively on CER and vice versa. The rest of the control variables produced similar coefficients in sign and significance compared with those on Table 6.4.

6.4.2 *Sensitivity Analysis*

In order to check the robustness of the main empirical results, we performed a number of sensitivity tests regarding the research design, and various definitions of the determinant variables. An important issue that has been considered in Chap. 5 as well and by several other studies (Testa and D'Amato 2017; Chen et al. 2018; Suto and Takehara 2018) is the existence of endogeneity meaning that CER performance might not only determine the firm's financial performance, but the causality mighty also go the other way round. Under this circumstance, the SUR estimation of models (6.1) and (6.2) may partially control for that by producing consistent estimators, yet not efficiently as the Generalized Method of Moments (GMM). For this reason, all regression models controlling for the impact of CER on CFP (model 6.2) and by including all CER components in the model concurrently, were estimated using the GMM method. The GMM method is selected over the Two-Stage Least Squares (2SLS) because in order to determine the 2SLS estimator we assume that the errors are i.i.d., meaning that they are identically and independently distributed over the observations. Having cross-sectional data the departure from independence may reflect neighboring effects in the sense that observations are similar in value (as in the case of board size, CEO duality, CER_COM existence etc.) and this way they share a correlation with the disturbances (Baum 2006). The instruments in the GMM estimation were those utilized in the previous chapter and specifically, a dummy variable COMMON receiving (1) if the firm is based on a common law country and (0) otherwise, and LISTED designating if the firm is listed in the stock market and (0) otherwise. In order to examine the relevance of the instruments we estimated the Anderson-Rubin likelihood ratio test, which has the null hypothesis that the smallest canonical correlation is zero and assumes that the regressors are distributed multivariate normal. A failure to reject the null hypothesis indicates that the status of the estimated equation is in question (Baum 2006). In our case the Anderson-Rubin test produced a chi-square value of 71.74 ($p > \chi^2 = 0.0001$) indicating that the null hypothesis is rejected at the 1 percent significance level at least. In addition the Wald χ^2 statistic of weak instrument produced a value of 57.33 ($p > \chi^2 = 0.0001$) thus our instruments are relevant and valid for our research setting. Also, the test of over-identifying restrictions (Hansen J-test) was performed, which examines the joint null hypothesis that the excluded instruments are uncorrelated with the error term and is correctly excluded from the second-stage equation. The results of this test (untabulated) corroborated findings in the previous tables and verified H1b of the substituting impact of R&D on the CER-CFP.

Moreover, we re-estimated all models by separating sample firms into two groups, R&D intensive firms and R&D non-intensive firms based on their decision to capitalize their R&D investments. Results are presented on the following Table 6.8. As we can see, the bidirectional association of CER and CFP is significant for the non R&D intensive firms since the coefficients of CER and ROE on models (6.1) and (6.2) are positive and highly significant. Again the results on Table 6.8 corroborate H1b and support inferences on Table 6.4. In addition, we decided to examine the validity of the main findings by extending the definition of R&D intensive firms by considering all firms that did not report R&D in their balance sheet (as assets) but they expensed it on the profit and loss account . Once again the result remained qualitatively unchanged and indicate that firms which expense their R&D activities are associated with lower CER performance. Also, we reached

Table 6.8 SUR Regression results on the association between CER, and CFP by separating R&D intensive firms

Variables	R&D intensive firms			Non-R&D intensive firms		
	Coef.	z-stat	P > |z|	Coef.	z-stat	P > |z|
Equation (6.1)—Dep. Variable: CER						
ROE	−0.044	−1.39	0.163	0.005***	4.67	0.001
SIZE	0.508***	6.93	0.001	−0.039	−1.08	0.280
LEV	0.661*	1.81	0.070	0.011	0.76	0.450
GROWTH	−0.002	−0.47	0.638	−0.002	−0.11	0.911
DUAL	−2.335***	−5.49	0.001	−2.929***	−14.43	0.001
BDSIZE	0.195***	4.40	0.001	0.059***	3.18	0.001
CER_COM	22.141***	60.36	0.001	21.641***	119.23	0.001
Constant	40.713***	23.76	0.001	54.649***	66.54	0.001
R-sq	0.282			0.261		
Chi^2 (p-value)	378.34*** (0.001)			146.08*** (0.001)		
Equation (6.2)—Dep. Variable: ROE						
CER	−0.004	−1.39	0.163	9.172***	4.67	0.001
SIZE	0.002	0.12	0.908	33.956**	2.31	0.021
LEV	0.161	1.39	0.166	0.636	0.11	0.912
GROWTH	0.002	0.26	0.793	−0.006	−0.08	0.938
DUAL	0.085	0.62	0.532	12.192	0.15	0.881
BDSIZE	0.025*	1.77	0.077	17.769**	2.39	0.017
CER_COM	−0.020	−0.15	0.884	−29.72***	−3.54	0.001
Constant	−0.093	−0.17	0.868	−12.38***	−3.59	0.001
R-sq	0.006			0.093		
Chi^2 (p-value)	7.98 (0.334)			35.99*** (0.001)		
Breusch-Pagan test of independence	0.485			5.444**		
Chi^2 (p-value)	(0.485)			(0.019)		

Note: **, *** indicate statistical significance at the 5% and 1% significance level respectively, z-statistics in the parenthesis

similar findings as those on Tables 6.5–6.7 after including the three CER components concurrently in the model. Finally, additional sensitivity testing was performed following different variables definitions such as LEV was estimated as the ratio of long term debt to total assets, growth was estimated as MV/BV ratio, profitability was estimated as return on assets (ROA). Results remained qualitatively unchanged compared with those on Tables 6.4–6.7.

6.5 Conclusion

Several studies on the environmental responsibility research field have focused on examining the impact of CER activities on financial performance (CFP). However, excluding R&D investments from the econometric models is problematic since according to McWilliams and Siegel (2000) R&D investments contribute to knowledge enhancement, product and process innovation and ultimately in financial performance. So if R&D has a positive impact on CFP then the coefficient of any variable correlated with R&D will be overestimated if R&D is absent from the regression model. The scope of this chapter is to shed further light on the CER-CFP relation documented in Chap. 5, by including the capitalized R&D in the regression equation. For this reason we selected firms which capitalize their R&D in the balance sheet, thus signaling to the market its future potential economic benefits. Moreover, we took into consideration the potential endogeneity between the examined factors and their bidirectional association, evidenced by previous studies on the field. We respond to the call for more research on the topic made by Zou et al. (2019), by covering a large multi-country sample (24 EU countries) of listed and unlisted European firms (7313 corporations) over a long period of time (2003–2018) and also by considering the bidirectional association between CER and CFP between R&D intensive and non-intensive firms.

Empirical analysis suggested that more profitable firms are associated with more CER related performance (Giannarakis 2014; Syed and Butt 2017). Moreover, the estimation of Eq. (6.2) provided positive and significant coefficients on the CER variable suggesting that CER impacts positively on ROE. However, the R&D_D variable and its interaction with ROE yield insignificant coefficients in both model estimations leading us to accept the substitution hypothesis. This result corroborates arguments in the literature that less innovative firms resort to increased CER activities in order to gain market differentiation (or competitive advantages) and enhance their financial performance. Consequently, R&D investments in our EU sample plays a substitution role on the CER-CFP relation and so it has an insignificant impact on CSR and corporate financial performance. Also, we incorporate in the analysis a decomposition of the CER variable and indicate which sub-category of CER performance contributes to CFP between R&D intensive and non-intensive firms, something that has not been considered extensively by previous researchers. All CER components produced positive and significant coefficients in almost all tables indicating that higher environmental innovation score, emissions score and resource use score lead to higher profitability, corroborating existing evidence in the literature and findings on Chap. 5. However R&D investments produced insignificant coefficients. Specifically, R&D intensive firms receive lower emissions score or invest less in emission reduction activities relative to their less R&D intensive counterparts. The results remained robust after several sensitivity tests regarding the estimation method, functional forms of the models and alternative variable definitions.

References

Ambec, S., and P. Lanoie. 2008. Does it pay to be green? A systematic overview. *Academy of Management Perspectives* 22 (4): 45–62.

Baum, C.F. 2006. *An introduction to econometrics using Stata*. College Station, TX: Stata Press Books, StataCorp LP.

Chen, F., T. Ngniatedema, and S. Li. 2018. A cross-country comparison of green initiatives, green performance and financial performance. *Management Decision* 56 (5): 1008–1032.

Cheng, S. 2008. Board size and the variability of corporate performance. *Journal of Financial Economics* 87 (1): 157–176.

Del Monte, A., and E. Papagni. 2003. R&D and the growth of firms: Empirical analysis of a panel of Italian firms. *Research Policy* 32: 1003–1014.

Dimitropoulos, P.E., and A. Tsagkanos. 2012. Financial performance and corporate governance in the European football industry. *International Journal of Sport Finance* 7 (4): 280–308.

Dinh, T., H. Kang, and W. Schultze. 2016. Capitalizing research and development: Signaling or earnings management? *European Accounting Review* 25 (2): 373–401.

El Ghoul, S., O. Guedhami, C. Kwok, and D. Mishra. 2011. Does corporate social responsibility affect the cost of capital? *Journal of Banking and Finance* 35: 2388–2406.

Emmanuel, C., D. Otley, and K. Merchant. 2012. *Accounting for management control*. 2nd ed. London: Cengage Learning EMEA.

Erhemjamts, O., Q. Li, and A. Venkateswaran. 2013. Corporate social responsibility and its impact on firms' investment policy, organizational structure and performance. *Journal of Business Ethics* 118: 395–412.

Giannarakis, G. 2014. Corporate governance, and financial characteristic effects on the extent of corporate social responsibility disclosure. *Social Responsibility Journal* 10 (4): 569–590.

Goodstein, L., K. Gautam, and W. Boeker. 1994. The effect of board size and diversity on strategic change. *Strategic Management Journal* 15: 241–250.

Hall, M. 2010. Accounting information and managerial work. *Accounting, Organizations and Society* 35 (3): 301–315.

Hull, C.E., and S. Rothenberg. 2008. Firm performance: The interactions of corporate social performance with innovation and industry differentiation. *Strategic Management Journal* 29 (7): 781–789.

Kim, W.C., and R. Mauborgne. 2005. Blue ocean strategy: From theory to practice. *California Management Review* 47 (3): 105–121.

Kreß, A., B. Eierle, and I. Tsalavoutas. 2019. Development costs capitalization and debt financing. *Journal of Business Finance and Accounting* 46: 636–685.

Lin, Y.C. 2017. Does R&D investment under corporate social responsibility increase firm performance? *Investment Management and Financial Innovations* 14 (1): 217–223.

Lin, C.-H., H.-L. Yang, and D.-Y. Liou. 2009. The impact of corporate social responsibility on financial performance: Evidence from business in Taiwan. *Technology in Society* 31: 56–63.

Lome, O., A.G. Heggeseth, and Ø. Moen. 2016. The effect of R&D on performance: Do R&D-intensive firms handle a financial crisis better? *Journal of High Technology Management Research* 27 (1): 65–77.

Luo, X., and C.B. Bhattacharya. 2006. Corporate social responsibility, customer satisfaction and market value. *Journal of Marketing* 70 (1): 1–18.

Mayhew, A. 1980. Schumpeterian capitalism versus the "Schumpeterian thesis". *Journal of Economic Issues* 14 (2): 583–592.

McGrath, R.G. 2010. Business models: A discovery driven approach. *Long Range Planning* 43 (2): 247–261.

McWilliams, A., and D. Siegel. 2000. Corporate social responsibility and financial performance: Correlation or misspecification? *Strategic Management Journal* 21: 603–609.

———. 2001. Corporate social responsibility: A theory of the firm perspective. *Academy of Management Review* 26 (1): 117–127.

Mitchell, L. 2005. Structural roles, CEOs and the missing link in corporate governance. *Brooklyn Law Review* 70: 1313–1368.

Nemlioglu, I., and S.K. Mallick. 2017. Do managerial practices matter in innovation and firm performance relations? Evidence from the UK. *European Financial Management* 23 (5): 1016–1061.

OECD/Eurostat. 2018. Oslo Manual 2018: Guidelines for collecting, reporting and using data on innovation [Online]. 4th ed. Paris: OECD publishing. Accessed June 13, 2019, from https://doi.org/10.1787/9789264304604-en

Orlitzky, M. 2001. Does firm size confound the relationship between corporate social performance and firm financial performance? *Journal of Business Ethics* 33: 167–180.

Otley, D. 1999. Performance management: A framework for management control systems research. *Management Accounting Research* 10 (4): 363–382.

Parcharidis, E.G., and N.C. Varsakelis. 2010. R&D and Tobin's q in an emerging financial market: The case of the Athens stock exchange. *Managerial and Decision Economics* 31: 353–361.

Rayna, T., and L. Striukova. 2016. 360° business model innovation: Toward an integrated view of business model innovation. *Research Technology Management* 59 (3): 21–28.

Reddy, K., S. Locke, and F. Srcimgeour. 2010. The efficacy of principle-based corporate governance practices and firm financial performance: An empirical investigation. *International Journal of Managerial Finance* 6 (3): 190–219.

Scapens, R.W. 2006. Understanding management accounting practices: A personal journey. *The British Accounting Review* 38 (1): 1–30.

Schumpeter, J., and A. Nichol. 1934. Robinson's economics of imperfect competition. *Journal of Political Economy* 42 (2): 249–259.

Suto, M., and H. Takehara. 2018. *Corporate social responsibility and corporate finance in Japan*. Singapore: Springer Nature.

Syed, M.A., and S.A. Butt. 2017. Financial and non-financial determinants of corporate social responsibility: Empirical evidence from Pakistan. *Social Responsibility Journal* 13 (4): 780–797.

Testa, M., and A. D'Amato. 2017. Corporate environmental responsibility and financial performance: Does bidirectional causality work? Empirical evidence from the manufacturing industry. *Social Responsibility Journal* 13 (2): 221–234.

Wang, J.-C., and K.-H. Tsai. 2004. *Productivity growth and R&D expenditure in Taiwan's manufacturing firms: Growth and productivity in East Asia. NBER-East Asia Seminar on Economics*. Chicago, IL: University of Chicago Press.

Wang, Y., R. Du, K.S. Koong, and W. Fan. 2017. Effects of R&D policy choice on accounting performance and market value. *R&D Management* 47 (4): 545–556.

Winston, A. 2009. Energize employees with green strategy. *Harvard Business Review* 87 (9): 24.

Zou, P., Q. Wang, J. Xie, and C. Zhou. 2019. Does doing good lead to doing better in emerging markets? Stock market performance to the SRI index announcements in Brazil, China, and South Africa. *Journal of the Academy of Marketing Science*. https://doi.org/10.1007/s11747-019-00651-z.

Chapter 7
Corporate Environmental Responsibility and Capital Structure

The debates on capital structure might originate from the irrelevance propositions provided by Modigliani and Miller (1958) claiming that the cost of capital of a firm is independent of its capital structure and the firm's value will not be affected by the extend of its debt financing, considering perfect markets. However, the Modigliani-Miller theorem was developed under the assumptions of the absence of taxes, bankruptcy costs, agency costs, and asymmetric information, at the same time, it requires an efficient market where securities are fairly priced and there are no transaction costs. CER related disclosures contribute to the reduction of information asymmetries between the firm and external stakeholders, a fact that may reduce firm risk and in turn affect capital structure decisions. Under this framework, several studies have examine the relationship between corporate environmental responsibility (CER) and capital structure decisions (CS) providing conflicting evidence of a differential impact of CER on long-term debt, short-term debt and equity issuance decisions. The purpose of this chapter is to study the impact of CER related activities on capital structure decisions taking into consideration the maturity of debt, potential endogeneity between the examined factors and their bidirectional association, evidenced by previous studies on the field.

7.1 Introduction

Capital structure is the capital composition of the firm by which the company finances its overall operations and growth. Debt is classified as short-term debt which consists of liabilities such as rent and bank loan that is due in less than 12 months and long-term debt comprising bonds, mortgages and capital leases. Equity comes in the form of common stock, preferred stock and retained earnings. The conception of capital structure varies among different contexts and is measured in different ways. While Besley and Brigham (2007) described capital it as a

P. Dimitropoulos, K. Koronios, *Corporate Environmental Responsibility, Accounting and Corporate Finance in the EU*, CSR, Sustainability, Ethics & Governance, https://doi.org/10.1007/978-3-030-72773-4_7

long-term financing method, Li et al. (2009) conceptualized it as the blend of common equity, preferred equity and long-term debt capital. Therefore, short-term debt is ignored in the definition of capital structure in traditional studies. An increasing number of studies incorporate short-term debt as a component of debt capital (Vătavu 2015; Zhang et al. 2016; Siddik et al. 2017). We thus define capital structure as the combination of short-term debt, long-term debt and equity capital. In fact, all capital that can be converted into cash is considered to be financial resources and thus a part of capital structure.

Corporate environmental responsibility (CER) could play a significant role in reducing the information asymmetries between the firm and its investors, since CER reporting firms tend to disclose more information regarding their activities, signaling their environmental awareness to the public. This fact can mitigate corporate risk thus allowing firms to raise capital or re-finance debt and even reducing cost of capital (Suto and Takehara 2018). According to Bae et al. (2019), CER activities are creating the so-called "halo" effect which enhances public and investors' trust towards the firm. This effect allows highly leveraged firms to mitigate negative actions from customers and competitors leading to higher ability to deal with financial difficulties. Also, high CER firms are perceived as less risky (lower litigation risk) allowing them better access to financing, lower cost of capital and less costs of predatory attacks from competitors (Bae et al. 2019; Suto and Takehara 2018). Specifically Bae et al. (2019) found that one standard deviation increase in CSR score reduces high-leveraged firms' losses in sales growth by 1.2%.

Moreover, an extensive number of studies have consider the impact of CER related activities and performance on cost of equity capital yielding a negative association. According to Xu et al. (2015), Harjoto (2017), Lee and Faff (2009) CER activities reduce risk and since the cost of equity is the required rate of return investors' desire according to their perception of firms' risk, CER reduce cost of equity, thus giving easier access to financing and with reduced cost (lower interest rates). Also, Dhaliwal et al. (2011), Cho et al. (2012) and Xu et al. (2015) empirically indicate that firms with better CSR performance disclose more information on corporate activities, achieving a positive image as responsible citizens, leading to lower information asymmetries which in fact reduces the perception of corporate risk. In addition, better CER firms could attract dedicated long-term investors prompting them to believe that they can achieve the desired rate of return easily and with less uncertainty. This fact could assist firms to obtain a wider range of investors and creditors.

Based on the above discussion, the fact that CER performing firms are associated with less cost of capital could have implications on the corporate decisions regarding the structure and maturity of debt and equity. According to Attig et al. (2014), Ge and Liu (2015) and Yang et al. (2018) the fact that CER activities improve credit ratings and reduce financing costs, high CER performing firms face less constraints on financing, leading to their decision to raise debt instead of capital. So this stream of research points towards a positive association between CER performance and debt. On the contrary, firms with more stakeholder oriented CER activities are associated with more risk aversion and ethical behavior, especially towards its employees. According to Benlemlih (2017), high CSR firms have higher stability, reduced perception of risk, and resort to equity instead of debt financing when they want to finance long term investments. So high CER firms could resort to less debt financing (leverage) and more common equity financing since higher leverage increases the bankruptcy risk. This means the high CER performing firms may be associated with lower levels of leverage so as to reduce stakeholders' concerns. However, this association can be dependent on the maturity of debt. Barnea and Rubin (2010), Benlemlih (2017) suggest that short-term debt might play a monitoring role on managerial behavior which may induce enhanced information disclosure and by reducing the free cash flows under managers' control and the potential for inefficient utilization of limited resources.

The scope of this chapter is to examine the impact of CER related performance on firm capital structure decisions, by distinguishing between short-term debt, long term debt and common equity capital and taking into consideration potential endogeneity between the examined factors and their bidirectional association, evidenced by previous studies on the field. We respond to the call for more research on the topic made by Xu et al. (2015), by covering a large multi-country sample of listed and unlisted European firms over a long period of time and also by considering the bidirectional association between CER and CS. In the following section we provide the theoretical background on the impact of CER on CS and the testable hypotheses. The third section discusses the data selection procedure and the research design. The fourth section provides the discussion of empirical results and the last section concludes the chapter.

7.2 CER-CS Association and Testable Hypothesis

Before discussing the empirical evidence on the association between CER performance and CS decisions it worth providing a brief discussion on the main capital structure theories and their association with the topic under research. There are many conditional theories of capital structure, and these diverse suggestions have aroused widespread discussion and high controversy. Modigliani and Miller (M-M) theory is the first proposition to formulate and demonstrate the effect of capital structure on firm value. In this view, under the perfect capital market, the mode to finance is irrelevant to firm value (Modigliani and Miller 1958). However, M-M model (1958) lacks of practicability in the real economies which is normally imperfect including tax expenses, agency costs and financial distress. Thus, Modigliani and Miller (1963) enhanced the theory by incorporating corporate tax, since interest payments can be deducted from tax and this tax benefits would lower the after-tax cost of capital and increase the value of the firm within a certain range. With the increase of debt, when firm is entirely debt financed and interest expenses reach the amount of earnings before income (EBIT), total value would be maximized. Thus, the levered firm value is greater than the unlevered because of the tax gain from leverage.

Agency theory, conceived by Jensen and Meckling (1976) and Jensen (1986), illustrates that debt serves as a disciplinary tool which helps to mitigate agency costs between managers and company owners. Jensen and Meckling (1976) stated that shareholders normally select managers to operate the corporation and act on their behalves. But as a result of separation between ownership and control, a conflict of interest arises, so managers normally work insufficiently and make decisions for their own utility like over-expanding firm size rather than maximizing shareholders' wealth. Debt financing, especially high debt ratios increases default risk, and thus the possibility of company's failure and the risk of managers losing their jobs. The consequence of such threatens impels and monitors managers to work hard, thereby reducing managerial cash flow waste. Therefore, debt financing is associated with the long-term profit increases (Grossman and Hart 1982; Jensen 1986; Harris and Raviv 1991). Nevertheless, another agency issue exists between equity holders and lenders which could deteriorate performance. Margaritis and Psillaki (2010) claimed that since the default risk is higher when there are debt investors, stockholders tend to engage in excessive risk-taking which provides them additional payments. Good investment normally enhances growth opportunities for the company and this would be only received if financed completely by equity. Although liability could protect against unnecessary wasteful investments, abandonment of projects with positive net present value contributes to the lack of investment and thus weakens firm value. This agency cost cannot be eliminated because the bondholders can perceive these incentives, even if they cannot identify whether non-payoffs are due to uncertainty or stockholders' decisions, so the debt capital cost would be higher (Jensen and Meckling 1976).

Although in M-M theory, 100% debt financing increases the total value to the maximum, irrespective of firm strategies, management or industry, there are a large body of theories criticizing this and attempting to find the drawback of debt financing. The trade-off theory is such an example where Myers (1984) considered bankruptcy costs in M-M model furtherly and claimed that too much leverage would increase financial distress costs and excessive risks, damaging the firm value. Especially in financial distress periods (or financial turmoil), the tax shield benefits are more likely to be wasted or offset by a variety of costs due to capital market imperfections. Hence, Brigham and Houston (2001) argued that the balance between advantages of getting a loan and costs of financial distress should be made when determining the optimal leverage ratio. Pecking order theory is another hypothesis challenging M-M theory and explaining drawbacks of debt and equity from the adverse selection perspective. It contends that because of information asymmetry between internal stakeholders and external providers, new issued shares may be undervalued and the interest of existing shareholders would be harmed. Internal financing requires the lowest transaction costs and issuing costs, followed by debt financing and equity financing (Myers and Majluf 1984). Additionally, the acceptable return of equity for stockholders is higher than that of debt because equity issuance has higher risk. Thus, Adair and Adaskou (2015) stated that the sequence should be retained earnings, debt issuance and equity financing when choosing a financial structure.

Because firm managers prefer financing new investments by internal sources and avoid outside funds if there are sufficient funding generated internally, a self-sufficient company tends to perform better, and an inverse association could be expected between debt ratio and firm's performance. Another major strand of theory, the market timing theory evolved to explore determinants of capital structure. It is premised on the idea that the optimized debt structure depends on the stock market situation. Managers have more information and knowledge about the operating performance and market conditions than those who are outside the organization (Sorin and Luigi 2009). They thus have tendencies to time the market by issuing new shares when there is overvaluation and repurchase shares once the capital cost of equity is high. Hence, changes in stock prices would affect capital structure of the firm continuously and accumulatively (Myers 1984; Baker and Wurgler 2002). Employing a different approach, incentive signaling theory claims that debt has a positive correlation with performance since it is issued to emphasize the optimistic prospect and the strong profit-making ability to the market and investors. Majumdar and Chhibber (1999) pointed out the reason why debt is interpreted as a positive signal, given that the increasing use of debt is associated with higher bankruptcy and liquidation costs. In contrast, issuing equity means that firms are aware of the lack of cash to meet legal obligations and are less confident about their future prospects.

Similarly, Jensen and Meckling (1976) developed the resource constraint theory to emphasize the significance of leverage. It indicates that when resources are limited in a firm, raising equity would dilute equity and sacrifice incentives intensity, while the issuance of debt ensures the concentration of equity and incentives, provides better monitoring and reduces owner manager's on-the-job consumption. But another alternative hypothesis says that because debt is riskier than equity and the constraint of firm resources, managers have to abandon high-risky but high-yield projects and cut back on research and development (R&D) expenditures, this would worsen performance in the long-run (Majumdar and Chhibber 1999). Free cash flow theory is for firms with overinvesting concern and problems. Myers (2001) indicated that the increase in liabilities is associated with the rise in company value provided that operating cash flow can successfully afford investment opportunities. In this circumstance, gains of employing debt significantly exceed potential financial distress costs.

Under the abovementioned framework, previous research on the impact of CSR on capital structure decisions has identified two streams of theoretical backgrounds and evidence pointing to a positive and a negative association between those variables. At first, Yang et al. (2018) document that CER activities reduce firms' idiosyncratic risk, information asymmetries, enhance corporate reputation and moral capital among stakeholders. The investment on CER activities could benefit stakeholders (creditors, customers, employees, lenders and investors) and through enhanced stakeholder relations firms can reduce their risk. Yang et al. (2018) suggest that creditors are less likely to provide long-term loans in firms with uncertain risk exposure. CER activities contribute to the reduction of long-term risk so under these circumstances creditors are willing to finance firms on a long-term basis. This fact may lead banks and credit rating agencies to reach a positive assessment on their future performance which is translated to an enhanced ability of firms to raise debt financing (especially long-term debt) with less strict terms and with lower cost (less interest rates). Thus, high CER performing firms tend to raise more debt relative to equity capital so a positive association is evidenced between CER and long-term debt.

On the contrary, agency cost theory suggests that debt maturity reduces the available resources under managerial control and so CER activities can be utilized in order managers and firm insiders to identify themselves to the public as socially responsible and enhance their reputation at the expense of shareholders. So Godfrey (2005) and Benlemlih (2017) propose that CSR investments reduce the cost of debt only when they are below an optimal level. So the use of less long-term debt can help shareholders to control potential overinvestment on CER activities by the managers. In addition, firms with more stakeholder oriented CER activities are associated with more risk aversion and ethical behavior, especially towards its employees and the environment. According to Benlemlih (2017), high CSR firms have higher stability, reduced perception of risk, and resort to short-term debt and equity instead of long-term debt financing when they want to finance long term investments. So high CER firms could resort to less debt financing (leverage) and more common equity financing since higher leverage increases the bankruptcy risk. This means the high CER performing firms may be associated with lower levels of leverage so as to reduce stakeholders' concerns. So under this framework, CER performance can have a negative association with long-term debt. So the first research hypotheses are stated as follows:

H1: CER performance will have a negative impact on long-term debt.

Furthermore, as previously mentioned high CER performing corporations tend to reduce the information asymmetries among their external stakeholders by increasing the information disclosure, which in turn reduces the cost of equity (Cohen et al. 2011; Benlemlih 2017). Suto and Takehara (2018) indicate that CSR activities reduce firms weighted average cost of capital and managerial forecast bias, thus increasing transparency and accountability. Consequently, if CER performance mitigates information asymmetries and reduces perceived firm risk, this will be reflected on the firm's choices between debt and equity financing. Sharfman and Fernando (2008), Girerd-Potin et al. (2014), El Ghoul et al. (2011) and Benlemlih (2017) provide evidence that investors require reduced returns by high CER performing firms and those firms could raise equity capital with lower costs. Also, CER and CS association can be dependent on the maturity of debt. Barnea and Rubin (2010), Benlemlih (2017) suggest that short-term debt might play a monitoring role on managerial behavior which may induce enhanced information disclosure and by reducing the free cash flows under managers' control and the potential for inefficient utilization of limited resources. So following this reasoning it is expected that high CER performing firms will tend to use more common equity financing and short term debt due to lower information asymmetries, reduced cost of equity and their intent to signal their high quality to the market and avoid CER overinvestment by the managers. Thus, the second research hypotheses are stated as follows:

H2a: CER performance will have a positive impact on shareholder's equity.
H2b: CER performance will have a positive impact on short-term debt.

7.3 Data Selection and Research Design

7.3.1 *Data Selection Procedure*

The current chapter utilizes a sample of corporations (both listed and unlisted) from 23 EU-member countries over the period 2003–2018. All financial, governance and environmental responsibility data have been extracted from Datastream database. The sample collection procedure started by including all corporations that have been covered by the environmental, social and governance (ESG) scores provided by Datastream during the sample period. 2003 has been selected as the initial sample year since that year was the first year with significant coverage (taking into consideration that this database officially was initiated in 2002). From a total of 656,734 firm-year observations we excluded those firms without environmental score data for at least five consecutive years in order to avoid bias in the empirical results and also to facilitate the estimation of differenced variables. Furthermore, we excluded firms with incomplete financial data related to short term debt lending, long-term lending and those that do not close their fiscal year on December. In addition, we winsorized the upper and down 1 per cent of the data distribution in order to mitigate the influence of significant outliers in the empirical analysis. After this procedure was finalized, we ended up with a final sample of 121,154 firm-year observations. The analysis and structure of the final sample is provided on Table 4.1 of Chap. 4.

7.3.2 *Research Design*

In order to examine the validity of the previously mentioned research hypotheses we will utilize panel regression models estimated with Generalized Least Squares (GLS) random effects. The panel random effect estimation method was chosen due to the fact that panel analysis contains more information with more variability and less collinearity among the variables, providing more efficient estimates and precise parameters of model estimation, allowing us to detect many effects that are not detectable in the cross-sectional data analysis. In order to choose the random effect estimation we performed the Breusch-Pagan Lagrange multiplier test for random effect which tests the null hypothesis that the residuals variance is equal to zero. The test produced a highly significant statistic (Chi2 682.47, $p < 0.00001$) leading us to reject the null hypothesis stating that the variance of residuals is equal to zero, and thus the random effects estimation was considered as the most appropriate method for our case. The functional form of the equations includes the average CER for each firm and year estimated as the annual average of the ESG resource use score, emissions reduction score and environmental innovation score, where i denotes the firm and t the year and e is the error term:

$$\begin{aligned} LTD_{it} = {} & a_0 + a_1 CER_{it} + a_2 SIZE_{it} + a_3 GROWTH_{it} + a_4 ROA_{it} \\ & + a_5 R\&D\_EXP_{it} + a_6 CFO_{it} + a_7 PPE_{it} + Year\ F.E \\ & + Industry\ F.E. + Country\ F.E. + e_{it} \end{aligned} \tag{7.1}$$

$$\begin{aligned} STD_{it} = {} & a_0 + a_1 CER_{it} + a_2 SIZE_{it} + a_3 GROWTH_{it} + a_4 ROA_{it} \\ & + a_5 R\&D\_EXP_{it} + a_6 CFO_{it} + a_7 PPE_{it} + Year\ F.E \\ & + Industry\ F.E. + Country\ F.E. + u_{it} \end{aligned} \tag{7.2}$$

$$\begin{aligned} CE\_TA_{it} = {} & a_0 + a_1 CER_{it} + a_2 SIZE_{it} + a_3 GROWTH_{it} + a_4 ROA_{it} \\ & + a_5 R\&D\_EXP_{it} + a_6 CFO_{it} + a_7 PPE_{it} + Year\ F.E \\ & + Industry\ F.E. + Country\ F.E. + z_{it} \end{aligned} \tag{7.3}$$

$$\begin{aligned} LTD_{it} = {} & a_0 + a_1 EN\_INNOV_{it} + a_2 SIZE_{it} + a_3 GROWTH_{it} + a_4 ROA_{it} \\ & + a_5 R\&D\_EXP_{it} + a_6 CFO_{it} + a_7 PPE_{it} + Year\ F.E \\ & + Industry\ F.E. + Country\ F.E. + e_{it} \end{aligned} \tag{7.1a}$$

$$\begin{aligned} STD_{it} = {} & a_0 + a_1 EN\_INNOV_{it} + a_2 SIZE_{it} + a_3 GROWTH_{it} + a_4 ROA_{it} \\ & + a_5 R\&D\_EXP_{it} + a_6 CFO_{it} + a_7 PPE_{it} + Year\ F.E \\ & + Industry\ F.E. + Country\ F.E. + u_{it} \end{aligned} \tag{7.2a}$$

$$\begin{aligned} CE\_TA_{it} = {} & a_0 + a_1 EN\_INNOV_{it} + a_2 SIZE_{it} + a_3 GROWTH_{it} + a_4 ROA_{it} \\ & + a_5 R\&D\_EXP_{it} + a_6 CFO_{it} + a_7 PPE_{it} + Year\ F.E \\ & + Industry\ F.E. + Country\ F.E. + z_{it} \end{aligned} \tag{7.3a}$$

$$\begin{aligned} LTD_{it} = {} & a_0 + a_1 EMM_{it} + a_2 SIZE_{it} + a_3 GROWTH_{it} + a_4 ROA_{it} \\ & + a_5 R\&D\_EXP_{it} + a_6 CFO_{it} + a_7 PPE_{it} + Year\ F.E \\ & + Industry\ F.E. + Country\ F.E. + e_{it} \end{aligned} \tag{7.1b}$$

$$\begin{aligned} STD_{it} = {} & a_0 + a_1 EMM_{it} + a_2 SIZE_{it} + a_3 GROWTH_{it} + a_4 ROA_{it} \\ & + a_5 R\&D\_EXP_{it} + a_6 CFO_{it} + a_7 PPE_{it} + Year\ F.E \\ & + Industry\ F.E. + Country\ F.E. + u_{it} \end{aligned} \tag{7.2b}$$

$$\begin{aligned} CE\_TA_{it} = {} & a_0 + a_1 EMM_{it} + a_2 SIZE_{it} + a_3 GROWTH_{it} + a_4 ROA_{it} \\ & + a_5 R\&D\_EXP_{it} + a_6 CFO_{it} + a_7 PPE_{it} + Year\ F.E \\ & + Industry\ F.E. + Country\ F.E. + z_{it} \end{aligned} \tag{7.3b}$$

$$\begin{aligned} LTD_{it} = {} & a_0 + a_1 RES\_USE_{it} + a_2 SIZE_{it} + a_3 GROWTH_{it} + a_4 ROA_{it} \\ & + a_5 R\&D\_EXP_{it} + a_6 CFO_{it} + a_7 PPE_{it} + Year\ F.E \\ & + Industry\ F.E. + Country\ F.E. + e_{it} \end{aligned} \tag{7.1c}$$

$$\begin{aligned} STD_{it} = {} & a_0 + a_1 RES\_USE_{it} + a_2 SIZE_{it} + a_3 GROWTH_{it} + a_4 ROA_{it} \\ & + a_5 R\&D\_EXP_{it} + a_6 CFO_{it} + a_7 PPE_{it} + Year\ F.E \\ & + Industry\ F.E. + Country\ F.E. + u_{it} \end{aligned} \tag{7.2c}$$

$$CE\_TA_{it} = a_0 + a_1 RES\_USE_{it} + a_2 SIZE_{it} + a_3 GROWTH_{it} + a_4 ROA_{it} + a_5 R\&D\_EXP_{it} + a_6 CFO_{it} + a_7 PPE_{it} + Year\ F.E + Industry\ F.E. + Country\ F.E. + z_{it} \tag{7.3c}$$

EN_INNOV denotes each company's environmental innovation score, which reflects the firm's ability to mitigate the environmental costs towards its customers or even creating new market and product opportunities via enhanced environmental technologies and processes. EMM denotes companies' emission reduction score estimated via their efforts and effectiveness in reducing emissions from their production and operation. RES_USE denotes the firms' resource use score and indicates firms' ability to reduce the usage of hazardous materials, energy, water and other valuable natural resources. Finally, CER is the firms' annual average of the aforementioned three scores. All environmental scores range between zero (0) and 100 so the higher the score the more environmentally responsible and efficient is the firm. Following Benlemlih (2017), we estimated the ratios of long-term debt to total debt (LTD), short-term debt to total debt (STD) in order to distinguish the differential impact of CER on debt maturity, and the ratio of shareholder's equity to total assets (CE_TA) so as to control for equity financing decisions. If H1 is valid and CER (and its components) impacts negatively on LTD we expect a negative coefficient in this variable. Respectively, if H2a and H2b are valid we expect a positive coefficient on CER and its components indicating that CER performing firms resort to short term debt and equity financing. Moreover, we have included in the models some additional controls that have proved significant determinants of capital structure by previous research.

SIZE is the natural logarithm of annual total assets capturing firm size. A positive relationship between firm size and leverage is expected since large firms are considered to be more diversified and less likely to go bankrupt resulting in lower bankruptcy cost (Ozkan 2001). Meanwhile, large firms seem to have a larger debt capacity, higher transparency and less information asymmetry, which means they might have easier access to the capital market and can borrow under relatively low interest rates (Ozkan 2001; Byoun 2008). At the same time, the transaction costs associated with debt issuing in large firms are expected to be lower compared to other firms. Thus, according to the trade-off theory, large firms might have a high leverage level (Jensen and Meckling 1976; Myers 1984), which complies with several empirical researches (Psillaki and Daskalakis 2009; Abor and Biekpe 2009). However, large firms may also prefer to issue shares because of their low information asymmetry, considering the pecking order theory, which may result in a low leverage level (Rajan and Zingales 1995). This opinion is supported by Titman and Wessels' (1988) empirical evidence.

GROWTH captures firms' growth opportunities and is estimated as the annual percentage change in sales revenues. According to the trade-off theory, firms with growth potentials are riskier since they have a higher bankruptcy cost and agency cost of debt, thus they might find it difficult to finance operations through debt and thus tend to have lower leverage (Myers 1977, 1984). Moreover, growth opportunities are intangible and cannot be collateralized, thus lenders might not be willing to borrow money to firms with high growth opportunities (O'Brien 2003). Nevertheless, the pecking order theory suggests a positive relationship between growth opportunities and leverage (Myers and Majluf 1984). Firms with high growth potentials might be more likely to exhaust internal funds, seeking external financing to meet their investment need. In this scenario, firms might prefer debt financing to equity financing due to the latter's high information cost result from asymmetric information, which will causing a lower leverage (Myers and Majluf 1984). At the same time, Yang et al. (2010) and Rajagopal (2011) argue that both negative and positive relationship exist between growth potential and leverage.

ROA is the ratio of return on assets capturing profitability. The relationship between profitability and leverage under different theories are conflicting. There might be a positive relationship between the variables according to the trade-off theory. Firms with high profitability are expected to have more capacity to benefit from the tax shield of debt because of their high profit and the corresponding low bankruptcy cost (Jensen and Meckling 1976; Myers 1984). Furthermore, debt can also reduce the agency cost of highly profitable firms by playing its disciplinary role since such firms often have high free cash flows, which might incur more agency problems and increase the agency cost of equity (Jensen and Meckling 1976; Myers 1984; Jensen 1986). However, the pecking order theory predict a negative relationship between profitability and leverage (Myers and Majluf 1984). Firms with high profitability seems to have adequate retained earnings to meet their investment needs and thus they will probably prefer financing through internal sources rather than seeking external financing due to latter's high information costs and transaction costs (Myers 1984; Myers and Majluf 1984).

R&D_EXP is the ratio of research and development expenses to total revenue. Highly leveraged firms are facing higher distress risk and this means that they face difficulties to re-finance their debt or access the capital market, thus they are inclined to cut back on R&D activities in order to reduce costs and maintain their sustainability. This decision can have a negative effect on corporate prospects leading to a deterioration on R&D expenses. This means a negative coefficient on the R&D_EXP variable (Harjoto 2017). CFO is the ratio of operating cash flows to total assets. Firms with enhanced operating cash flows resort less on financing since they can utilize internally generated funds instead of raising debt. So a negative coefficient is also expected on this variable.

Finally, PPE is the ratio of property, plant and equipment (tangible fixed assets) to total assets. Asset tangibility seems to be positively related to debt according to the trade-off theory since tangible assets can be used as collateral for creditors to borrow or issue debt (Miller 1977; Myers 1984). Meanwhile, the more tangible assets the firm has, the lower the possibility of bankruptcy of the firm, making it possible for the firm to issue more debt to gain on the tax benefit of debt (Myers 1984). However, the correlation between asset tangibility and leverage might be the opposite when considering the pecking order theory. This is because the tangible assets are considered as stable sources of return generating more internal funds, which allow firms to rely on internally generated funds to invest and do not need to seek external financing (Myers and Majluf 1984; Psillaki and Daskalakis 2009; Bevan and Danbolt 2002).

7.4 Empirical Results

7.4.1 Descriptive Statistics and Correlations

Table 7.1 presents the descriptive statistics of the variables utilized in the regression models. First of all, the mean CER is 64.5 (with a standard deviation of 21.1) indicating that the sample firms have a satisfactory environmental performance score. The highest average score is that of resources usage (66.5) followed by emission reduction activities (65.9) while environmental innovation activities lag behind with an average of 61.3. Also, the capital structure is mainly comprised by long term debt (0.601) and less by short term debt (0.399), while the average common equity is negative indicating that in the sample there are some firms with financial difficulties and increased bankruptcy potential. The sample firms are highly profitable (mean ROA 21.28) and present significant growth opportunities (mean GROWTH at 58.9). In addition, the asset tangibility of the sample firms is almost 60 per cent of total assets and spent almost 2 per cent of their revenues on R&D expenses and activities.

Table 7.2 presents the differences between the main capital structure variables after separating firms within low, medium and high CER performance groups in order to examine the extent to which CER impacts on CFP. Following Suto and Takehara (2018) we followed the portfolio formation approach and estimated the average of the CER variable per year and country and assigned each firm as low CER firm if its annual score is below the 1st quartile, high CER firms are those that their

Table 7.1 Descriptive statistics of the sample variables

Variables	Mean	Standard deviation	Min	Max
CER	64.597	21.167	3.161	99.429
EN_INNOV	61.333	25.962	0.202	99.919
EMM	65.915	26.189	0.161	99.919
RES_USE	66.542	25.730	0.805	99.844
LTD	0.601	0.338	0.004	1.00
STD	0.399	0.338	0.001	1.00
CE_TA	−3.627	4.319	−25.69	1.00
SIZE	20.496	2.848	2.302	30.311
GROWTH	58.915	141.720	−69.711	409.071
ROA	21.281	48.257	−27.861	174.17
R&D_EXP	0.018	0.017	−0.025	0.082
CFO	0.030	5.042	−29.42	39.88
PPE	0.596	0.245	0.001	1.00

Table 7.2 Capital structure based on CER ranked portfolios

Variables	P1—High CER	P2—Medium CER	P3—Low CER	Difference (P1 − P3)	p-value
LTD	0.592	0.646	0.657	−0.065***	0.001
STD	0.407	0.354	0.342	0.065***	0.001
CE_TA	−4.051	−1.569	−0.034	−4.017	0.161

CER score is above the 3rd quartile while all other firm observations lying between the 1st and 3rd quartiles are assigned as medium CER firms. The column under the title "difference" reports the average CER difference between the high CER portfolio firms and the low CER portfolio. The "p-value" column presents the corresponding probability values from estimating the Welch two-sample t-test and asterisks indicate statistical significance at the 1 per cent significance level (Suto and Takehara 2018). As we can see from the last column, high CER performance firms are characterized by statistically higher short term debt and less long term debt. The ratio of CE_TA did not present a significant difference between the low and high CER groups. These findings corroborate arguments by Benlemlih (2017) that more socially responsible firms make different choices regarding their capital structure and seem to resort to more short term debt to finance their activities, while on the contrary less CER performing firms tend to choose long term debt as the more prevalent form of financing. This provides a first indication for the support of our main research hypotheses.

Table 7.3 presents the Pearson correlation coefficients between the sample variables. The majority of the correlation coefficients are significant with economic meaning, and are not very high indicating the absence of multicollinearity on the data. The VIF estimation of the sample variables provided values that were far below the threshold of 10 indicating no multicollinearity. As we can see, CER is positively associated with STD, while is negatively and significantly associated with LTD, suggesting that higher CER performance is positively associated with short term debt and negatively with long term debt. Also CE_TA is not significantly correlated to CER so this findings goes along with the results on the Table 7.2. This type of association is evidenced for the emission score and product responsibility score. Environmental innovation score is negatively and significantly correlated with CE_TA but no significant association is evidenced with any type of lending. Also, more profitable firms are positively associated with SIZE and LTD. The same arguments stand for the rest three environmental performance scores. The only difference is observed in the resource use score (RES_USE) which is negatively associated with ROA and with SIZE. Finally, growth opportunities did not provide any significant correlation coefficients with the CER related and financial performance. Of course correlations do not indicate causality or direction of association so we will proceed with the panel regression analysis results in order to extract more specific inferences.

Table 7.3 Pearson correlation coefficients of sample variables

Variables	1	2	3	4	5	6	7	8	9	10	11	12	13
1. LTD	1												
2. STD	**−1.00**	1											
3. CE_TA	**0.009**	**−0.008**	1										
4. SIZE	**0.291**	**−0.291**	**0.046**	1									
5. GROWTH	0.001	−0.001	0.000	0.001	1								
6. ROA	**0.019**	**−0.019**	**0.070**	**0.127**	−0.002	1							
7. R&D_EXP	**−0.037**	**0.037**	**0.019**	**−0.025**	−0.001	**−0.055**	1						
8. CFO	**0.005**	**−0.005**	**0.626**	**0.018**	0.001	**0.062**	**−0.062**	1					
9. PPE	**0.352**	**−0.352**	**0.023**	**0.279**	0.002	**0.019**	**−0.026**	**0.011**	1				
10. EN_INNOV	0.004	−0.004	**−0.007**	**0.013**	−0.005	**0.018**	**−0.045**	0.001	**−0.010**	1			
11. EMM	**−0.006**	**0.006**	−0.001	−0.002	−0.001	**0.011**	−0.018	−0.002	0.004	**0.377**	1		
12. RES_USE	**−0.021**	**0.021**	**−0.006**	**−0.040**	0.002	**−0.011**	−0.005	−0.005	0.000	**0.405**	**0.710**	1	
13. CER	**−0.010**	**0.010**	−0.006	**−0.012**	−0.002	**0.007**	**−0.028**	**−0.002**	−0.001	**0.728**	**0.854**	**0.863**	1

Note: Correlation coefficients in bold indicate statistical significance at least at the 5% significance level

7.4.2 Regression Results

Table 7.4 presents the GLS panel regression results from the estimation from models (7.1)–(7.1c) using the LTD variable as the capital structure measure. The first column uses the CER variable as the environmental performance measure and the other three columns the components of CER namely EN_INNOV, EMM and RES_USE. The Chi-square values are highly significant and the R^2s are very satisfactory considering the size of the sample and other relative studies published in the field. The estimation of model (7.1) provided a negative and significant coefficients (−0.005) on the CER verifying evidence provided on the literature by Godfrey (2005) and Benlemlih (2017) that CSR investments reduce the cost of debt only when they are below an optimal level. In addition, firms with more stakeholder oriented CER activities are associated with more risk aversion and ethical behavior, especially towards its employees and the environment. According to Benlemlih (2017), high CSR firms have higher stability, reduced perception of risk, and resort to short-term debt and equity instead of long-term debt financing when they want to

Table 7.4 Panel regression results on the association between CER and LTD

Dep. Variable: LTD	Model (7.1)	Model (7.1a)	Model (7.1b)	Model (7.1c)	Overall
CER	−0.005** (−2.02)				
EN_INNOV		−0.006*** (−2.62)			−0.005** (−2.31)
EMM			−0.001 (−0.66)		−0.001 (−0.32)
RES_USE				−0.002 (−1.36)	−0.001 (−0.49)
SIZE	0.004 (1.16)	0.003 (1.11)	0.003 (1.06)	0.004 (1.10)	0.037 (1.10)
GROWTH	0.001 (1.15)	0.002 (1.15)	0.003 (1.11)	0.001 (1.13)	0.001 (1.15)
ROA	0.151*** (2.59)	0.148*** (2.57)	0.151*** (2.59)	0.150*** (2.59)	0.148*** (2.58)
R&D_EXP	−1.524*** (−4.05)	−1.850*** (−4.61)	−1.388*** (−3.76)	−1.361*** (−3.71)	−1.822*** (−4.54)
CFO	−0.202*** (−2.72)	−0.199*** (−2.69)	−0.201*** (−2.71)	−0.202*** (−2.71)	−0.199*** (−2.70)
PPE	0.335*** (8.61)	0.336*** (8.59)	0.334*** (8.56)	0.335*** (8.57)	0.335*** (8.59)
Constant	0.548*** (7.90)	0.553*** (7.92)	0.532*** (7.68)	0.538*** (7.69)	0.554*** (7.93)
R-sq	0.178	0.188	0.185	0.086	0.200
Wald Chi^2 (p-value)	168.99*** (0.001)	171.36*** (0.001)	165.15*** (0.001)	166.70*** (0.001)	171.87*** (0.001)

Note: **, *** indicate statistical significance at the 5% and 1% significance level respectively, z-statistics in the parenthesis

finance long term investments. This finding provides support to H1 and suggests that high CER performing firms could resort to less debt financing since higher leverage increases the bankruptcy risk and stakeholders' concerns.

The same argument is valid for sample firms with increased environmental innovation score. The coefficient on the EN_INNOV variable is also negative and significant indicating that firms investing funds on environmental friendly activities rely less on long term debt supporting H1. Nevertheless, the coefficients on the EMM and RES_USE variables had the expected sign but were not statistically significant within conventional levels.

Regarding the rest of the control variables, profitability (ROA) and asset tangibility (PPE) produced positive and highly significant coefficients in all models estimations, corroborating arguments based on the trade-off theory of capital structure. Practically, firms with high profitability have more capacity to benefit from the tax shield of debt because of their high profit and the corresponding low bankruptcy cost (Jensen and Meckling 1976; Myers 1984). Also, debt reduces the agency cost within highly profitable firms by playing its disciplinary role since such firms often have high free cash flows, which might incur more agency problems and increase the agency cost of equity (Jensen and Meckling 1976; Myers 1984; Jensen 1986). Under the same theory, tangible assets can be used as collateral for creditors to borrow or issue debt (Miller 1977; Myers 1984) thus the more tangible assets the firm has, the lower the possibility of bankruptcy of the firm, making it possible for the firm to issue more debt to gain on the tax benefit of debt (Myers 1984). On the contrary, R&D expenses and operating cash flows (CFO) produced negative and significant coefficients in all model estimations, a result which is according to our expectations. Highly leveraged firms are facing higher distress risk and this means that they may face difficulties to re-finance their debt or access the capital market, thus they are inclined to cut back on R&D activities in order to reduce costs and maintain their sustainability, leading to a negative impact on R&D expenses (Harjoto 2017). Finally, firms with enhanced operating cash flows resort less on long term financing since they can utilize internally generated funds instead of raising long term debt.

Tables 7.5 and 7.6 present the results from the estimation of the GLS panel regression models using the short term debt (STD) and shareholders' equity (CE_TA) as the dependent variables. As we can see on Table 6.4, the sign of regression coefficients is exactly the opposite of those on Table 6.3, indicating that short-term debt plays a substitute role on our sample firms' capital structure decisions. The CER variable produced positive and significant coefficients in both Tables 7.5 and 7.6 verifying hypotheses H2a and H2b and corroborate arguments in the literature by Sharfman and Fernando (2008), Girerd-Potin et al. (2014), El Ghoul et al. (2011) and Benlemlih (2017) that investors require reduced returns by high CER performing firms and those firms could raise equity capital with lower costs. Also, our evidence support arguments by Barnea and Rubin (2010) and Benlemlih (2017) that high CER performing firms will tend to use more common equity financing and short term debt due to lower information asymmetries, reduced cost of equity and their intent to signal their high quality to the market and avoid CER overinvestment by the managers. The rest of the control variables produced the

Table 7.5 Panel regression results on the association between CER and STD

Dep. Variable: STD	Model (7.2)	Model (7.2a)	Model (7.2b)	Model (7.2c)	Overall
CER	0.005** (2.02)				
EN_INNOV		0.006*** (2.62)			0.005** (2.31)
EMM			0.001 (0.66)		0.001 (0.32)
RES_USE				0.002 (1.36)	0.001 (0.49)
SIZE	−0.004 (−1.16)	−0.003 (−1.11)	−0.003 (−1.06)	−0.004 (−1.10)	−0.037 (−1.10)
GROWTH	−0.001 (−1.15)	−0.002 (−1.15)	−0.003 (−1.11)	−0.001 (−1.13)	−0.001 (−1.15)
ROA	−0.151*** (−2.59)	−0.148*** (−2.57)	−0.151*** (−2.59)	−0.150*** (−2.59)	−0.148*** (−2.58)
R&D_EXP	1.525*** (4.05)	1.850*** (4.61)	1.388*** (3.76)	1.361*** (3.71)	1.822*** (4.54)
CFO	0.202*** (2.72)	0.199*** (2.69)	0.201*** (2.71)	0.202*** (2.71)	0.199*** (2.70)
PPE	−0.335*** (−8.61)	−0.336*** (−8.59)	−0.334*** (−8.56)	−0.335*** (−8.57)	−0.335*** (−8.59)
Constant	0.451*** (6.49)	0.446*** (6.38)	0.467*** (6.74)	0.461*** (6.60)	0.445*** (6.36)
R-sq	0.188	0.188	0.186	0.086	0.189
Wald Chi^2 (p-value)	168.99*** (0.001)	171.36*** (0.001)	165.15*** (0.001)	166.70*** (0.001)	171.87*** (0.001)

Note: **, *** indicate statistical significance at the 5% and 1% significance level respectively, z-statistics in the parenthesis

opposite sign compared with those on Table 7.4. Specifically, less profitable firms and those with less tangible assets resort more on short term debt and shareholders' capital, while the opposite is true for firms with higher R&D expenses and operating cash flows. Interpreting Tables 7.4–7.6 together we can argue the CER performing firms rely more on their own shareholders' equity and short term debt to finance their activities, and less on long term debt. This finding is in line with evidence by Benlemlih (2017) in the US market and provide support to the arguments about the different motives behind short term and long term financing decisions.

7.4.3 Sensitivity Analysis

In order to check the robustness of the main empirical results, we performed a number of sensitivity tests regarding the research design, and various definitions of the determinant variables. An important issue that has been raised by several

Table 7.6 Panel regression results on the association between CER and CE_TA

Dep. Variable: CE_TA	Model (7.3)	Model (7.3a)	Model (7.3b)	Model (7.3c)	Overall
CER	0.006*** (2.76)				
EN_INNOV		0.002* (1.65)			0.001 (0.65)
EMM			0.003** (2.51)		0.002 (1.32)
RES_USE				0.004*** (2.56)	0.002 (1.64)
SIZE	−0.013*** (−2.91)	−0.013*** (−2.77)	−0.013*** (−2.88)	−0.013*** (−2.88)	−0.013** (−2.53)
GROWTH	−0.002** (−2.54)	−0.002** (−2.55)	−0.001** (−2.51)	−0.001** (−2.52)	−0.002** (−2.53)
ROA	0.332*** (5.88)	0.333*** (5.90)	0.331*** (5.87)	0.332*** (5.89)	0.331*** (5.87)
R&D_EXP	5.955*** (76.86)	5.996*** (43.63)	5.814*** (129.09)	5.737*** (122.56)	5.855*** (40.87)
CFO	−0.089 (−0.96)	−0.092 (−0.99)	−0.089 (−0.96)	−0.089 (−0.97)	−0.089 (−0.96)
PPE	0.002 (0.07)	0.004 (0.12)	0.004 (0.12)	0.003 (0.09)	0.002 (0.07)
Constant	0.711*** (7.41)	0.722*** (7.50)	0.722*** (7.58)	0.719*** (7.51)	0.714*** (7.43)
R-sq	0.168	0.169	0.170	0.067	0.167
Wald Chi^2 (p-value)	26423.78*** (0.001)	28116.34*** (0.001)	165.15*** (0.001)	25894.70*** (0.001)	26181.26*** (0.001)

Note: **, *** indicate statistical significance at the 5% and 1% significance level respectively, z-statistics in the parenthesis

studies (Benlemlih 2017; Harjoto 2017; Yang et al. 2018; Suto and Takehara 2018) is the existence of endogeneity meaning that CER performance might not only determine the capital structure, but the causality mighty also go the other way round. Under this circumstance, the GLS panel estimation of models may partially control for that by producing consistent estimators, yet not efficiently as the Generalized Method of Moments (GMM). For this reason, all regression models controlling for the impact of CER on CS were estimated using the GMM method. The GMM method is selected over the Two-Stage Least Squares (2SLS) because in order to determine the 2SLS estimator we assume that the errors are i.i.d., meaning that they are identically and independently distributed over the observations. Having cross-sectional data the departure from independence may reflect neighboring effects in the sense that observations are similar in value and this way they share a correlation with the disturbances (Baum 2006). Also, observations that are close in time may be correlated with the strength of the correlation increasing with proximity. Thus, in the case that the errors are not i.i.d., 2SLS estimators can be consistent but inefficient. In other words, the 2SLS estimator is a special case of the GMM estimator. The GMM method allows us to estimate efficient and consistent estimators in the case of non-i.i.d. errors.

The next step to the estimation of the models is to select proper instruments, meaning variables that are correlated with the CER and its components, but uncorrelated with CFP. In this case, we selected three instrumental variables that have the abovementioned characteristic. The first is a dummy variable COMMON receiving (1) if the firm is based on a common law country and (0) otherwise. Evidence on Chap. 4 indicate that firms' legal origin is a significant determinant of CER performance. The second instrument is the firm listing status dummy receiving (1) if the firms is listed in the stock market and (0) otherwise. Listed firms receive higher scrutiny from regulators and investors and usually are subjected to various reporting standards imposed by securities commissions. Listed firms are facing more pressures from internal and external stakeholders to engage in social and environmentally responsible investments in order to improve their image, gain their legitimacy on their operations and even enhance their stock market performance (Monteiro and Aibar-Guzmán 2010). The third instrument is the 1 year lag of CER performance since according to Benlemlih (2017), current CER performance is affected by previous years CER related activities. In order to examine the relevance of the instruments we estimated the Anderson-Rubin likelihood ratio test, which has the null hypothesis that the smallest canonical correlation is zero and assumes that the regressors are distributed multivariate normal. A failure to reject the null hypothesis indicates that the status of the estimated equation is in question (Baum 2006). In our case the Anderson-Rubin test produced a chi-square value of 74.45 ($p > \chi^2 = 0.0001$) indicating that the null hypothesis is rejected at the 1 percent significance level at least. In addition the Wald χ^2 statistic of weak instrument produced a value of 59.21 ($p > \chi^2 = 0.0001$) thus our instruments are relevant and valid for our research setting. Also, the test of over-identifying restrictions (Hansen J-test) was performed, which examines the joint null hypothesis that the excluded instruments are uncorrelated with the error term and is correctly excluded from the second-stage equation. The results of this tests suggested that the used instruments were statistically meaningful, as the Hansen statistic was insignificant in conventional levels.

Table 7.7 presents the GMM empirical results from the estimation of model (7.1) and by including the LTD, STD and CE_TA as the dependent variables. We also provide the first stage regression results which indicate that 1 year lagged CER performance and COMMON_LAW variables are highly significant justifying their use as instruments. The results of the second stage regression corroborate the main findings on the previous tables and verify the main research hypotheses. CER produced positive and highly significant coefficients for both STD and CE_TA dependent variables and a negative and significant coefficient on the LTD model. Again this finding indicates that our sample firms high enhanced CER performance resort less on long term debt and more on common equity and short term debt in order to finance their activities. Consequently all of our research hypotheses are again verified. Also, the sign and significance of the control variables is qualitatively similar on those on Tables 7.4–7.6. Another sensitivity test regarding the selection of

Table 7.7 IV GMM regression results on the association between CER and measures of capital structure

Dep. Variable	LTD	STD	CE_TA
First stage results			
SIZE	0.600*** (2.59)	0.600*** (2.59)	0.532** (2.37)
GROWTH	0.006** (2.29)	0.006** (2.29)	0.006** (2.49)
ROA	1.135 (0.26)	1.135 (0.26)	1.173 (0.28)
R&D_EXP	−18.476*** (−5.54)	−18.476*** (−5.54)	−17.363*** (−5.29)
CFO	−3.704 (−0.60)	−3.705 (−0.60)	−3.633 (−0.61)
PPE	−0.540 (−0.21)	−0.540 (−0.21)	1.454 (0.59)
COMMON_LAW	−6.948*** (−6.19)	−6.948*** (−6.19)	−6.903*** (−6.51)
LISTED	0.604 (0.56)	0.604 (0.56)	1.189 (1.12)
CER_{it-1}	0.292*** (14.04)	0.292*** (14.04)	0.287*** (14.26)
Constant	31.30*** (6.89)	31.30*** (6.89)	31.94*** (7.24)
R^2-adjusted	0.111	0.111	0.109
F-stat (p-value)	40.60*** (0.001)	40.60*** (0.001)	41.68*** (0.001)
Second stage results			
CER	−0.003*** (−4.57)	0.003*** (4.57)	0.001* (1.66)
SIZE	0.004 (1.51)	−0.004 (−1.51)	−0.016*** (−6.14)
GROWTH	0.001 (1.56)	−0.001 (−1.56)	−0.003*** (−3.08)
ROA	0.185*** (2.57)	−0.185*** (−2.57)	0.354*** (3.59)
R&D_EXP	−8.907*** (−12.32)	8.907*** (12.32)	7.250*** (8.76)
CFO	−0.280*** (−2.57)	0.280*** (2.57)	−0.169 (−1.50)
PPE	0.415*** (10.79)	−0.415*** (−10.79)	−0.016 (−0.56)
Constant	0.688*** (11.25)	0.311*** (5.09)	0.743*** (12.01)
R-sq	0.141	0.142	0.093
Wald Chi^2 (p-value)	680.31*** (0.001)	680.31*** (0.001)	589.48*** (0.001)

Note: **, *** indicate statistical significance at the 5% and 1% significance level respectively, t-tests in the parentheses on the first stage results and z-statistics in the parenthesis on the second stage results the instruments was to re-estimate models of Table 6.6 by using the 1 and 2 year lags of the CER and its components as proposed by Suto and Takehara (2018) and Testa and D'Amato (2017). Despite the fact that Larcker and Rusticus (2010) argue that the use of lagged endogenous regressors as the instrument assumes that the exogenous part of the regressor persists over time, but on the contrary, the endogenous part does not persist over time, suggesting that the use of lagged endogenous variables as instruments does not solve the problem of endogeneity. Empirical evidence provided qualitatively similar arguments based on those presented on Table 7.7, so the control for endogeneity and simultaneity should be controlled when examining the association between CER and CS.

Furthermore, additional sensitivity testing was performed following different definitions of the dependent variables. Specifically, we estimated LTD as the ratio of long term debt to total assets, STD as the ratio of short term debt to total assets and also utilized leverage (total debt to total assets) as the dependent variables. Results are presented on Table 7.8 and indicate that CER has a negative coefficient on the LTD model while a positive coefficient on the STD model. So under this test high CER performing firms are associated with less long term debt and more short term debt, corroborating our hypotheses and verifying the differential impact of CER performance on capital structure decisions. Also, we excluded from the sample all firms having a negative ratio of shareholder's equity to total assets (CE_TA) in order to avoid any bias in the results arising from the negative common equity. Untabulated findings remained robust compared with those on Tables 7.4–7.6. Also, all models were re-estimated by substituting the average CER variable with a factor extracted using a principal component analysis from the EN_INNOV, EMM and RES_USE scores. Again the results remained qualitatively unchanged from this modification. Finally, we conducted additional robustness checks by changing the definition of GROWTH (using MV/BV), ROA (using ROE and profit margin) and SIZE (using the natural logarithm of sale revenues) but the results again remained qualitatively unchanged compared with those on Tables 7.4–7.6.

7.5 Conclusion

Several studies on the environmental responsibility research field have focused on examining the impact of CER activities on capital structure (CS). Researchers have expressed different views on how a company's social and environmental performance affects its capital structure decisions and their selection between short term and long term debt (Xu et al. 2015; Benlemlih 2017; Yang et al. 2018). The scope of this chapter is to examine the impact of CER related performance on firm capital structure decisions, by distinguish between short-term debt, long term debt and common equity capital and taking into consideration potential endogeneity between the examined factors and their bidirectional association, evidenced by previous studies on the field. We respond to the call for more research on the topic made by Xu et al. (2015), by covering a large multi-country sample (24 EU countries) of

Table 7.8 Panel regression results on the association between CER and alternative measures of capital structure

Dep. Variable	TD/TA	LTD/TA	STD/TA
CER	−0.006*** (−2.76)	−0.005*** (−2.80)	0.008* (1.69)
SIZE	0.013*** (2.91)	0.004* (1.66)	0.006 (0.71)
GROWTH	0.002** (2.54)	0.002*** (3.85)	0.003 (0.31)
ROA	−0.332*** (−5.88)	−0.075 (−1.34)	0.009 (1.07)
R&D_EXP	−5.955*** (−76.86)	−0.357*** (−4.91)	−0.093*** (−3.67)
CFO	0.089 (0.96)	−0.037 (−0.57)	−0.024** (−2.19)
PPE	−0.002 (−0.07)	0.085*** (2.98)	0.007 (0.87)
Constant	0.288*** (3.00)	0.100* (1.86)	0.014 (0.86)
R-sq	0.169	0.171	0.097
Wald Chi^2 (p-value)	26423.79*** (0.001)	144.96*** (0.001)	73.12*** (0.001)

Note: **, *** indicate statistical significance at the 5% and 1% significance level respectively, z-statistics in the parenthesis

listed and unlisted European firms (7313 corporations) over a long period of time (2003–2018) and also by considering the bidirectional association between CER and CS.

Empirical analysis suggested that CER performance is negatively impacting long term debt but positively on short term debt and shareholders' equity indicating a differential impact of CER on capital structure decisions. Practically, we can argue the CER performing firms rely more on their own shareholders' equity and short term debt to finance their activities, and less on long term debt. This finding is in line with evidence by Benlemlih (2017) in the US market and provide support to the arguments about the different motives behind short term and long term financing decisions. The empirical results remain robust after several sensitivity tests regarding the definition of the dependent variables and after controlling for the impact of endogeneity on the association between CER and CS. The findings of this chapter contribute to the ongoing debate about the association between CER and CS is several merits. At first we complement previous studies by Yang et al. (2018), Harjoto (2017) and Suto and Takehara (2018), by incorporating a multi-country instead of a single country research setting. Also, we incorporate in the analysis a decomposition of the CER variable and indicate which sub-category of CER performance contributes to capital structure decisions, something that has not been considered extensively by previous researchers.

References

Abor, J., and N. Biekpe. 2009. How do we explain the capital structure of SMEs in sub-Saharan Africa? *Journal of Economic Studies* 36 (1): 83–97.

Adair, P., and M. Adaskou. 2015. Trade-off-theory vs. pecking order theory and the determinants of corporate leverage: Evidence from a panel data analysis upon French SMEs (2002-2010). *Cogent Economics & Finance* 3 (1): 1–12.

Attig, N., S.W. Cleary, S. El Ghoul, and O. Guedhami. 2014. Corporate legitimacy and investment – Cash flow sensitivity. *Journal of Business Ethics* 121 (2): 297–314.

Bae, K.-H., S. El Ghoul, O. Guedhami, C.C.Y. Kwok, and Y. Zheng. 2019. Does corporate social responsibility reduce the costs of high leverage? Evidence from capital structure and product market interactions. *Journal of Banking and Finance* 100: 165–150.

Baker, M., and J. Wurgler. 2002. Market timing and capital structure. *The Journal of Finance* 57 (1): 1–32.

Barnea, A., and A. Rubin. 2010. Corporate social responsibility as a conflict between shareholders. *Journal of Business Ethics* 97: 71–86.

Baum, C.F. 2006. *An introduction to econometrics using Stata*. College Station, TX: Stata Press Books, StataCorp LP.

Benlemlih, M. 2017. Corporate social responsibility and firm debt maturity. *Journal of Business Ethics* 144 (3): 491–517.

Besley, S., and E. Brigham. 2007. *Essentials of managerial finance*. Boston, MA: Cengage Learning.

Bevan, A., and J. Danbolt. 2002. Capital structure and its determinants in the UK – A decompositional analysis. *Applied Financial Economics* 12 (3): 159–170.

Brigham, E.F., and J.F. Houston. 2001. *Fundamentals of financial management.* Boston, MA: Cengage Learning.

Byoun, S. 2008. How and when do firms adjust their capital structures toward targets? *The Journal of Finance* 63 (6): 3069–3096.

Cho, S.Y., C. Lee, and R.J. Pfeiffer. 2012. Corporate social responsibility performance and information asymmetry. *Journal of Accounting and Public Policy* 32: 71–83.

Cohen, J., L. Holder-Webb, L. Nath, and D. Wood. 2011. Retail investors' perceptions of the decision-usefulness of economic performance, governance, and corporate social responsibility disclosures. *Behavioral Research in Accounting* 23: 109–129.

Dhaliwal, D., O. Li, A. Zhang, and Y. Yang. 2011. Voluntary nonfinancial disclosure and the cost of equity capital: The initiations of corporate social responsibility reporting. *The Accounting Review* 86: 59–100.

El Ghoul, S., O. Guedhami, C. Kwok, and D. Mishra. 2011. Does corporate social responsibility affect the cost of capital? *Journal of Banking and Finance* 35: 2388–2406.

Ge, W., and M. Liu. 2015. Corporate social responsibility and the cost of corporate bonds. *Journal of Accounting and Public Policy* 34 (6): 597–624.

Girerd-Potin, I., S. Jimenez-Garces, and P. Louvet. 2014. Which dimensions of social responsibility concern financial investors? *Journal of Business Ethics* 121 (4): 559–576.

Godfrey, P.C. 2005. The relationship between corporate philanthropy and shareholder wealth: A risk management perspective. *Academy of Management Review* 30 (4): 777–798.

Grossman, S.J., and O.D. Hart. 1982. *Corporate financial structure and managerial incentive. The economics of information and uncertainty*, 107–140. Chicago, IL: University of Chicago Press.

Harjoto, M.A. 2017. Corporate social responsibility and degrees of operating and financial leverage. *Review of Quantitative Finance and Accounting* 49: 487–513.

Harris, M., and A. Raviv. 1991. The theory of capital structure. *Journal of Finance* 46 (1): 297–356.

Jensen, M.C., and W.H. Meckling. 1976. Theory of the firm: Managerial behavior, agency costs and ownership structure. *Journal of Financial Economics* 3 (4): 305–360.

Jensen, M.C. 1986. Agency costs of free cash flow, corporate finance and takeovers. *American Economic Review* 76 (2): 323–339.

Larcker, D.F., and T.O. Rusticus. 2010. On the use of instrumental variables in accounting research. *Journal of Accounting and Economics* 49 (3): 186–205.

Lee, D.D., and R.W. Faff. 2009. Corporate sustainability performance and idiosyncratic risk: A global perspective. *Financial Review* 44: 213–237.

Li, K., H. Yue, and L. Zhao. 2009. Ownership, institutions, and capital structure: Evidence from China. *Journal of Comparative Economics* 37 (3): 471–490.

Majumdar, S.K., and P. Chhibber. 1999. Capital structure and performance: Evidence from a transition economy on an aspect of corporate governance. *Public Choice* 98 (3/4): 287–305.

Margaritis, D., and M. Psillaki. 2010. Capital structure, equity ownership and firm performance. *Journal of Banking and Finance* 34 (3): 621–632.

Miller, M. 1977. Debt and taxes. *The Journal of Finance* 32 (2): 261–275.

Modigliani, F., and M. Miller. 1958. The cost of capital, corporation finance and the theory of investment. *American Economic Review* 48 (3): 261–297.

———. 1963. Corporate income taxes and the cost of capital: A correction. *American Economic Review* 53 (3): 433–443.

Monteiro, S., and B. Aibar-Guzmán. 2010. Determinants of environmental disclosure in the annual reports of large companies operating in Portugal. *Corporate Social Responsibility and Environmental Management* 17: 185–204.

Myers, S. 1977. Determinants of corporate borrowing. *Journal of Financial Economics* 5 (2): 147–175.

———. 1984. The capital structure puzzle. *The Journal of Finance* 39 (3): 574–592.

Myers, S.C. 2001. Capital structure. *The Journal of Economic Perspectives* 15 (2): 81–102.

Myers, S., and N. Majluf. 1984. Corporate financing and investment decisions when firms have information that investors do not have. *Journal of Financial Economics* 13 (2): 187–221.

O'Brien, J. 2003. The capital structure implications of pursuing a strategy of innovation. *Strategic Management Journal* 24 (5): 415–431.

Ozkan, A. 2001. Determinants of capital structure and adjustment to long run target: Evidence from UK company panel data. *Journal of Business Finance & Accounting* 28 (1–2): 175–198.

Psillaki, M., and N. Daskalakis. 2009. Are the determinants of capital structure country or firm specific? *Small Business Economics* 33 (3): 319–333.

Rajagopal, S. 2011. The portability of capital structure theory: Do traditional models fit in an emerging economy? *Journal of Finance and Accountancy* 5: 1–17.

Rajan, R., and L. Zingales. 1995. What do we know about capital structure? Some evidence from international data. *The Journal of Finance* 50 (5): 1421–1460.

Sharfman, M.P., and C.S. Fernando. 2008. Environmental risk management and the cost of capital. *Strategic Management Journal* 29: 569–592.

Siddik, M., S. Kabiraj, and S. Joghee. 2017. Impacts of capital structure on performance of banks in a developing economy: Evidence from Bangladesh. *International Journal of Financial Studies* 5 (2): 1–13.

Sorin, V., and P. Luigi. 2009. A review of the capital structure theories. *Annals of the University of Oradea: Economic Science* 3 (1): 315–320.

Suto, M., and H. Takehara. 2018. *Corporate social responsibility and corporate finance in Japan*. Singapore: Springer Nature.

Testa, M., and A. D'Amato. 2017. Corporate environmental responsibility and financial performance: Does bidirectional causality work? Empirical evidence from the manufacturing industry. *Social Responsibility Journal* 13 (2): 221–234.

Titman, S., and R. Wessels. 1988. The determinants of capital structure choice. *The Journal of Finance* 43 (1): 1–19.

Vătavu, S. 2015. The impact of capital structure on financial performance in Romanian listed companies. *Procedia Economics and Finance* 32: 1314–1322.

Xu, S., D. Liu, and J. Huang. 2015. Corporate social responsibility, the cost of equity capital and ownership structure: An analysis of Chinese listed firms. *Australian Journal of Management* 40 (2): 245–276.

Yang, C., C. Lee, Y. Gu, and Y. Lee. 2010. Co-determination of capital structure and stock returns—A LISREL approach. *The Quarterly Review of Economics and Finance* 50 (2): 222–233.

Yang, S., F. He, Q. Zhu, and S. Li. 2018. How does corporate social responsibility change capital structure? *Asia-Pacific Journal of Accounting and Economics* 25 (3–4): 352–387.

Zhang, D., H. Cao, and P. Zou. 2016. Exuberance in China's renewable energy investment: Rationality, capital structure and implications with firm level evidence. *Energy Policy* 95: 468–478.

Chapter 8
Corporate Environmental Responsibility and Corporate Risk

Corporate Environmental Responsibility (CER) (within the context of Corporate Social Responsibility—CSR) has been considered as a significant factor related with the corporate risk. The ethics surrounding this mission should be considered self-evident, as should the fact that the contribution of each business/organization should reflect both the size of the organization and its carbon footprint. Nevertheless, significant research on the topic have found contradictory evidence regarding the sign of the CER impact (positive or negative) on corporate systematic risk within different countries and stages of economic development. The scope of this chapter is to examine the direct association between corporate environmental responsibility and corporate systematic risk within a multi-country setting, dealing with the potential endogenous relation between CER and risk and also controlling for the potential impact of countries legal origin.

8.1 Introduction

Today it is well-known and understood that business actions contribute significantly to environmental degradation. In order for businesses to continue to exist, they must commit to implementing a voluntary environmental policy that is one of the most important criteria for business judgment by society, consumers and even their shareholders. Corporate environmental responsibility (CER) refers to the actions of businesses, which aim at addressing environmental and social issues. Businesses, especially today, are inextricably linked to the society in which they operate, and therefore have an obligation to take appropriate responsibility towards society and the environment (Albuquerque et al. 2019; Chabowski et al. 2019). According to Chakraborty et al. (2019), there are two main streams on the literature that justify the adoption of CER activities by corporations. The first is the agency theory which focuses on the maximization of shareholders' wealth and considers CER as an

P. Dimitropoulos, K. Koronios, *Corporate Environmental Responsibility, Accounting and Corporate Finance in the EU*, CSR, Sustainability, Ethics & Governance, https://doi.org/10.1007/978-3-030-72773-4_8

agency problem contributing to a misallocation of scarce corporate resources (Barnea and Roubin 2010). In other words CER is undertaken at the expense of shareholders' leading to lower firm value and increase in firms' systematic risk. The other stream of literature resides on stakeholder theory which posits that CER enhances corporate and shareholder wealth because it enhances the stakeholder motivation to provide resources to the firm, contributing significantly to corporate value and the reduction of systematic risk (Cai et al. 2016).

In recent years businesses and companies have simply spent on donations and other forms of social activities. However, with the shift of stakeholder's attention on environmental issues, they are seeking to increase their long-term performance through practices that will satisfy all stakeholders (i.e. individuals and groups affected by the firms). Empirical evidence document the benefits of corporate environmental responsibility on several merits such as improving the working environment, enhance motivation and greater engagement of employees, acquiring competitive advantages, creating a growing trend in the company's reputation and customer base, even increase interest from customers and greater consumer response. All these create the so called "moral capital" or "relational capital" as argued by Godfrey (2005). This moral capital enhances firms' reputation and goodwill, providing an insurance-type of protection in cases of poor financial performance or adverse financial conditions. This fact mitigates negative stakeholder perceptions of firm risk which is the financial uncertainty faced by an investor who holds securities or invests in a particular business.

The majority of empirical studies so far point towards a negative association between CER and risk (Nguyen and Nguyen 2015; Cai et al. 2016; Chakraborty et al. 2019; Albuquerque et al. 2019). Even a recent study by Chabowski et al. (2019) provide evidence of a negative association between bond credit ratings and environmental inefficiency, arguing that firms which are more environmentally sensitive (are associated with less polluting production techniques and policies leading to a significant reduction of emissions) receive higher bond credit ratings, thus are perceived by the market and investors as carrying less litigation risk and lower uncertainty regarding their future operations. Nevertheless, there are some studies which point to the opposite direction. Cespa and Cestone (2007), Barnea and Roubin (2010) points towards a positive association between CER and corporate risk because as argued, CER diverts valuable corporate resources (exacerbating managerial entrenchment) making it more vulnerable to external pressures, which contributes to enhanced systematic risk. The reason for this inconsistency of empirical findings is attributed to different estimations of risk and CER, different stages of economic development and even the legal origin within each firm operates, distinguishing between stakeholder—oriented countries (code law) and shareholder—oriented countries (common law) (Benlemlih and Girerd-Potin 2017).

The scope of this chapter is to examine the impact of CER related performance on firms' systematic risk, by controlling for potential endogeneity between CER and systematic risk and by distinguishing sample firms between common law and code law countries, evidenced as significant factors for the inconsistency of empirical results by previous studies on the field. We respond to the call for more research on the topic made by Cai et al. (2016), by covering a large multi-country sample of listed European firms over a long period of time and also by considering the bidirectional association between CER and risk, consider each firm's legal origin. In the following section we provide the theoretical background on the impact of CER on CS and the testable hypotheses. The third section discusses the data selection procedure and the research design. The fourth section provides the discussion of empirical results and the last section concludes the chapter.

8.2 CER: Risk Association and Testable Hypothesis

The scope of this section is to discuss the CER-risk literature and to establish an association between firm's systematic risk and the environmental performance. As discussed above, the main theoretical traits on the CER-risk association are split into two streams which consider CER on the one hand as value enhancing for shareholders and stakeholders (stakeholder theory and risk reduction hypothesis) while on the other hand CER is considered as a value destroying activity (agency theory). Starting from the first stream of the literature, Godfrey (2005) argues that CER engagement creates a moral capital which can provide to shareholders a protection against bad reputation and adverse effects of economic activities. In other words, this moral capital is considered as a form of intangible capital which can assist the firm to reduce systematic risk by improving customer loyalty and employee commitment (Godfrey et al. 2009; Melo 2012). Moreover, Cai et al. (2016) argue that ethical managers can utilize CER as a communication mechanism in order to improve information transparency and dissemination of details regarding strategic focus and responsible investments. Accordingly, enhanced levels of transparency contribute to reduced information asymmetries between the firm and its main stakeholders, mitigating the investor's perception of firm risk (Dhaliwal et al. 2011; Suto and Takehara 2018).

Furthermore, Cheng et al. (2014) point that CER engagement reduces firms' financial constraints, allowing them to gain easier access to capital and financial markets, leading to improved trust and cooperation between the firms and its external stakeholders. This in turn reduces agency problems within the firm and the costs associated with those problems, through forcing managers to focus on the long-term performance (viability and sustainability) instead of the short-term financial focus. Under this framework, firms can avoid conflicts of interests between managers and stakeholders, establish solid and trustworthy relations and even mitigate the costs associated with polluting activities, a fact that contributes to lower perception of risk (Sen et al. 2006). Several studies support this argument by providing evidence of a reduction of firm's cost of capital (Sharfman and Fernando 2008; El Ghoul et al. 2011) and cost of debt (Chava 2011; Chabowski et al. 2019) for firms with enhanced CER performance. In other words, managerial decisions to engage in CER activities can contribute towards a reduction of firm risk by providing an insurance type of protection to stakeholders, making the firm more appealing to customers, creditors and investors, reducing their perception of future uncertainty (Cai et al. 2016; Nguyen and Nguyen 2015).

However, the opposite scenario could also be true, since some stakeholders (especially shareholders and creditors) could perceive CER investments and activities as a waste of valuable firm resources, enhancing firm's uncertainty about future operations and potential cash flows, contributing to enhance systematic risk (agency theory). As pointed out by Barnea and Roubin (2010), investment in CER diverts firm's resources from investment activities with positive NPVs such as building new products or services, improving production lines, or even taking advantage of profitable investment opportunities. This can reduce firm's future competitiveness making it more vulnerable to financial frictions or takeover activities by the competitors. Under this situation, CER engagement could be translated into higher risk (or uncertainty) by the stakeholders. Also, because CER involves a trade-off between the interests of various stakeholders, focusing in some of those groups could yield a cost to other groups. Cespa and Cestone (2007) argue that firms which systematically engage in environmentally responsible investments as a way to secure social and local acceptance for their operations, could be associated with higher managerial entrenchment which in turn increases the perceived risk of the firm (Nguyen and Nguyen 2015). Overall, under the agency theory of the firm (resource constraint hypothesis), managers pursuing CER activities reduce firm's ability to maximize shareholders' wealth and this could lead to a positive association between CER performance and systematic risk (Barnett and Salomon 2012). So considering the opposing theoretical arguments and the contradictory empirical evidence in the literature, we state the first research hypotheses as follows:

H1a: Under the risk-reduction hypothesis, CER performance will have a negative impact on systematic risk.

H1b: Under the resource-constraint hypothesis CER performance will have a positive impact on systematic risk.

Legal origin is another determinant factor for firm's engagement on CER and has implications for systematic risk. La Porta et al. (1998, 2008) argued that firms in code law countries are oriented towards stakeholder engagement since they are characterized by more concentrated ownership, higher legal protection of outside investors or minority shareholders and even more managers are associated with the firms through managerial shareholdings. All these characteristics may motivate managers in code law firms to engage in CER related activities so as to improve firms' long term investments, via improving the interests' alignment between owners, managers and external stakeholders (Kim et al. 2012). On the contrary, firms from common law countries are more shareholder-oriented organizations having as top priority the protection of investor rights. Under this framework, common law firms tend to focus on short-term profitability and performance and avoid to harm the wealth of outside investors. This fact creates less incentives for managers to engage in CER related activities since short-term investors may resist managerial decisions focusing on long-term perspectives and require the reduction of short-term financial performance so as to achieve higher future sustainability.

Empirical evidence in Chap. 4 indicate that code law firms invest more on CER activities. This result is attributed to the fact that managers of such firms tend to adhere to stakeholder-oriented ideologies and face less pressure when they decide to invest in CER projects. Consequently, the country's legal framework creates different motives towards business risk and investments and different stances against various issues relating to internal and external stakeholders, such as environmental responsibility.

This fact could have direct implications for corporate systematic risk. According to Benlemlih and Girerd-Potin (2017), laws and market regulations in common law countries are characterized by higher protection of shareholders' and investors' interests leading to the maximization of stakeholders interests only through shareholders value maximization. So CER activities are considered as a waste on company's resources and under this culture they may be associated with increased financial risk (since investors and shareholders could ask for a higher premium in order to safeguard their investments from the risk that CER activities create). On the contrary, Liang and Renneboog (2017) and Allen et al. (2009) argue that within code law countries (which are more stakeholder oriented), employees show enhanced loyalty to the firm, consumers often use social responsibility criteria for their purchase choices and show significant patience under critical or unexpected events. So CER performance could improve firm's value via better reputation, customer and employee loyalty and product quality confidence. All these can contribute into a reduction of systematic risk (beta) for firms engaging in CER activities within code law countries. Consequently, in code law countries the business environment stimulates CER policies via protection of the environment, reduction of waste and emissions and efficient use of scarce resources. So CER activities could be perceived as a risk reduction investment leading to lower financial risk. Thus, based on the above discussion the second research hypothesis is stated as follows:

H2: The association between CER performance and systematic risk will differ between code law and common law firms.

8.3 Data Selection and Research Design

8.3.1 Data Selection Procedure

The current chapter utilizes a sample of listed only corporations from 23 EU-member countries over the period 2003–2018. All financial, governance and environmental responsibility data have been extracted from Datastream database. The sample collection procedure started by including all corporations that have been covered by the environmental, social and governance (ESG) scores provided by Datastream during the sample period. 2003 has been selected as the initial sample year since that year was the first year with significant coverage (taking into consideration that this database officially was initiated in 2002). From a total of 656,734 firm-year observations we excluded those firms without environmental score data for at least five consecutive years in order to avoid bias in the empirical results and also to facilitate the estimation of differenced variables. Furthermore, we excluded firms with incomplete financial and stock market data related to stock prices and returns, and those that do not close their fiscal year on December. In addition, we winsorized the upper and down 1 per cent of the data distribution in order to mitigate the influence of significant outliers in the empirical analysis. After this procedure was finalized, we ended up with a final sample of 62,672 firm-year observations.

8.3.2 Research Design

Contemporary portfolio theory argues that corporate systematic risk is the only type of risk that matters for financial decision making since idiosyncratic risk could be diversified by investors, thus is not priced by them. Following previous studies by Cai et al. (2016) and Albuquerque et al. (2019), we estimate firm systematic risk using the CAPM's beta, which captures firm's systematic risk relative to the risk of the stock market. So we run the following time—series regression for every firm i and year t utilizing monthly stock price data:

$$R_{i,s} - R_f = a_i + \beta_i(R_{M,s} - R_f) + \varepsilon_{i,s} \tag{8.1}$$

Where $R_{i,s}$ is the stock return of firm i in month s, R_f is the one-month T-bill rate in month s for each country the firm is belonging to and $R_{M,s}$ is the return of each country's general stock price index in month s. The annual value of each firm's systematic risk is the estimated value of β_i. In order to capture the impact of CER on systematic risk under both negative and positive association of firms' stock returns with the market returns we will utilize the absolute value of beta (ABS_BETA) in the regression models.

For examining the validity of the previously mentioned research hypotheses we will consider an important issue that has been raised by several studies (Cai et al. 2016; Albuquerque et al. 2019) which is the existence of endogeneity meaning that CER performance might not only determine the firm systematic risk, but the causality mighty also go the other way round. Hong et al. (2012), provide empirical evidence that financially constrained firms are less likely to invest in socially responsible activities, something which reverts when the financial constraints are relaxed. Also, Albuquerque et al. (2019) point that firms with lower levels of systematic risk receive higher valuations by market participants and can devote more resources to CER activities. In addition, reverse causality could be generated by the fact that firms which have built strong relationships with customers have lower systematic risk and so can invest more in CER activities or firms with lower systematic risk could be operating within sectors that are more incentivized to engage in CER policies.

Under this circumstance, the GLS panel estimation of models may partially control for that by producing consistent estimators, yet not efficiently as the Generalized Method of Moments (GMM). For this reason, all regression models controlling for the impact of CER on risk will be estimated using the GMM method. The GMM method is selected over the Two-Stage Least Squares (2SLS) because in order to determine the 2SLS estimator we assume that the errors are i.i.d., meaning that they are identically and independently distributed over the observations. Having cross-sectional data the departure from independence may reflect neighboring effects in the sense that observations are similar in value and this way they share a correlation with the disturbances (Baum 2006). Also, observations that are close in time may be correlated with the strength of the correlation increasing with proximity. Thus, in the case that the errors are not i.i.d., 2SLS estimators can be consistent but inefficient. In other words, the 2SLS estimator is a special case of the GMM estimator. The GMM method allows us to estimate efficient and consistent estimators in the case of non-i.i.d. errors.

Following the study by Cai et al. (2016) we will estimate a system of simultaneous equations where the first stage equation includes CER performance as the dependent variable and other controls as the independent variables, while the second stage regression includes firm's systematic risk (beta) as the dependent variable and the CER variable with the same controls as in the first equation. The next step to the estimation of the models is to select proper instruments, meaning variables that are correlated with the CER and its components, but uncorrelated with risk. For this reason we will use as instruments the two-year lags of CER and its components as in Cai et al. (2016). In order to check the validity of the instruments all result tables report AR(1) and AR(2) tests which represent the first and second-order serial correlation test in the first differenced residuals, under the null hypothesis of no serial correlation. All AR tests rejected the null hypothesis suggesting that the instruments are valid for our research setting. Moreover, we performed the Hansen J-statistic of over-identifying restrictions which tests the joint null hypothesis that instrumental variables are valid (uncorrelated with the error term). All Hansen statistics were insignificant within conventional levels suggesting that the instruments were completely exogenous. The functional form of the equations includes the average CER for each firm and year estimated as the annual average of the ESG resource use score, emissions reduction score and environmental innovation score, where i denotes the firm and t the year and e is the error term:

First stage:

$$\begin{aligned} CER_{it} = {} & a_0 + a_1 ROE_{it} + a_2 LEV_{it} + a_3 SIZE_{it} + a_4 GROWTH_{it} \\ & + a_5 DUAL_{it} + a_6 BDSIZE_{it} + a_7 CER\_COM_{it} + Year\ F.E \\ & + Industry\ F.E. + Country\ F.E. + e_{it} \end{aligned} \tag{8.1a}$$

$$
\begin{aligned}
EN\_INNOV_{it} = {} & a_0 + a_1 ROE_{it} + a_2 LEV_{it} + a_3 SIZE_{it} + a_4 GROWTH_{it} \\
& + a_5 DUAL_{it} + a_6 BDSIZE_{it} + a_7 CER\_COM_{it} + Year\ F.E \\
& + Industry\ F.E. + Country\ F.E. + e_{it}
\end{aligned} \tag{8.1b}
$$

$$
\begin{aligned}
EMM_{it} = {} & a_0 + a_1 ROE_{it} + a_2 LEV_{it} + a_3 SIZE_{it} + a_4 GROWTH_{it} \\
& + a_5 DUAL_{it} + a_6 BDSIZE_{it} + a_7 CER\_COM_{it} + Year\ F.E \\
& + Industry\ F.E. + Country\ F.E. + e_{it}
\end{aligned} \tag{8.1c}
$$

$$
\begin{aligned}
RES\_USE_{it} = {} & a_0 + a_1 ROE_{it} + a_2 LEV_{it} + a_3 SIZE_{it} + a_4 GROWTH_{it} \\
& + a_5 DUAL_{it} + a_6 BDSIZE_{it} + a_7 CER\_COM_{it} + Year\ F.E \\
& + Industry\ F.E. + Country\ F.E. + e_{it}
\end{aligned} \tag{8.1d}
$$

Second stage:

$$
\begin{aligned}
ABS\_BETA_{it} = {} & a_1 CER_{it} + a_2 ROE_{it} + a_3 LEV_{it} + a_4 SIZE_{it} \\
& + a_5 GROWTH_{it} + a_6 DUAL_{it} + a_7 BDSIZE_{it} \\
& + a_8 CER\_COM_{it} + Year\ F.E + Industry\ F.E. \\
& + Country\ F.E. + u_{it}
\end{aligned} \tag{8.2a}
$$

$$
\begin{aligned}
ABS\_BETA_{it} = {} & a_1 EN\_INNOV_{it} + a_2 ROE_{it} + a_3 LEV_{it} + a_4 SIZE_{it} \\
& + a_5 GROWTH_{it} + a_6 DUAL_{it} + a_7 BDSIZE_{it} \\
& + a_8 CER\_COM_{it} + Year\ F.E + Industry\ F.E. \\
& + Country\ F.E. + u_{it}
\end{aligned} \tag{8.2b}
$$

$$
\begin{aligned}
ABS\_BETA_{it} = {} & a_1 EMM_{it} + a_2 ROE_{it} + a_3 LEV_{it} + a_4 SIZE_{it} \\
& + a_5 GROWTH_{it} + a_6 DUAL_{it} + a_7 BDSIZE_{it} \\
& + a_8 CER\_COM_{it} + Year\ F.E + Industry\ F.E. \\
& + Country\ F.E. + u_{it}
\end{aligned} \tag{8.2c}
$$

$$
\begin{aligned}
ABS\_BETA_{it} = {} & a_1 RES\_USE_{it} + a_2 ROE_{it} + a_3 LEV_{it} + a_4 SIZE_{it} \\
& + a_5 GROWTH_{it} + a_6 DUAL_{it} + a_7 BDSIZE_{it} \\
& + a_8 CER\_COM_{it} + Year\ F.E + Industry\ F.E. \\
& + Country\ F.E. + u_{it}
\end{aligned} \tag{8.2d}
$$

EN_INNOV denotes each company's environmental innovation score, which reflects the firm's ability to mitigate the environmental costs towards its customers or even creating new market and product opportunities via enhanced environmental technologies and processes. EMM denotes companies' emission reduction score estimated via their efforts and effectiveness in reducing emissions from their production and operation. RES_USE denotes the firms' resource use score and indicates firms' ability to reduce the usage of hazardous materials, energy, water and other valuable natural resources. Finally, CER is the firms' annual average of the aforementioned three scores. All environmental scores range between zero (0) and 100 so the higher the score the more environmentally responsible and efficient is the firm. ABS_BETA is the absolute value of firm's beta estimated from model (8.1). If the risk reduction hypothesis is valid we expect coefficient a_1 to be negative and statistically significant for all model estimations. Moreover, we have included in the models some additional controls that have proved significant determinants of firm systematic risk by previous research (Cai et al. 2016; Albuquerque et al. 2019; Benlemlih and Girerd-Potin 2017).

SIZE is the natural logarithm of annual total assets capturing firm size. A negative relationship between firm size and systematic risk is expected since large firms are considered to be more diversified and less likely to go bankrupt resulting in lower bankruptcy cost (Suto and Takehara 2018). However, large firms may also invest in higher risk investment projects and so they may be associated with more volatile future cash flows which increases their market valuation and potential, leading to increased systematic risk (beta). LEV is financial leverage estimated as the ratio of total debt to total assets. Highly leveraged firms are associated with higher systematic risk so a positive coefficient is expected in this variable (Albuquerque et al. 2019). GROWTH captures firms' growth opportunities and is estimated as the annual percentage change in sales revenues. According to Cai et al. (2016), firms with growth potentials are riskier since they have a higher bankruptcy cost and agency cost of debt, thus they might have enhanced betas. Thus a positive coefficient is expected on that variable. ROE is the ratio of return on equity capturing profitability. Firms with high profitability seems to have adequate retained earnings to meet their investment needs and thus they will probably face less systematic risk and will be associated with lower beta values (Cai et al. 2016). Finally. We include some governance variables as in the fourth and fifth chapters and specifically CEO duality, board size and the existence of a sustainability committee. Firms with improved and efficient governance mechanisms are expected to be associated with lower systematic risk so a negative coefficient is expected in the governance variables.

Moving on to examine the second research hypothesis, we introduce in model (8.2a) a dummy variable COMMON_LAW which receives unity (1) for firms established in a common law country and zero (0) if a firm belongs to a code law country. As discussed in Chap. 4, firms from common law countries are more shareholder-oriented organizations having as top priority the protection of investor rights. Under this framework, common law firms tend to focus on short-term profitability and performance and avoid to harm the wealth of outside investors. This fact creates less incentives for managers to engage in CER related activities since short-term investors may resist managerial decisions focusing on long-term perspectives and require the reduction of short-term financial performance so as to achieve higher future sustainability. Empirical evidence by Kim et al. (2012) and from Chap. 4 indicate that code law firms invest more on environmental protection activities and projects compared to their common law counterparts. This result may have implication for firms' systematic risk as argued by Benlemlih and Girerd-Potin (2017). So model (8.2a) will be adjusted as follows:

First stage:

$$\begin{aligned} CER_{it} = a_0 + a_1ROE_{it} + a_2LEV_{it} + a_3SIZE_{it} + a_4GROWTH_{it} \\ + a_5COMMON\_LAW_{it} + a_6DUAL_{it} + a_7BDSIZE_{it} \\ + a_8CER\_COM_{it} + Year\ F.E + Industry\ F.E. + Country\ F.E. \\ + e_{it} \end{aligned} \tag{8.3a}$$

Second stage:

$$\begin{aligned} ABS\_BETA_{it} = {} & a_1CER_{it} + a_2COMMON\_LAW_{it} * CER_{it} + a_2ROE_{it} \\ & + a_3LEV_{it} + a_4SIZE_{it} + a_5GROWTH_{it} + a_6DUAL_{it} \\ & + a_7BDSIZE_{it} + a_8CER\_COM_{it} + Year\ F.E \\ & + Industry\ F.E. + Country\ F.E. + u_{it} \end{aligned} \tag{8.3b}$$

So if a country's legal origin affects the association between CER and the second research hypothesis is valid and firm's systematic risk we expect coefficient a_2 to be statistically significant and with the opposite sign relative to coefficient a_1.

8.4 Empirical Results

8.4.1 Descriptive Statistics and Correlations

Table 8.1 presents the descriptive statistics of the variables utilized in the regression models. First of all, the mean CER is 64.5 (with a standard deviation of 21.1) indicating that the sample firms have a satisfactory environmental performance score. The highest average score is that of resources usage (66.5) followed by emission reduction activities (65.9) while environmental innovation activities lag behind with an average of 61.3. Also, unsigned BETA has an average value of 0.78

Table 8.1 Descriptive statistics of the sample variables

Variables	Mean	Standard deviation	Min	Max
CER	64.597	21.167	3.161	99.429
EN_INNOV	61.333	25.962	0.202	99.919
EMM	65.915	26.189	0.161	99.919
RES_USE	66.542	25.730	0.805	99.844
BETA	0.789	0.584	−9.744	9.159
ABS_BETA	0.818	0.542	0.000	9.744
ROE	79.511	34.815	−33.315	118.315
LEV	4.625	4.319	−0.682	25.691
SIZE	20.496	2.848	2.302	30.311
GROWTH	58.915	141.720	−69.711	409.071
DUAL	0.717	0.450	0.000	1.000
BDSIZE	12.271	4.703	1	38
CER_COM	0.552	0.497	0.000	1.000

Table 8.2 Firm systematic risk based on CER ranked portfolios

Variables	P1—High CER	P2—Medium CER	P3—Low CER	Difference (P1 − P3)	p-value
BETA	0.776	0.815	0.842	−0.066***	0.001
ABS_BETA	0.807	0.842	0.860	−0.053**	0.021

indicating that the sample firms are underperforming relative to the market and are characterized as low risk firms. The absolute value of BETA is close to the unsigned average and is up to 0.82. This value of EU listed firms beta is lower relative to US firms as depicted in the studies of Cai et al. (2016) and Albuquerque et al. (2019), where in both studies the estimate beta was above unity (1). The sample firms are highly profitable (mean ROA 21.28) and present significant growth opportunities (mean GROWTH at 58.9). Finally, almost the 72 per cent of the sample listed firms have a dual structure where the CEO serves also as the board chairman, average board size comprises of 13 directors and the slight majority of the firms have established a sustainability committee.

Table 8.2 presents the differences between the firm's systematic risk after separating firms within low, medium and high CER performance groups in order to examine the extent to which CER impacts on firm beta. Following Suto and Takehara (2018) we followed the portfolio formation approach and estimated the average of the CER variable per year and country and assigned each firm as low CER firm if its annual score is below the 1st quartile, high CER firms are those that their CER score is above the 3rd quartile while all other firm observations lying between the 1st and 3rd quartiles are assigned as medium CER firms. The column under the title "difference" reports the average CER difference between the high CER portfolio firms and the low CER portfolio. The "p-value" column presents the corresponding probability values from estimating the Welch two-sample t-test and asterisks indicate statistical significance at the 1 per cent significance level (Suto and Takehara 2018). As we can see from the last column, high CER performance firms are characterized by statistically lower beta (both in absolute value and unsigned value). Systematic risk seems to increase as we move from the high CER performance group to the low CER performance group, reaching to 0.86 for the low CER group of firms. These findings corroborate arguments by Albuquerque et al. (2019) and Suto and Takehara (2018) that more socially responsible firms are associated with lower systematic risk since CER related activities reduce information uncertainty and the investors' perception of firm risk. This provides a first indication for the support of our main research hypotheses, but since this association does not point towards any form of causality between the variables under interest, we proceed with the GMM regression analysis.

Table 8.3 GMM regression results on the association between CER and firm systematic risk

Variables			
First stage equation—Dep. Variable: CER	**Coef.**	**z-stat**	**P > |z|**
ROE	−0.001	−0.92	0.359
LEV	0.457	1.07	0.285
SIZE	0.315***	4.24	0.001
GROWTH	0.002	1.20	0.229
DUAL	−2.949***	−7.33	0.001
BDSIZE	0.121***	3.33	0.001
CER_COM	21.175***	57.76	0.001
Constant	45.88***	27.13	0.001
R-sq	0.265		
Chi^2 (p-value)	346.96*** (0.001)		
Second stage equation—Dep. Variable: ABS_BETA	**Coef.**	**z-stat**	**P > |z|**
CER	−0.005**	−1.91	0.050
ROE	−0.021***	−4.68	0.001
LEV	0.035***	2.83	0.005
SIZE	0.021***	9.71	0.001
GROWTH	−0.001	−0.19	0.848
DUAL	−0.016	−1.37	0.171
BDSIZE	−0.003***	−2.64	0.008
CER_COM	0.041***	3.29	0.001
Constant	0.471***	9.20	0.001
R-sq	0.136		
Chi^2 (p-value)	133.23*** (0.001)		
AR(1) test (p-value)	0.014		
AR(2) test (p-value)	0.076		
Hansen test of over-identification (p-value)	0.248		

Note: **, *** indicate statistical significance at the 5% and 1% significance level respectively

8.4.2 Regression Results

Table 8.3 presents the GMM regression results from the estimation from models (8.1a) and (8.2a) using the absolute value of beta as the dependent variable. The Chi-square values are highly significant and the R^2s are very satisfactory considering the size of the sample and other relative studies published in the field. All Tables present both the first stage and second stage results so the discussion will focus on the second stage results. The estimation of model (8.2a) provide a negative and significant coefficient (−0.005) on the CER verifying evidence provided on the literature by Cai et al. (2016), Benlemlih and Girerd-Potin (2017), Suto and Takehara (2018) and Albuquerque et al. (2019) that CER investments reduce the firm's systematic risk, so the risk reduction hypothesis is verified and so we can accept hypothesis H1a. In addition, firms with more stakeholder oriented CER activities are associated with more risk aversion and ethical behavior, especially towards its employees and the environment could be perceived as having less uncertainty and thus risk, so investors require a smaller risk premium to invest in such corporations.

Regarding the rest of the control variables, profitability (ROE) produced a negative and highly significant coefficient, corroborating arguments on the literature that highly profitable firms are associated with higher future cash flows and thus lower systematic risk (Cai et al. 2016). Also, LEV impact positively on systematic risk, which is also evidenced in the US market by Albuquerque et al. (2019) and Cai et al. (2016), because debt reduces the free cash flows available to the firm mitigating their ability to take advantage of future growth opportunities thus leading to increased systematic risk. In addition, SIZE produced a positive and significant coefficient suggesting that large firms may invest in higher risk investment projects and so they may be associated with more volatile future cash flows which increases their market valuation and potential, leading to increased beta. Finally, all governance variables produced highly significant coefficients suggesting that those firms with CEO duality, smaller board size and the non-existence of a sustainability committee are associated with lower systematic risk. It seems that investors in the EU market perceive CEO duality as a beneficial governance structure because when the CEO serves also as board chairman firms may face lower information asymmetries and achieve improved and fast decision making, taking acute decisions when necessary. Also, smaller boards are considered as more able to communicate and take more efficient decisions relative to larger boards so less directors on the board could reduce free-rider and other agency problems and further reduce firm beta (Jensen and Meckling 1976). Finally, while CER performance reduces firm beta, the existence of a sustainability committee is perceived as a waste of valuable resources (including human and financial resources) leading to increased systematic risk.

Tables 8.4, 8.5, and 8.6 present the results from the estimation of the GMM regression models using the CER components as the dependent variables in the second stage models. As we can see on Table 8.4, environmental innovation score (EN_INNOV) produced a negative and significant coefficient indicating the firms which invest in environmentally innovative projects are associated with lower systematic risk. On Table 8.6 the emissions score (EMM) also has a negative and significant coefficient indicating that firms engaging in emission reduction activities are facing less uncertainty on their future operations and potential litigation risk so investors require lower risk premiums to invest in such corporations. Nevertheless, the resource use score (RES_USE) on Table 8.6 had a negative sigh but was not significant within conventional levels, so resource use score does not seem to affect firm beta. Overall, both Tables 8.4 and 8.5 support the risk reduction hypothesis and so we can accept H1a. The coefficients on the control variables have the same sign and significance as on Table 8.3 so the same inferences are made for the controls.

Table 8.7 presents the GMM regression results after considering the country's legal origin on the association between CER performance and systematic risk. Again the CER coefficient (capturing the CER-BETA association for the code law countries) is negative and significant corroborating the main findings on the previous tables. However, the interaction variable COMMON_LAW*CER produced a

Table 8.4 GMM regression results on the association between EN_INNOV and firm systematic risk

Variables			
First stage equation—Dep. Variable: EN_INNOV	**Coef.**	**z-stat**	**P > |z|**
ROE	−0.001	−0.57	0.571
LEV	0.553	0.96	0.339
SIZE	0.370***	3.69	0.001
GROWTH	0.002	0.79	0.428
DUAL	−1.564***	−2.88	0.004
BDSIZE	0.210***	4.29	0.001
CER_COM	16.491***	33.30	0.001
Constant	45.41***	18.56	0.001
R-sq	0.108		
Chi^2 (p-value)	116.93*** (0.001)		
Second stage equation—Dep. Variable: ABS_BETA	**Coef.**	**z-stat**	**P > |z|**
EN_INNOV	−0.004**	−2.02	0.043
ROE	−0.002***	−4.67	0.001
LEV	0.035***	2.83	0.005
SIZE	0.021***	9.70	0.001
GROWTH	−0.001	−0.20	0.843
DUAL	−0.015	−1.29	0.198
BDSIZE	−0.002***	−2.61	0.009
CER_COM	0.036***	3.20	0.001
Constant	0.464***	9.23	0.001
R-sq	0.137		
Chi^2 (p-value)	133.66*** (0.001)		
AR(1) test (p-value)	0.023		
AR(2) test (p-value)	0.045		
Hansen test of over-identification (p-value)	0.247		

Note: **, *** indicate statistical significance at the 5% and 1% significance level respectively

positive and highly significant coefficient indicating that CER performing firms in common law countries are associated with higher systematic risk. This result verifies H2 and suggests that CER activities in common law countries (having more shareholder-oriented firms) are considered as a waste on company's resources and under this may be associated with increased financial risk, since investors and shareholders could ask for a higher premium in order to safeguard their investments from the risk that CER activities create. The control variables have again the same sign and significance as in Table 8.3.

Table 8.5 GMM regression results on the association between EMM and firm systematic risk

Variables			
First stage equation—Dep. Variable: EMM	**Coef.**	**z-stat**	**P > |z|**
ROE	−0.001	−0.49	0.621
LEV	0.134	0.29	0.809
SIZE	0.513***	5.33	0.001
GROWTH	0.004	1.49	0.137
DUAL	−3.486***	−6.69	0.004
BDSIZE	0.097**	2.07	0.039
CER_COM	22.57***	47.54	0.001
Constant	43.12***	19.69	0.001
R-sq	0.196		
Chi^2 (p-value)	236.66*** (0.001)		
Second stage equation—Dep. Variable: ABS_BETA	**Coef.**	**z-stat**	**P > |z|**
EMM	−0.005**	−1.99	0.047
ROE	−0.002***	−4.67	0.001
LEV	0.035***	2.81	0.005
SIZE	0.021***	9.73	0.001
GROWTH	−0.001	−0.18	0.853
DUAL	−0.016	−1.36	0.173
BDSIZE	−0.003***	−2.66	0.008
CER_COM	0.039***	3.29	0.001
Constant	0.465***	9.23	0.001
R-sq	0.136		
Chi^2 (p-value)	133.51*** (0.001)		
AR(1) test (p-value)	0.002		
AR(2) test (p-value)	0.065		
Hansen test of over-identification (p-value)	0.336		

Note: **, *** indicate statistical significance at the 5% and 1% significance level respectively

8.4.3 *Sensitivity Analysis*

In order to check the robustness of the main empirical results, we performed a number of sensitivity tests regarding the research design, and various definitions of the determinant variables. At first, following Cai et al. (2016) we replaced the CER variable definition (which is the simple average of the three sub-scores) by estimating a principal component from the three sub-scores. So we constructed the CER_PCA variable which equals the first principal component from the abovementioned scores. The principal component was determined using eigen values greater than 1.0. The principal component analysis was used since the solutions generated from principal component analysis do not have significant differences from the factor analysis techniques. In this context, we utilized the orthogonal method of extraction and particularly Varimax which tries to maximize the dispersion of loadings within factors. Eventually, for including the factor scores

Table 8.6 GMM regression results on the association between RES_USE and firm systematic risk

Variables			
First stage equation—Dep. Variable: RES_USE	**Coef.**	**z-stat**	**P > |z|**
ROE	−0.002	−1.08	0.280
LEV	0.685	1.29	0.198
SIZE	0.062	0.67	0.501
GROWTH	0.001	0.49	0.623
DUAL	−3.796***	−7.59	0.004
BDSIZE	0.055	1.23	0.221
CER_COM	24.46***	53.67	0.001
Constant	52.12***	24.79	0.001
R-sq	0.236		
Chi^2 (p-value)	298.54*** (0.001)		
Second stage equation—Dep. Variable: ABS_BETA	**Coef.**	**z-stat**	**P > |z|**
RES_USE	−0.001	−0.35	0.280
ROE	−0.002***	−4.66	0.001
LEV	0.035***	2.81	0.005
SIZE	0.020***	9.64	0.001
GROWTH	−0.001	−0.21	0.831
DUAL	−0.014	−1.25	0.210
BDSIZE	−0.003***	−2.70	0.007
CER_COM	0.031**	2.53	0.011
Constant	0.449***	8.82	0.001
R-sq	0.132		
Chi^2 (p-value)	129.64*** (0.001)		
AR(1) test (p-value)	0.001		
AR(2) test (p-value)	0.088		
Hansen test of over-identification (p-value)	0.472		

Note: **, *** indicate statistical significance at the 5% and 1% significance level respectively

as an independent variable in models (8.1a) and (8.2a) we employed the Anderson–Rubin method. The outcome of the principal component analysis supported that data do not suffer from multicollinearity since the KMO statistic of sampling adequacy had a value up to 0.93 indicating that this kind of analysis is appropriate for our data. Eventually the extracted communalities are close to 1 proposing that the factor explains the original data adequately, and the factor loadings are above the critical value of 0.70 proposing that the estimated PCA is of statistical significance. Table 8.8 presents the GMM empirical results from the estimation of models (8.1a) and (8.2a) by using the CER_PCA variable instead of the average CER. As we can see CER_PCA produced a negative and significant coefficient (at the 10% per cent level) corroborating H1a and the inferences drawn from Table 8.3, so our main results remain robust.

Another sensitivity test dealt with the functional form of the main models by including 1 year lags of the independent variables in the main models as in Cai et al. (2016) and Albuquerque et al. (2019). The results remained qualitatively unchanged

Table 8.7 GMM regression results on the association between CER-systematic risk considering each country's legal origin

Variables			
First stage equation—Dep. Variable: CER	**Coef.**	**z-stat**	**P > |z|**
ROE	−0.001	−0.90	0.367
LEV	0.338	0.79	0.429
SIZE	0.273***	3.67	0.001
GROWTH	0.003	1.30	0.194
COMMON_LAW	−4.627***	−6.58	0.001
DUAL	−3.021***	−7.52	0.001
BDSIZE	0.102***	2.79	0.005
CER_COM	21.18***	57.91	0.001
Constant	47.46***	27.85	0.001
R-sq	0.267		
Chi^2 (p-value)	352.84*** (0.001)		
Second stage equation—Dep. Variable: ABS_BETA	**Coef.**	**z-stat**	**P > |z|**
CER	−0.006**	−2.03	0.042
COMMON_LAW*CER	0.001***	3.50	0.001
ROE	−0.001***	−4.68	0.001
LEV	0.037***	2.97	0.003
SIZE	0.021***	9.93	0.001
GROWTH	−0.001	−0.24	0.813
DUAL	−0.014	−1.21	0.226
BDSIZE	−0.002**	−2.43	0.015
CER_COM	0.039***	3.21	0.001
Constant	0.453***	8.81	0.001
R-sq	0.149		
Chi^2 (p-value)	145.85*** (0.001)		
AR(1) test (p-value)	0.044		
AR(2) test (p-value)	0.101		
Hansen test of over-identification (p-value)	0.410		

Note: **, *** indicate statistical significance at the 5% and 1% significance level respectively

relative to those on Tables 8.3, 8.4, 8.5, and 8.6. Furthermore, additional sensitivity testing was performed following different definitions of the independent variables. Specifically, we estimated LEV as the ratio of long term debt to total assets, GROWTH as the ratio of market value to book value of equity (MV/BV) and ROE was replaced by ROA but the results remain unchanged after these modifications. Finally, we conducted additional robustness checks by changing the definition of BETA by estimating the Fama-French three factor model as in Cai et al. (2016) but once again the results remained qualitatively unchanged compared with those on Tables 8.3, 8.4, 8.5, and 8.6.

Table 8.8 GMM regression results with a principal component of CER and systematic risk

Variables			
First stage equation—Dep. Variable: CER_PCA	**Coef.**	**z-stat**	**P > |z|**
ROE	−0.001	−0.92	0.355
LEV	0.030	1.05	0.295
SIZE	0.020***	4.11	0.001
GROWTH	0.001	1.20	0.232
DUAL	−0.203***	−7.55	0.001
BDSIZE	0.007***	3.10	0.002
CER_COM	1.429***	58.34	0.001
Constant	−1.235***	−10.93	0.001
R-sq	0.268		
Chi^2 (p-value)	353.84*** (0.001)		
Second stage equation—Dep. Variable: ABS_BETA	**Coef.**	**z-stat**	**P > |z|**
CER_PCA	−0.008*	−1.82	0.069
ROE	−0.001***	−4.68	0.001
LEV	0.035***	2.82	0.005
SIZE	0.021***	9.70	0.001
GROWTH	−0.001	−0.19	0.847
DUAL	−0.016	−1.37	0.172
BDSIZE	−0.003**	−2.64	0.008
CER_COM	0.041***	3.24	0.001
Constant	0.435***	8.76	0.001
R-sq	0.136		
Chi^2 (p-value)	132.87*** (0.001)		
AR(1) test (p-value)	0.006		
AR(2) test (p-value)	0.065		
Hansen test of over-identification (p-value)	0.394		

Note: **, *** indicate statistical significance at the 5% and 1% significance level respectively

8.5 Conclusion

Several studies on the environmental responsibility research field have focused on examining the impact of CER activities on firm risk. Researchers have expressed different views on how a company's social and environmental performance affects its systematic risk with contradictory findings. The risk-reduction hypothesis posits that CER performance contributes to a reduction of firm's perceived risk while the resource constraint hypothesis argues that CER leads to a reduction of future investment opportunities (by misallocating valuable firm resources) increasing firms' systematic risk (Cai et al. 2016; Albuquerque et al. 2019; Chabowski et al. 2019). The scope of this chapter is to examine the impact of CER related performance on firm systematic risk, by taking into consideration potential endogeneity between the examined factors and their bidirectional association, evidenced by previous studies on the field and by separating sample firms according to their legal origin. We respond to the call for more research on the topic made by Cai et al. (2016), by covering a large multi-country sample (24 EU countries) of listed European firms (62,672 firm-year observations) over a long period of time (2003–2018) and also by considering the bidirectional association between CER and systematic risk (beta).

Empirical analysis suggested that CER performance provide a negative and significant impact on firm systematic risk (beta) verifying existing evidence on the literature by Cai et al. (2016), Benlemlih and Girerd-Potin (2017), Suto and Takehara (2018) and Albuquerque et al. (2019) that CER investments reduce the firm's systematic risk, so leading us to accept the risk reduction hypothesis. In addition, firms with more stakeholder oriented CER activities are associated with more risk aversion and ethical behavior, especially towards its employees and the environment, thus perceived as having less uncertainty and risk, so investors require a smaller risk premium to invest in such corporations. Moreover, after separating sample firms based on their country's legal origin we found that CER performing firms in common law countries are associated with higher systematic risk. This result verifies the resource constraint hypothesis and suggests that CER activities in common law countries are considered as a waste on company's resources and under this may be associated with increased financial risk, since investors and shareholders could ask for a higher premium in order to safeguard their investments from the risk that CER activities create. The findings of this chapter contribute to the ongoing debate about the association between CER and firm risk is several merits. At first we complement previous studies by Suto and Takehara (2018), Cai et al. (2016) and Albuquerque et al. (2019) by incorporating a multi-country instead of a single country research setting, considering at eh same time the legal origin of firms. Also, we incorporate in the analysis a decomposition of the CER variable and indicate which sub-category of CER performance contributes to capital structure decisions, something that has not been considered extensively by previous researchers.

References

Albuquerque, R., Y. Koskinen, and C. Zhang. 2019. Corporate social responsibility and firm risk: Theory and empirical evidence. *Management Science* 65 (10): 4451–4469.

Allen, F., E. Carletti, and R. Marquez. 2009. Stakeholder capitalism, Corporate governance, and firm value. Wharton Finance Working Paper.

Barnea, A., and A. Roubin. 2010. Corporate social responsibility as a conflict between shareholders. *Journal of Business Ethics* 97 (1): 71–86.

Barnett, M.L., and R.M. Salomon. 2012. Does it pay to be really good? Addressing the shape of the relationship between social and financial performance. *Strategic Management Journal* 33: 1304–1320.

Baum, C.F. 2006. *An introduction to econometrics using Stata*. College Station, TX: Stata Press Books, StataCorp LP.

Benlemlih, M., and I. Girerd-Potin. 2017. Corporate social responsibility and firm financial risk reduction: On the moderating role of the legal environment. *Journal of Business Finance and Accounting* 44 (7–8): 1137–1166.

Cai, L., J. Cui, and H. Jo. 2016. Corporate environmental responsibility and firm risk. *Journal of Business Ethics* 139: 563–594.

Cespa, G., and G. Cestone. 2007. Corporate social responsibility and managerial entrenchment. *Journal of Economics and Management Strategy* 16 (3): 741–771.

Chabowski, B., W.-C. Chiang, K. Deng, and L. Sun. 2019. Environmental inefficiency and bond credit rating. *Journal of Economics and Business* 101: 17–37.

Chakraborty, A., L.S. Gao, and S. Sheikh. 2019. Managerial risk taking incentives, corporate social responsibility and firm risk. *Journal of Economic and Business* 101: 58–72.

Chava, S. 2011. Environmental externalities and cost of capital. Working paper. Georgia Institute of Technology. Accessed from http://papers.ssrn.com/sol3/papers.cfm?abstract_id=1677653

Cheng, B., I. Ioannou, and G. Serafeim. 2014. Corporate social responsibility and access to finance. *Strategic Management Journal* 35 (1): 1–23.

Dhaliwal, D., O. Li, A. Zhang, and Y. Yang. 2011. Voluntary nonfinancial disclosure and the cost of equity capital: The initiations of corporate social responsibility reporting. *The Accounting Review* 86 (1): 59–100.

El Ghoul, S., O. Guedhami, C. Kwok, and D. Mishra. 2011. Does corporate social responsibility affect the cost of capital? *Journal of Banking & Finance* 35 (9): 2388–2406.

Godfrey, P.C. 2005. The relationship between corporate philanthropy and shareholder wealth: A risk management perspective. *Academy of Management Review* 30: 777–798.

Godfrey, P.C., C. Merrill, and J. Hansen. 2009. The relationship between corporate social responsibility and shareholder value: An empirical test of the risk management hypothesis. *Strategic Management Journal* 30 (4): 425–445.

Hong, H., J.D. Kubik, and J.A. Scheinkman. 2012. *Financial constraints on corporate goodness. NBER working paper, 18476*. Cambridge, MA: National Bureau of Economic Research.

Jensen, M.C., and W.H. Meckling. 1976. Theory of the firm: Managerial behavior, agency costs and ownership structure. *Journal of Financial Economics* 3 (4): 305–360.

Kim, Y., M. Park, and B. Wier. 2012. Is earnings quality associated with corporate social responsibility? *The Accounting Review* 87: 761–796.

La Porta, R., F. Lopez-de-Silanes, A. Shleifer, and R.W. Vishny. 1998. Law and finance. *Journal of Political Economy* 106 (6): 1134–1155.

La Porta, R., F. Lopez-de-Silanes, and A. Shleifer. 2008. The economic consequence of legal origins. *Journal of Economic Literature* 46: 285–332.

Liang, H., and L. Renneboog. 2017. On the foundations of corporate social responsibility. *The Journal of Finance* 72 (2): 853–910.

Melo, T. 2012. Determinants of corporate social performance: The influence of organizational culture, management tenure and financial performance. *Social Responsibility Journal* 8 (1): 33–47.

Nguyen, P., and A. Nguyen. 2015. The effect of corporate social responsibility on firm risk. *Social Responsibility Journal* 11 (2): 324–339.

Sen, S., C.B. Bhattacharya, and D. Korschun. 2006. The role of corporate social responsibility in strengthening multiple stakeholder relationships. *Journal of the Academy of Marketing Science* 34 (2): 158–166.

Sharfman, M.P., and C.S. Fernando. 2008. Environmental risk management and the cost of capital. *Strategic Management Journal* 29: 569–592.

Suto, M., and H. Takehara. 2018. *Corporate social responsibility and corporate finance in Japan*. Singapore: Springer Nature.

Chapter 9
Corporate Environmental Responsibility, Cash Holding and Dividend Policy Decisions

Cash is a very important item on the firms' balance sheet since it determines its liquidity and ability to finance daily operations, pay liabilities on time and take advantage of investment opportunities. On the same merit, cash dividend payments are the part of earnings distributed to shareholders and is closely connected with the liquidity of the firm. Several studies have examined the determinants of cash holdings and dividend payments and CER related disclosures have been considered as information which contribute to the reduction of information asymmetries between the firm and external stakeholders, a fact that affects the decisions of firms to distribute dividends and hold cash. Nevertheless, empirical findings have been quite contradictory suggesting a differential impact of CER on cash holdings and cash dividend payments. The purpose of this chapter is to study the impact of CER related activities on dividend and cash holding decisions taking into consideration the potential endogeneity between the examined factors and their bidirectional association, evidenced by previous studies on the field.

9.1 Introduction

The literature on the topic of cash holding decisions provide two main theoretical underpinnings in order to justify managerial behavior on that issue. The first theoretical argument refers to the precautionary motive for cash holdings which designates the goal of firms to adhere to business uncertainty, meet unexpected events and reduce the risk of illiquidity (Dimitropoulos 2020; Dimitropoulos et al. 2020). The second theoretical argument refers to the transactional motive for holding cash, which dictates that firms hold excessive cash amounts in order to reduce transactional costs of daily operations. This means that firms decide to awash cash so as to meet overdue payables without having to raise external financing. This decision of course can have implications for the investment decisions of the firm,

P. Dimitropoulos, K. Koronios, *Corporate Environmental Responsibility, Accounting and Corporate Finance in the EU*, CSR, Sustainability, Ethics & Governance, https://doi.org/10.1007/978-3-030-72773-4_9

since those firms with increased cash assets can take advantage of investment opportunities that could be rejected by less cash holding firms (Martínez-Sola et al. 2018; Dimitropoulos 2020). Moreover, dividend payments is another corporate strategic decision which affects firms' market valuation, uncertainty and investor behavior, while is closely connected to cash flows. According to Hasan and Habib (2020) and Benlemlih (2019), dividend payouts contribute significantly on the control of agency costs of free cash flows by reducing the available resources under managerial control and the inefficient usage of scarce corporate resources.

CER activities have been considered by several researchers as a determinant of cash holding and dividend payment decisions. Specifically, the CER motivation for holding more cash amounts on the balance sheet is derived from managers' goal to maximize shareholder value and adhere to stakeholders' needs, therefore it suggests a positive association between CER and cash holdings. So in order to satisfy those needs managers hold more cash for social and environmental activities instead of increasing the payouts to shareholders. Empirical evidence by Cheung (2016) and Huang et al. (2019) indicate that socially responsible firms hold more cash in their attempt to satisfy not only their shareholders, but also the majority of their key stakeholders. In addition, CER activities create value for their shareholders because CER strategic focus creates trust and reciprocity with stakeholders by showing to them the firms' dedication for meeting its implicit commitments. This fact enhances stakeholders' willingness to support firms operations. According to Chang et al. (2019), higher cash holdings allows environmentally responsible firms to signal to their stakeholders, their ability to fulfil those imbedded commitments.

On the contrary the precautionary motive for cash holdings (as already mentioned) leads to higher cash holdings in order to deal with business uncertainty of future payments and risks. Empirical evidence provided by Boutin-Dufesne and Savaria (2004) and Huang et al. (2019) suggest that environmentally responsible firms are associated with lower risks and cost of equity, and easier access to financing, implying that those firms are associated with diminished operational costs and so they need to hold less cash for covering such costs. Thus the precautionary theory argues for a negative association between CER and cash holdings.

Moreover, there is an extensive literature on the impact of CER performance on cash dividend payments with two contrasting theories which try to provide theoretical and empirical explanations. The first theory is based on the free cash flow argument (Jensen 1986) which argues that when managers have increased free cash flows on their hands they tend to be reckless and overinvest and so they use inefficiently the firm's resources. Thus, higher dividends reduce the free cash flows available to managers' discretion. From a CER perspective, if managers use environmental responsibilities for deriving private benefits by being identified as socially or environmentally responsible, dividend payments can be used as a controlling or discipline mechanism of managerial overinvestment and shareholders' wealth appropriation (Barnea and Rubin 2010; Benlemlih 2019). Consequently, in the case where dividends are used as a discipline mechanism for preventing managers from wasting cash and firm resources, a positive association is expected between CER performance and cash dividends. In the same direction, signaling theory proposed by Bhattacharya (1979) posits that dividends are used as a signal to the market about the firm's future prospects and profitability. So CER performing firms who are devoted in honoring implicit and explicit stakeholder claims will tend to distribute more cash dividends in order to be able to sustain such activities in the future and signal that such CER activities do not destroy company resources and cash assets (Benlemlih 2019). However, the substitution theory argues for a negative association between CER and cash dividend payments. This is attributed to the fact that CER related activities and disclosures improve the information environment of the firm and so there is less need on behalf of the managers to pay cash dividends in order to mitigate information asymmetries between minority investors and stakeholders. So this fact reduces the need to resolve to costly dividend payouts (Hail et al. 2014; Ni and Zhang 2019).

Following the abovementioned discussion, the scope of this chapter is to examine the impact of CER related performance on firm dividend and cash holding decisions, by taking into consideration potential endogeneity between the examined factors, evidenced by previous studies on the field. We respond to the call for more research on the topic made by Benlemlih (2019) and Huang et al. (2019), by covering a large multi-country sample of listed and unlisted European firms over a long period of time and also by considering different environmental performance ratings. In the following section we provide the theoretical background on the impact of CER on cash holdings and dividend payments along with the testable hypotheses. The third section discusses the data selection procedure and the research design. The fourth section provides the discussion of empirical results and the last section concludes the chapter.

9.2 CER: Cash—Dividends Association and Testable Hypothesis

Based on the prior findings on the literature, the CER motive and the precautionary motive of cash holdings are the main theoretical frameworks that we will discuss on this section. Under the CER motive of cash holdings, Arouri and Pijorlet (2017), Huang et al. (2019) and Cheung (2016) argue that environmentally responsible firms are devoted in adhering to stakeholders' needs and demands, and in order to achieve such strategy they are more in need to hold cash and cash equivalents in their balance sheets. Furthermore, according to Jiraporn and Chintrakarn (2013) and Dittmar et al. (2003), CER activities may also suggest enhanced managerial entrenchment leading to several agency problems and costs within the firms. Logically, holding more cash allows firms to overcome those agency costs. In the same token, environmental responsible activities by firms indicate that they are governed by responsible and competent managers, suggesting a low intensity of agency conflicts. If this argument is valid, we can infer that managers of environmentally responsible firms are focusing more on maximizing stakeholder value (and not only shareholders' value) leading to more cash holdings for being able to achieve that goal (Vafeas and Nikolaou 2001; Chang et al. 2019).

On the contrary, following the precautionary argument for cash holdings, firms tend to hold more cash when they face higher environmental uncertainty and risks, so as to be able to adhere to those difficulties without liquidating assets or raising external financing. As evidenced by the previous chapter, CER performing firms are associated with lower systematic risk, which reduces the firm's perceived risk by the public (Feldman et al. 1997). Also Suto and Takehara (2018) indicated that socially responsible firms have lower cost of equity, so are facing less costly access to the capital markets. So following the discussion on the precautionary motive, firms with enhanced CER performance are expected to hold less cash assets since they have to face less risks relative to their low CER performing counterparts. So due to the conflicting theories, arguments and findings on the literature regarding the impact of CER on cash holding decisions, we state first research hypothesis (and the alternative H1a) as follows:

H1: CER performance will have a positive impact on cash holding decisions.
H1a: CER performance will have a negative impact on cash holding decisions.

Moving on to the association between CER and dividend payments, two theoretical perspectives are offered to the literature, which point to a positive association between the environmental responsibility and dividend payments. Firstly, the agency theory of the firm argues that dividends are useful in controlling agency costs associated with free cash flows (Jensen 1986). Cash dividend payments reduced the cash available to managers' discretion and so control for wealth and resources appropriation by managers for their private benefits. CER investments and activities moderate this agency problem by the fact that CER performing firms finance activities that interest different stakeholder groups. This fact creates benefits for both inside and outside parties by reducing information asymmetries and enhances monitoring capabilities (Benlemlih 2019). Under the reduction of information asymmetries between stakeholders, firms which have incorporated more detailed CER rules and mechanisms tend to be more scrutinized by stakeholders, they can establish more trustful relationships with them and so they can enhance their performance and profitability on the long-term (Ni and Zhang 2019). This means that CER performing firms are in better position to pay cash dividends than low CER performing firms.

Secondly, from the perspective of signaling theory firms devoted to the protection of the environment are keen on creating their wealth within ethical and sustainable limits, and so they are focused in satisfying the interests of financial and non-financial stakeholders. A prosperous dividend payout policy can create reputational benefits by communicating the firm's fairness and ethical stance on wealth distribution between the key players which contribute to its creation (shareholders, customers, lenders, creditors etc.) (He et al. 2012; Benlemlih 2019; Ni and Zhang 2019). Moreover, increased cash dividend payments can signal to the market that CER activities do not simply consume firm's cash assets and can end to more efficient allocation of resources and improved stakeholder (and shareholder) satisfaction. According to Hasan and Habib (2020), firms that are headquarter in higher social capital regions pay more dividends because they cash access funds from various sources at lower costs and so they can distribute more cash dividends relative to firms operating in low social capital regions. Also, firms operating in high social capital areas tend to be more ethical and distribute excess cash as dividends. Consequently, from the one side by paying more dividends firms can reduce agency costs and information asymmetries between the stakeholders by encouraging managers to invest in CER activities, while simultaneously high dividends serve as signals for corporate reputation and the interest of firms for the needs of stakeholders.

On the contrary, the substitution view on payout decisions argues for a negative association between CER and dividend payments. Specifically, the substitutive view considers dividends as tools of reputation building through signals to the market about the decision of the firm not to waste free cash flows (Hasan and Habib 2020). So higher social capital (with improved social and environmental behavior) can mitigate the free cash flow problem and thus substitute the need to build corporate reputation through cash dividend payments. In other words, managers of high CER performing firms may consider dividend payments as having marginal benefit, thus they will be less incentivized to distribute cash dividends. Moreover, since the firm's information environment is improved via the firm's implication of CER activities into its daily operations, the necessity for demonstrating firm's commitment via costly dividend payments is even lower. Consequently, under the different arguments and finding of the substitution theory, the agency and signaling theories regarding the impact of CER on payout policies we state the second research hypothesis (and the alternative H2a hypothesis) in the following form:

H2: CER performance will impact positively on dividend payments.
H2a: CER performance will impact negatively on dividend payments.

9.3 Data Selection and Research Design

9.3.1 Data Selection Procedure

The current chapter utilizes a sample of listed only corporations from 23 EU-member countries over the period 2003–2018. All financial, governance and environmental responsibility data have been extracted from Datastream database. The sample collection procedure started by including all corporations that have been covered by the environmental, social and governance (ESG) scores provided by Datastream during the sample period. 2003 has been selected as the initial sample year since that year was the first year with significant coverage (taking into consideration that this database officially was initiated in 2002). From a total of 656,734 firm-year observations we excluded those firms without environmental score data for at least five consecutive years in order to avoid bias in the empirical results and also to facilitate the estimation of differenced variables. Furthermore, we excluded firms with incomplete financial data related cash and cash equivalents and dividend payments, as well as those that do not close their fiscal year on December. In addition, we winsorized the upper and down 1 per cent of the data distribution in order to mitigate the influence of significant outliers in the empirical analysis. After this procedure was finalized, we ended up with a final sample of 1288 distinct firms summing up to 115,748 firm-year observations.

9.3.2 Research Design

For examining the validity of the first research hypotheses we will follow previous studies and estimate a linear regression model of cash holdings as the dependent variable and CER performance along with other controls as independent variables (Chang et al. 2019). Also, in order to mitigate concerns of any omitted correlated variables we will utilize a panel fixed effects regression analysis as in Huang et al. (2019) including year, industry and country fixed effects. The functional form of the equations includes the average CER for each firm and year estimated as the annual average of the ESG resource use score, emissions reduction score and environmental innovation score, where i denotes the firm and t the year and e is the error term:

$$\begin{aligned} CASH_{it} = {} & a_0 + a_1 CER_{it} + a_2 SIZE_{it} + a_3 LEV_{it} + a_4 CFO_{it} + a_5 PPE_{it} \\ & + a_6 GROWTH_{it} + a_7 RnD\_EXP_{it} + a_8 ROA_{it} + a_9 NWC_{it} \\ & + a_8 DIV\_TA_{it} + Year\ F.E + Industry\ F.E. + Country\ F.E. \\ & + u_{it} \end{aligned} \tag{9.1a}$$

$$\begin{aligned} CASH_{it} = {} & a_0 + a_1 EN\_INNOV_{it} + a_2 SIZE_{it} + a_3 LEV_{it} + a_4 CFO_{it} \\ & + a_5 PPE_{it} + a_6 GROWTH_{it} + a_7 RnD\_EXP_{it} + a_8 ROA_{it} \\ & + a_9 NWC_{it} + a_8 DIV\_TA_{it} + Year\ F.E + Industry\ F.E. \\ & + Country\ F.E. + u_{it} \end{aligned} \tag{9.1b}$$

$$\begin{aligned} CASH_{it} = {} & a_0 + a_1 EMM_{it} + a_2 SIZE_{it} + a_3 LEV_{it} + a_4 CFO_{it} + a_5 PPE_{it} \\ & + a_6 GROWTH_{it} + a_7 RnD\_EXP_{it} + a_8 ROA_{it} + a_9 NWC_{it} \\ & + a_8 DIV\_TA_{it} + Year\ F.E + Industry\ F.E. + Country\ F.E. \\ & + u_{it} \end{aligned} \tag{9.1c}$$

$$\begin{aligned} CASH_{it} = {} & a_0 + a_1 RES\_USE_{it} + a_2 SIZE_{it} + a_3 LEV_{it} + a_4 CFO_{it} \\ & + a_5 PPE_{it} + a_6 GROWTH_{it} + a_7 RnD\_EXP_{it} + a_8 ROA_{it} \\ & + a_9 NWC_{it} + a_8 DIV\_TA_{it} + Year\ F.E + Industry\ F.E. \\ & + Country\ F.E. + u_{it} \end{aligned} \tag{9.1d}$$

CASH is the ratio of cash and cash equivalents to total assets at the end of the fiscal year (Huang et al. 2019). EN_INNOV denotes each company's environmental innovation score, which reflects the firm's ability to mitigate the environmental costs towards its customers or even creating new market and product opportunities via enhanced environmental technologies and processes. EMM denotes companies' emission reduction score estimated via their efforts and effectiveness in reducing emissions from their production and operation. RES_USE denotes the firms' resource use score and indicates firms' ability to reduce the usage of hazardous materials, energy, water and other valuable natural resources. Finally, CER is the firms' annual average of the aforementioned three scores. All environmental scores range between zero (0) and 100 so the higher the score the more environmentally responsible and efficient is the firm. If the CSR motive hypothesis is valid we expect coefficient a_1 to be positive and statistically significant for all model estimations, while on the contrary if the precautionary motive of cash holdings is true coefficient a_1 will be negative and statistically significant. Moreover, we have included in the models some additional controls that have proved significant determinants of firm systematic risk by previous research (Huang et al. 2019; Chang et al. 2019; Ghaly et al. 2015).

SIZE captures firm size and is measured as the natural logarithm of total assets. As argued by Chang et al. (2019), larger firms have easier access to financial markets since they have lower likelihood of financial distress. So smaller firms are expected to hold more cash and this may assist them in improving their financial performance and sustaining their viability in troubled times. So we expect a negative coefficient on the SIZE variable. LEV captures firm leverage and is estimated as the ratio of total debt to total assets. Higher leverage may lead to less information asymmetries leading to increased capital market access and thus a reduced need to hold more cash (Kim et al. 2013; Chang et al. 2019). So a negative coefficient is expected on this variable. Firms with a greater ability to generate internal funds (operating cash flows—CFO) are less in need of external financing. Kim et al. (2013) indicate that firms with higher cash flows were associated with lower cash holdings since they can substitute cash, so a negative coefficient is expected for this variable. The ratio of property, plant and equipment to total assets (PPE) is an additional control included in models (9.1a–9.1d). Firms with more tangible assets can have easier access to financial markets since those assets could be used as collateral thus reducing the cost of lending borrowing. This means that firms with higher tangibility have less need to hold cash to support their financial performance, thus we expect a negative coefficient on this variable.

Another significant determinant of cash holdings is firms' growth opportunities (GROWTH), which are measured as the annual change in sale revenues. Firms with high growth opportunities need more cash in order to finance them. The precautionary motive for high growth firms is more intense so a positive coefficient is expected for the GROWTH coefficient. Research and Development expenses (RnD_EXP) to sales is another determinant of cash holdings. According to Chang et al. (2019), RnD intensive firms are associated with enhanced future investment opportunities and so they will tend to hold more cash for precautionary reasons. Thus, a positive coefficient is expected on RnD_EXP. Profitability is expected to be associated with higher cash holdings since firms with higher ROA are more able to retain earnings and generate higher cash amounts relative to less profitable firms. Net working capital (NWC) is another determinant of cash holding decisions and is estimated as the ratio of (inventories + debtors − creditors)/total assets. Firms holding more liquid assets substitutes have a reduced need for cash. So firms with higher NWC are expected to hold less cash. Finally the last control variable is the cash dividends paid divided to total assets (DIV_TA). Firms which pay higher dividends are thought to hold more cash in order to sustain that policy in the near future and this in turn is expected to have a positive impact on cash holdings decisions. Thus, a positive coefficient is also expected for that variable.

Moving on to examine the second research hypotheses, we will follow previous studies and estimate a linear regression model of cash dividend payments to total assets as the dependent variable and CER performance along with other controls as independent variables (Benlemlih 2019; Ni and Zhang 2019; Hasan and Habib 2020). As in Benlemlih (2019), we scale dividend payments with total assets (DIV_TA) instead of earnings or market value because earnings can easily be manipulated and so this fact can create a false picture of the true earnings paid as dividends. In addition, by scaling dividends with assets we avoid any inconsistency of data arising from negative earnings ratios, and by avoiding the use of market capitalization in the denominator we control for pricing problems due to low share prices. Also, in order to mitigate concerns of any omitted correlated variables we will utilize a panel fixed effects regression analysis as in Benlemlih (2019) including year, industry and country fixed effects. The functional form of the equations includes the average CER for each firm and year estimated as the annual average of the ESG resource use score, emissions reduction score and environmental innovation score, where i denotes the firm and t the year and e is the error term:

$$\begin{aligned} DIV\_TA_{it} = {} & a_0 + a_1 CER_{it} + a_2 SIZE_{it} + a_3 LEV_{it} + a_4 GROWTH_{it} \\ & + a_5 ROA_{it} + a_6 PPE_{it} + a_7 CFO_{it} + a_8 RnD\_EXP_{it} \\ & + a_9 CASH_{it} + Year\ F.E + Industry\ F.E. + Country\ F.E. \\ & + u_{it} \end{aligned} \tag{9.2a}$$

$$\begin{aligned} DIV\_TA_{it} = {} & a_0 + a_1 EN\_INNOV_{it} + a_2 SIZE_{it} + a_3 LEV_{it} + a_4 GROWTH_{it} \\ & + a_5 ROA_{it} + a_6 PPE_{it} + a_7 CFO_{it} + a_8 RnD\_EXP_{it} \\ & + a_9 CASH_{it} + Year\ F.E + Industry\ F.E. + Country\ F.E. \\ & + u_{it} \end{aligned} \tag{9.2b}$$

$$\begin{aligned} DIV\_TA_{t} = {} & a_0 + a_1 EMM_{it} + a_2 SIZE_{it} + a_3 LEV_{it} + a_4 GROWTH_{it} \\ & + a_5 ROA_{it} + a_6 PPE_{it} + a_7 CFO_{it} + a_8 RnD\_EXP_{it} \\ & + a_9 CASH_{it} + Year\ F.E + Industry\ F.E. + Country\ F.E. \\ & + u_{it} \end{aligned} \tag{9.2c}$$

$$\begin{aligned} DIV\_TA_{it} = {} & a_0 + a_1 RES\_USE_{it} + a_2 SIZE_{it} + a_3 LEV_{it} + a_4 GROWTH_{it} \\ & + a_5 ROA_{it} + a_6 PPE_{it} + a_7 CFO_{it} + a_8 RnD\_EXP_{it} \\ & + a_9 CASH_{it} + Year\ F.E + Industry\ F.E. + Country\ F.E. \\ & + u_{it} \end{aligned} \tag{9.2d}$$

If H2 is valid we expect a positive and statistically significant a_1 coefficient suggesting that a high CER performing firms utilize environmental investments as a good way to signal to the market and its stakeholders of a firm's reputation and ability to pay dividends to its shareholders. Also, following Benlemlih (2019), Ni and Zhang (2019) we added additional control variables which impact firms' dividend policy decisions. The definition of the control variables is the same as in models (9.1a)–(9.1d). As argued by Benlemlih (2019), larger firms are more mature and less risky and so they have an enhanced ability to pay dividends. So we expect a positive coefficient on the SIZE variable. Higher leverage firms are more financial constrained by their lenders and so they may face more restrictions on dividend payments. So a negative coefficient is expected on this variable.

Firms with high growth opportunities will probably retain more earnings to finance such opportunities in the future and are less probable to pay dividends to their shareholders. Thus, a negative coefficient is expected for the GROWTH coefficient. Also, profitability is expected to be associated with higher dividend payments since firms with higher ROA are more able to generate free cash flows and distribute profits relative to less profitable firms. Following that argument it is expected that firms with a greater ability to generate operating cash flows (CFO) are less probable to pay cash dividends, so a negative coefficient is expected on the CFO variable. Furthermore, to control for investment opportunities we include the ratio of property, plant and equipment to total assets (PPE) and RnD expenses to sales revenues (RnD_EXP). Firms with more tangible assets and RnD activities tend to distribute less earnings to their shareholders in order to have the required free cash flow to finance such investment activities in the future. Thus a negative coefficient is expected on the PPE and RnD_EXP variable. Finally, we include the ratio of cash and cash equivalents to total assets (CASH). In the case that managers want to reduce the agency costs of free cash flows they will distribute more earnings as dividends, so a positive association is expected on that case. On the contrary, if agency conflicts are evident within the firm managers can increase short term investments and cash holdings leading to less dividend payments. So a negative relation is expected between dividends and cash on that occasion.

9.4 Empirical Results

9.4.1 Descriptive Statistics and Correlations

Table 9.1 presents the descriptive statistics of the variables utilized in the regression models. First of all, the mean CER is 64.5 (with a standard deviation of 21.1) indicating that the sample firms have a satisfactory environmental performance score. The highest average score is that of resources usage (66.5) followed by emission reduction activities (65.9) while environmental innovation activities lag behind with an average of 61.3. Also, the cash ratio (CASH) has an average value of 0.283 indicating that the sample firms are holding almost the one third of their assets

Table 9.1 Descriptive statistics of the sample variables

Variables	Mean	Standard deviation	Min	Max
CER	64.597	21.167	3.161	99.429
EN_INNOV	61.333	25.962	0.202	99.919
EMM	65.915	26.189	0.161	99.919
RES_USE	66.542	25.730	0.805	99.844
CASH	0.283	0.090	0.069	0.360
DIV_TA	0.337	0.022	0.001	0.434
ROA	21.28	4.825	−27.86	174.14
LEV	4.625	4.319	−0.682	25.691
SIZE	20.496	2.848	2.302	30.311
GROWTH	58.915	141.720	−69.711	409.071
CFO	0.030	5.042	−294.25	389.94
PPE	0.596	0.245	0.001	0.999
RnD_EXP	0.018	0.017	0.001	0.008
NWC	−2.974	95.87	−359.93	0.999

Table 9.2 Firm cash ratio and dividend payout ratio based on CER ranked portfolios

Variables	P1—High CER	P2—Medium CER	P3—Low CER	Difference (P1 − P3)	p-value
CASH	0.302	0.150	0.197	0.105***	0.001
DIV_TA	0.345	0.044	0.029	0.316***	0.001

in cash and cash equivalents. This number is higher relative to the studies by Chang et al. (2019) and Huang et al. (2019) in the US market. Thus European firms tend to hold more cash on their balance sheets. In addition, the ratio of dividends to assets (DIV_TA) has an average value of 0.337 which is again significantly higher relative to US and Chinese firms evidenced in prior studies on the field (Benlemlih 2019; Ni and Zhang 2019). The sample firms are highly profitable (mean ROA 21.28) and present significant growth opportunities (mean GROWTH at 58.9). The sample firms CFO cover on average the 3 per cent of total assets and are characterized by enhanced asset tangibility (PPE average is up to 60 per cent of total assets) However, sample firms have negative net working capital (−2.97) indicating that current liabilities are more than their current assets.

Table 9.2 presents the differences between the firm's cash and dividend payout ratios after separating firms within low, medium and high CER performance groups as in Suto and Takehara (2018). We followed the portfolio formation approach and estimated the average of the CER variable per year and country and assigned each firm as low CER firm if its annual score is below the 1st quartile, high CER firms are those that their CER score is above the 3rd quartile while all other firm observations lying between the 1st and 3rd quartiles are assigned as medium CER firms. The column under the title "difference" reports the average CER difference between the high CER portfolio firms and the low CER portfolio. The "p-value" column presents the corresponding probability values from estimating the Welch two-sample t-test and asterisks indicate statistical significance at the 1 per cent significance level (Suto and Takehara 2018). As we can see from the last column, high CER performance firms are characterized by statistically higher cash holdings and dividend payments relative to low CER performing firms. Both ratios seem to increase as we move from the low CER performance group to the high CER performance group. These findings corroborate arguments by Benlemlih (2019) and Huang et al. (2019) that more environmentally responsible firms are associated with higher dividend payments and cash holdings.

Table 9.3 presents the Pearson correlation coefficients of the sample variables. Correlation coefficients in bold indicate statistical significance at the 5 per cent level at least. First, CASH is insignificantly correlated with the environmental performance variables, yet is positively correlated with firm size (SIZE), PPE and ROA, but negatively correlated with NWC. It seems that more profitable firms are more able to generate higher free cash flows and eventually to hold more cash. On the contrary, firms with higher net working capital tend to hold less cash. In addition, cash dividends are positively and significantly correlated with environmental innovation performance (EN_INNOV), SIZE, CFO and PPE but negatively correlated with R&D expenses. Thus, cash dividend paying firms have higher size and asset tangibility, generate more operating cash flows and engage more in environmentally responsible investments. The rest of the statistically significant correlation coefficients have economic meaning, and are not very high indicating the absence of multicollinearity on the data. The VIF estimation of the sample variables provided values that were far below the threshold of 10 indicating no multicollinearity. However, the correlation coefficients do not suggest causality, so we will proceed with the fixed effect regression analysis in the following sub-section.

9.4.2 Regression Results

Table 9.4 presents the fixed effects regression results from the estimation from models (9.1a)–(9.1d) using the CASH ratio as the dependent variable. The F-stat values are highly significant and the R^2s are very high considering the size of the sample and other relative studies published in the field. The estimation of model (9.1a) provide a positive and significant coefficient (0.01) on the CER verifying evidence provided on the literature by Huang et al. (2019) and Chang et al. (2019) that CER performance lead firms to hold more cash on hand so the CSR motive of cash holdings is verified suggesting that managers of high CER performing firms focus more on maximizing stakeholder value leading to enhanced cash holdings to achieve such goals. Consequently, we can accept hypothesis H1. In addition, firms with higher emission (EMM) and resource use scores (RES_USE) are significantly associated with higher cash holdings. The coefficient on environmental innovation activities (EN_INNOV) although is positive is not statistically significant. Thus, overall evidence on Table 9.4 suggest a positive impact of environmental responsibility on cash holdings so we can accept H1.

Table 9.3 Pearson correlation coefficients of sample variables

Variables	1	2	3	4	5	6	7	8	9	10	11	12	13	14
1. Cash	1													
2. CER	0.001	1												
3. EN_INNOV	0.001	**0.728**	1											
4. EMM	0.007	**0.854**	**0.377**	1										
5. RES_USE	−0.006	**0.863**	**0.405**	**0.710**	1									
6. SIZE	**0.102**	**−0.012**	**0.013**	−0.002	**−0.040**	1								
7. LEV	−0.001	0.005	**0.007**	0.001	**0.006**	**−0.046**	1							
8. CFO	0.001	−0.002	0.001	−0.002	−0.005	**0.018**	**−0.265**	1						
9. PPE	**0.012**	−0.002	**0.010**	0.004	0.001	**0.279**	**−0.024**	**0.011**	1					
10. Growth	−0.002	−0.002	−0.005	−0.001	0.002	0.001	−0.001	0.001	0.002	1				
11. R&D_EXP	−0.006	−0.028	**−0.045**	−0.018	−0.005	**−0.026**	**−0.019**	**−0.062**	**−0.026**	−0.001	1			
12. ROA	**0.656**	**0.007**	**0.018**	**0.011**	**−0.011**	**0.127**	**−0.070**	**0.062**	**0.019**	−0.001	**−0.055**	1		
13. NWC	**−0.688**	−0.001	−0.002	−0.007	0.006	**−0.099**	**−0.151**	**−0.012**	**−0.010**	0.001	0.005	**−0.656**	1	
14. DIV_TA	−0.002	0.005	**0.009**	0.006	−0.001	**0.047**	−0.001	**0.019**	**0.013**	−0.001	**−0.150**	−0.001	0.002	1

Note: Correlation coefficients in bold indicate statistical significance at the 5 per cent significance level at least

Table 9.4 Fixed effects regression results on the association between environmental performance and cash holdings

Model (9.1a–d)—Dep. Variable: CASH	CER	EN_INNOV	EMM	RES_USE
ER	0.010** (2.06)			
EN_INNOV		0.004 (1.05)		
EMM			0.005** (2.33)	
RES_USE				0.011*** (2.61)
SIZE	−0.002*** (−2.71)	−0.001*** (−2.52)	−0.001*** (−2.62)	−0.001*** (−2.72)
LEV	−0.078*** (−9.64)	−0.077*** (−9.58)	−0.077*** (−9.59)	−0.079*** (−9.74)
CFO	−0.296*** (−7.90)	−0.293*** (−7.80)	−0.295*** (−7.87)	−0.296*** (−7.95)
PPE	−0.729*** (−44.63)	−0.731*** (−45.12)	−0.729*** (−44.72)	−0.729*** (−44.99)
GROWTH	0.004*** (5.71)	0.004*** (5.71)	0.003*** (5.72)	0.003*** (5.72)
RnD_EXP	−2.893*** (−8.71)	−2.904*** (−8.72)	−2.942*** (−8.75)	−2.902*** (−8.75)
ROA	0.105*** (3.65)	0.103*** (3.35)	0.104*** (3.60)	0.107*** (3.67)
NWC	−0.678*** (−31.26)	−0.681*** (−31.47)	−0.679*** (−31.32)	−0.679*** (−31.47)
DIV_TA	0.252*** (7.56)	0.252*** (7.57)	0.254*** (7.62)	0.249*** (7.51)
Constant	0.744*** (40.69)	0.747*** (41.71)	0.746*** (40.92)	0.745*** (40.91)
R-sq	0.854	0.852	0.854	0.855
F-Stat (p-value)	634.89*** (0.001)	603.65*** (0.001)	619.77*** (0.001)	642.12*** (0.001)

Note: **, *** indicate statistical significance at the 5% and 1% significance level respectively. T-statistics in the parentheses

Regarding the rest of the control variables, all of them are highly significant and with the expected sign except for the RnD_EXP variable. Specifically, SIZE LEV, CFO, NWC and PPE produced negative and significant coefficients in all four models. This corroborates evidence in the literature the larger, more leveraged firms which have higher asset tangibility and net working capital, and those that generate more operating cash flows hold less cash on their balance sheets (Huang et al. 2019). On the contrary, firms with higher growth opportunities (GROWTH), dividend payments (DIV_TA) and enhanced profitability hold more cash firms since the coefficients on those variables produced positive and highly significant coefficients, corroborating arguments on the literature that firms with higher ROA, more

Table 9.5 Fixed effects regression results on the association between environmental performance and dividends

Model (9.2a–d)—Dep. Variable: DIV_TA	CER	EN_INNOV	EMM	RES_USE
ER	0.002*** (4.79)			
EN_INNOV		0.003** (2.09)		
EMM			0.001 (0.26)	
RES_USE				0.003** (2.08)
SIZE	0.006*** (3.35)	0.001*** (2.87)	0.002*** (2.74)	0.001*** (2.96)
LEV	−0.003 (−1.05)	−0.004 (−0.77)	−0.004 (−0.70)	−0.005 (−0.81)
GROWTH	−0.002*** (−4.99)	−0.001** (−2.30)	−0.002** (−2.28)	−0.001** (−2.27)
ROA	0.275*** (7.99)	0.434*** (6.49)	0.432*** (6.46)	0.436*** (6.46)
PPE	−0.031*** (−6.39)	−0.041*** (−3.38)	−0.040*** (−3.34)	−0.040*** (−3.34)
CFO	0.226*** (5.54)	0.118** (2.40)	0.120** (2.43)	0.116** (2.30)
RnD_EXP	−0.220*** (−9.20)	−1.996*** (−5.67)	−2.021*** (−5.80)	−2.011*** (−5.78)
CASH	0.028 (1.80)	0.003 (0.12)	0.005 (0.18)	0.002 (0.10)
Constant	0.021*** (3.05)	0.033** (1.98)	0.041*** (2.59)	0.031 (1.85)
R-sq	0.638	0.697	0.696	0.697
F-Stat (p-value)	449.43*** (0.001)	435.28*** (0.001)	438.93*** (0.001)	431.72*** (0.001)

Note: **, *** indicate statistical significance at the 5% and 1% significance level respectively. T-statistics in the parentheses

dividend payments and positive changes in revenues (GROWTH) are more able to generate free cash flows and hold more cash relative to less profitable and growing firms (Chang et al. 2019). Finally, RnD_EXP produced negative and significant coefficients which does not verify the precautionary motive of cash holdings for RnD intensive firms, contradicting existing evidence in the literature. This result could be attributed to the fact that RnD intensive firms in EU pay such expenses through cash leading to higher cash consumption and lower cash reserves.

Table 9.5 presents the results from the estimation of the panel fixed effects regression models (9.2a)–(9.2d) capturing the impact of CER performance on dividend cash payments The F-stat values are highly significant and the R^2s are very high considering the size of the sample and other relative studies published in the field. As we can see, all environmental performance scores (except for the emission score—EMM) produced positive and significant coefficients indicating the firms which invest in environmentally innovative projects, utilize resources more efficiently and have enhanced overall environmental performance, pay more cash dividends to their shareholders. Practically, 1 percentage point increase in CER performance score is associates with 0.2 per cent of dividend to asset ratio, ceteris paribus. These evidence verify the existing evidence in the literature that a high dividend payout ratio can signal to the market that CER related activities are not a waste of firms resources and in fact could lead to a better allocation of those resources and enhanced shareholders' satisfaction (Benlemlih 2019; Ni and Zhang 2019; Hasan and Habib 2020). Consequently, evidence on Table 9.5 lead us to accept H2.

The coefficients on the majority of the control variables (except for LEV and CASH) have the expected sign and are highly significant. Being more specific, SIZE, ROA and CFO produced positive coefficients indicating that larger, more profitable and firms which can generate higher operating cash flows pay more cash dividends to their shareholders, since they are more able to generate free cash flows and distribute their profits (Benlemlih 2019). On the contrary, firms with enhanced growth opportunities (more positive changes in revenues), higher asset tangibility (PPE) and RnD expenses pay less dividends to their shareholders. Again, these results are in accordance with existing evidence by Ni and Zhang (2019) arguing that those firms with more tangible assets, growth opportunities and RnD activities tend to distribute less earnings to their shareholders in order to have the required free cash flow to finance and even enhance such investment activities in the future.

9.4.3 Sensitivity Analysis

In order to check the robustness of the main empirical results, we performed a number of sensitivity tests regarding the research design, and various definitions of the determinant variables. At first, following Cai et al. (2016) we replaced the CER variable definition (which is the simple average of the three sub-scores) by estimating a principal component from the three sub-scores. So we constructed the CER_PCA variable which equals the first principal component from the abovementioned scores. The principal component was determined using eigen values greater than 1.0. The principal component analysis was used since the solutions generated from principal component analysis do not have significant differences from the factor analysis techniques. In this context, we utilized the orthogonal method of extraction and particularly Varimax which tries to maximize the dispersion of loadings within factors. Eventually, for including the factor scores as an independent variable in model (9.1a) we employed the Anderson–Rubin method. The outcome of the principal component analysis supported that data do not suffer from multicollinearity since the KMO statistic of sampling adequacy had a value up to 0.91 indicating that this kind of analysis is appropriate for our data. Eventually the extracted communalities are close to 1 proposing that the factor

Table 9.6 Fixed effects regression results on the association between alternative CER measures and cash holdings

Model (9.1a)—Dep. Variable: CASH	LnCER	CER_PCA
LnCER	0.005** (2.06)	
CER_PCA		0.001** (2.10)
SIZE	−0.001*** (−2.66)	−0.001*** (−2.72)
LEV	−0.078*** (−9.64)	−0.078*** (−9.64)
CFO	−0.296*** (−7.92)	−0.296*** (−7.91)
PPE	−0.729*** (−44.82)	−0.729*** (−44.63)
GROWTH	0.003*** (5.69)	0.003*** (5.72)
RnD_EXP	−2.896*** (−8.70)	−2.895*** (−8.72)
ROA	0.106*** (3.68)	0.105*** (3.65)
NWC	−0.679*** (−31.41)	−0.678*** (−31.26)
DIV_TA	0.251*** (7.55)	0.252*** (7.57)
Constant	0.730*** (33.88)	0.751*** (42.79)
R-sq	0.854	0.854
F-Stat (p-value)	628.04*** (0.001)	636.38*** (0.001)

Note: **, *** indicate statistical significance at the 5% and 1% significance level respectively. T-statistics in the parentheses

explains the original data adequately, and the factor loadings are above the critical value of 0.70 proposing that the estimated PCA is of statistical significance. Table 9.6 presents the fixed effects regression results from the estimation of model (9.1a) by using the CER_PCA variable instead of the average CER, as well as the natural logarithm of CER. As we can see both LnCER and CER_PCA produced a positive and significant coefficient (at the 5% per cent level) corroborating H1 and the inferences drawn from Table 9.4, suggesting that more environmentally responsible firms hold more cash on their balance sheets, so our main results remain robust.

Another important issue that has been raised by previous studies (Huang et al. 2019) is the existence of endogeneity meaning that CER performance might not only determine the cash holdings, but the causality mighty also go the other way round. Under this circumstance, the fixed effect panel estimation of models may partially control for that by producing consistent estimators, yet not efficiently as the 2SLS method. For this reason, regression model (9.1a) was re-estimated by applying the

Table 9.7 2SLS regression results of CER on cash holdings

Variables			
First stage equation—Dep. Variable: CER	**Coef.**	**t-stat**	**P > |t|**
SIZE	0.879***	2.52	0.012
LEV	1.067***	2.60	0.009
CFO	3.921**	2.39	0.017
PPE	−2.265***	−4.04	0.001
GROWTH	−0.575	−1.31	0.192
RnD_EXP	−3.581**	−2.06	0.040
ROA	−4.251***	−2.78	0.006
NWC	−1.994***	−2.72	0.007
DIV_TA	2.676**	2.03	0.042
Lag_CER	0.331***	11.95	0.001
Constant	3.204***	3.60	0.001
R-sq	0.141		
F-Stat (p-value)	20.80*** (0.001)		
Second stage equation—Dep. Variable: CASH	**Coef.**	**z-stat**	**P > |z|**
CER	0.002**	2.64	0.009
SIZE	−0.003	−0.54	0.588
LEV	−0.102***	−9.10	0.001
CFO	−0.313***	−7.82	0.001
PPE	−0.725***	−38.41	0.001
GROWTH	0.004***	5.01	0.001
RnD_EXP	−2.891***	−8.54	0.001
ROA	0.122***	3.86	0.001
NWC	−0.663***	−26.08	0.001
DIV_TA	0.239***	6.44	0.001
Constant	0.717***	29.64	0.001
R-sq	0.848		
Wald Chi^2 (p-value)	622.92*** (0.001)		
AR(1) test (p-value)	0.002		
Hansen test of over-identification (p-value)	0.421		

Note: **, *** indicate statistical significance at the 5% and 1% significance level respectively

2SLS method using the one-year lags of CER and its components as instruments. In order to check the validity of the instruments we performed the AR(1) test which represent the first-order serial correlation test in the first differenced residuals, under the null hypothesis of no serial correlation. The AR tests rejected the null hypothesis suggesting that the instruments are valid for our research setting. Moreover, we performed the Hansen J-statistic of over-identifying restrictions which tests the joint null hypothesis of that instrumental variables are valid (uncorrelated with the error term). The Hansen statistic was insignificant within conventional levels suggesting that the instruments were completely exogenous. The second stage results presented on Table 9.7 indicate that CER produced a positive and significant coefficient

Table 9.8 Fixed effects regression results on the association between CER and alternative dividend ratios

Model (9.2a)—Dep. Variable:	DIV_CFO	DIV_SALES
LnCER		0.003*** (5.20)
CER_PCA	0.015** (1.96)	
SIZE	0.021*** (3.46)	0.001** (2.47)
LEV	0.123 (1.58)	−0.044*** (−12.28)
CFO	−0.081 (−0.25)	0.099*** (5.85)
PPE	−0.362*** (−3.90)	−0.048*** (−10.45)
GROWTH	0.003*** (5.69)	−0.001*** (−3.17)
RnD_EXP	4.096 (0.81)	1.791 (0.64)
ROA	1.325*** (4.75)	0.282*** (19.72)
CASH	0.653*** (3.65)	0.120*** (12.79)
Constant	−0.107 (−0.66)	−0.016** (−2.03)
R-sq	0.141	0.415
F-Stat (p-value)	24.59*** (0.001)	354.02*** (0.001)

Note: **, *** indicate statistical significance at the 5% and 1% significance level respectively. T-statistics in the parentheses

corroborating H1 and the findings on Table 9.4. So our main results are unaffected by endogeneity issues.

Another sensitivity test dealt with the definition of the variables in model (9.2a) by substituting the dependent variable (DIV_TA) with alternative payout ratios as in Hasan and Habib (2020). Specifically on Table 9.8 we estimated model (9.2a) using the ratios of dividends to operating cash flows (DIV_CFO) and dividends to sale revenues (DIV_SALES). The results remained qualitatively unchanged relative to those on Table 9.5 indicating that high CER performing firms pay more cash dividends to their shareholders, so again H2 is verified. Furthermore, additional sensitivity testing was performed following different definitions of the independent variables. Specifically, we estimated LEV as the ratio of long term debt to total assets, GROWTH as the ratio of market value to book value of equity (MV/BV) and ROA was replaced by ROE but the results remain unchanged after these modifications. Finally, in order to mitigate any omitted variable bias we performed a change analysis in order to extract more efficient inferences regarding the impact of CER on the differences of cash and dividend ratios. So we re-estimated models (9.1a) and (9.2a) by substituting the dependent variables with the annual change of cash and dividends from year (t-1) to year (t) (Huang et al. 2019; Benlemlih 2019). Once again the results remained qualitatively unchanged compared with those on Tables 9.4 and 9.5 verifying H1 and H2.

9.5 Conclusion

Several studies on the environmental responsibility research field have focused on examining the impact of CER activities on cash holding and dividend payments decisions. Researchers have expressed different views on how a company's social and environmental performance affects firms' cash holding decisions (Chang et al. 2019; Huang et al. 2019) and cash dividend distributions (Benlemlih 2019; Ni and Zhang 2019; Hasan and Habib 2020). The scope of this chapter was to examine the impact of CER related performance on firm dividend and cash holding decisions, by taking into consideration potential endogeneity between the examined factors, evidenced by previous studies on the field. We respond to the call for more research on the topic made by Benlemlih (2019) and Huang et al. (2019), by covering a large multi-country sample of listed and unlisted European firms over a long period of time and also by considering different environmental performance ratings.

Empirical analysis suggested that CER performance is positively and significantly impacting on cash holdings, supporting evidence provided on the literature by Huang et al. (2019) and Chang et al. (2019) that CER performance lead firms to hold more cash, suggesting that managers of high CER performing firms focus more on maximizing stakeholder value leading to enhanced cash holdings to achieve such goals. Furthermore, we found firms with higher emission (EMM) and resource use scores (RES_USE) are significantly associated with higher cash holdings. Additionally, all environmental performance scores (except for the emission score—EMM) proved to have a positive and significant impact on dividends payments. Practically, 1 percentage point increase in CER performance score is associated with 0.2 per cent of dividend to asset ratio, ceteris paribus. These evidence verify the existing arguments in the literature that a high dividend payout ratio can signal to the market that CER related activities are not a waste of firms resources and in fact could lead to a better allocation of those resources and to enhanced shareholders' satisfaction (Benlemlih 2019; Ni and Zhang 2019; Hasan and Habib 2020). The results remain robust after controlling for endogeneity and several other sensitivity tests. The findings of this chapter contribute to the ongoing debate about the association between CER and corporate finance decisions is several merits. At first we complement previous studies by Benlemlih (2019), Chang et al. (2019) and Hasan and Habib (2020), by incorporating a multi-country instead of a single country research setting. Also, we incorporate in the analysis a decomposition of the CER variable and indicate which sub-category of CER performance contributes to dividend and cash holding decisions, something that has not been considered extensively by previous researchers.

References

Arouri, M., and G. Pijorlet. 2017. CSR performance and the value of cash holdings: International evidence. *Journal of Business Ethics* 140: 263–284.

Barnea, A., and A. Rubin. 2010. Corporate social responsibility as a conflict between shareholders. *Journal of Business Ethics* 97: 71–86.

Benlemlih, M. 2019. Corporate social responsibility and dividend policy. *Research in International Business and Finance* 47: 114–138.

Bhattacharya, S. 1979. Imperfect information, dividend policy and the "bird in the hand" fallacy. *Bell Journal of Economics* 10 (1): 259–270.

Boutin-Dufesne, F., and P. Savaria. 2004. Corporate social responsibility and financial risk. *Journal of Investing* 13 (1): 57–66.

Cai, L., J. Cui, and H. Jo. 2016. Corporate environmental responsibility and firm risk. *Journal of Business Ethics* 139: 563–594.

Chang, C.-H., S.-S. Chen, Y.-S. Chen, and S.-C. Peng. 2019. Commitment to built trust by socially responsible firms: Evidence from cash holdings. *Journal of Corporate Finance* 56: 364–387.

Cheung, A. 2016. Corporate social responsibility and corporate cash holdings. *Journal of Corporate Finance* 37: 412–430.

Dimitropoulos, P.E. 2020. Cash holding determinants in the Greek hotel industry. In *Cultural and tourism innovation in the digital Era*, Springer Proceedings in Business and Economics, ed. V. Katsoni and T. Spyriadis, 585–596. New York: Springer.

Dimitropoulos, P.E., K. Koronios, A. Thrassou, and D. Vrontis. 2020. Cash holdings, corporate performance and viability of Greek SMEs: Implications for stakeholder relationship management. *EuroMed Journal of Business*. https://doi.org/10.1108/EMJB-08-2019-0104.

Dittmar, A., J. Mahrt-Smith, and H. Servaes. 2003. International corporate governance and corporate cash holdings. *Journal of Financial and Quantitative Analysis* 38: 111–133.

Feldman, S.J., P.A. Soyka, and P. Ameer. 1997. Does improving a firm's environmental management system and environmental performance result in a higher stock price? *Journal of Investing* 6 (4): 87–97.

Ghaly, M., V. Dang, and K. Stathopoulos. 2015. Cash holdings and employee welfare. *Journal of Corporate Finance* 33: 53–70.

Hail, L., A. Tahoun, and C. Wang. 2014. Dividend payouts and information shocks. *Journal of Accounting Research* 52: 403–456.

Hasan, M.M., and A. Habib. 2020. Social capital and payout policies. *Journal of Contemporary Accounting and Economics* 16: 100183.

He, T.T., W.X.B. Li, and G.Y.N. Tang. 2012. Dividends behavior in state-versus family-controlled firms: Evidence from Hong Kong. *Journal of Business Ethics* 110: 97–112.

Huang, H.H., C. Liu, and L. Sun. 2019. Chemical releases and corporate cash holdings. *International Review of Financial Analysis* 64: 159–173.

Jensen, M. 1986. Agency costs of free cash flow, corporate finance and takeovers. *American Economic Review* 76 (2): 323–329.

Jiraporn, P., and P. Chintrakarn. 2013. Hoe do powerful CEOs view corporate social responsibility? An empirical note. *Economics Letters* 119 (3): 344–347.

Kim, J., D. Woods, and H. Kim. 2013. Identifying the financial characteristics of cash-rich and cash-poor restaurant firms: A logistic regression analysis. *Tourism Economics* 19 (3): 583–598.

Martínez-Sola, C., P.J. García-Teruel, and P. Martínez-Solano. 2018. Cash holdings in SMEs: Speed of adjustment, growth and financing. *Small Business Economics* 51: 823–842.

Ni, X., and H. Zhang. 2019. Mandatory corporate social responsibility disclosure and dividend payouts: Evidence from a quasi-natural experiment. *Accounting and Finance* 58: 1581–1612.

Suto, M., and H. Takehara. 2018. *Corporate social responsibility and corporate finance in Japan*. Singapore: Springer Nature.

Vafeas, N., and V. Nikolaou. 2001. The association between corporate environmental and financial performance. *Advances in Public Interest Accounting* 8: 195–214.

Chapter 10
Corporate Environmental Responsibility and Earnings Value Relevance

The significant evolution and growth of the environmentally responsible corporate behavior internationally, revealed that the capital markets have developed environmental criteria in assessing the firm risk and valuation, which operate as supplements on the traditional financial accounting information. Thus investors, consider along with accounting reports, the firms' sustainability reports in order to make informed investment decisions. Under this framework, several studies have examine the relationship between corporate environmental responsibility (CER) and accounting value relevance providing conflicting evidence of a differential impact of CER on stock prices and returns, earnings persistence and predictability. The purpose of this chapter is to study the impact of CER related activities on earnings value relevance and persistence by taking into consideration the moderating role of CER on the value relevance of traditional accounting information. We predict and find that CER performing firms are more likely to report financial numbers which have more favorite attributes than less CER performing firms, such as more value relevance, higher persistence and predictability.

10.1 Introduction

The issue of accounting value relevance has received significant attention by market participants namely investors, corporate managers and investment analysts. The main concern of managers is to create a stable profitability growth in the near future because their remuneration and even their position may by closely tied to reporting financial numbers published in the firm's financial statements (Dimitropoulos and Asteriou 2009). On the other hand, investment analysts' main job is to assess firms' future growth prospects, risk and sustainability so as to be able to give transparent and reliable investment information to their clients. Several empirical and theoretical studies in the past have provided concrete evidence that accounting numbers and

P. Dimitropoulos, K. Koronios, *Corporate Environmental Responsibility, Accounting and Corporate Finance in the EU*, CSR, Sustainability, Ethics & Governance, https://doi.org/10.1007/978-3-030-72773-4_10

specifically income and common equity are associated with positive returns, thus are value relevant for market participants by providing acute signals to the market about firms' future cash flows (Latane and Jones 1979; Foster et al. 1984; Bernard and Thomas 1989). However, this relation is only dependent on the reliability and quality of financial numbers and whether those numbers are depicting the actual economic condition of the firm.

The financial accounting reporting systems serve as information providers to the market by signaling firm's past performance and future sustainability (Kang et al. 2012; Dimitropoulos and Koronios 2018). Earnings persistence and predictability are two additional characteristics (or proxies) for earnings quality since according to Kang et al. (2012), Dimitropoulos and Koronios (2018) and Dimitropoulos et al. (2019), the role of accounting reports is to provide useful information to investors for making their assessments and judgements regarding the firms' future earnings and cash flows. Consequently, they are an important attribute for financial decision making. In practice, earnings persistence express the extent that current earnings may be sustained in the future. This is estimated by a regression of future earnings on current earnings and the closer the regression coefficient is on unity (1) the higher the earnings persistence (Sloan 1996; Richardson et al. 2005; Oei et al. 2008).

Beyond the contribution of financial accounting information on its assistance towards investors and managers, environmentally responsible activities and investments have been emerged as an additional stream of information which completes and even complements the classic financial accounting reports published by the firms. According to Wang (2016), the disclosure of environmental information improves the reliability and transparency of accounting information and even argue about the necessity of regulating the dissemination of non-financial information (social and environmental responsibility disclosures) on a permanent basis. Moreover, Laksmana and Yang (2009) document that the engagement in CER related activities provides an indication that firms are more prone to provide an insurance-type of protection towards their main stakeholders, against the costs that would be possibly created from negligence to social matters. So under this idea, CER performing firms are considered as more transparent and with a more stable earnings growth and higher earnings predictability compared to their low CER performing counterparts.

Instrumental theory has been utilized as the main theoretical framework on this research issue because it regards CER as a mechanism of wealth creation. CER is utilized by firms as a strategic tool in order to maximize shareholders' wealth and improve profitability in the long-term. The reason for that is because CER activities safeguard shareholders' wealth from litigation costs and reputation damages which can diminish stock returns and future prospects. So since high CER performing firms are less probable to experience litigation costs they should have more stable earnings and less downgrading volatility (Laksmana and Yang 2009). Empirical evidence by Hassel et al. (2005), Lourenço et al. (2012) and Clarkson et al. (2013) support the argument that CER related disclosures provide incremental value relevance and explain the stock prices in a significant extent. In the same context, Al-Tuwaijri et al. (2004), Moneva and Cuellar (2009), and Schadewitz and Niskala (2010) provide empirical evidence for an increase in share prices for firms disclosing environmental information in different international markets. Finally, evidence by Alipour et al. (2019), Gregory et al. (2016), Reverte (2016), Wang (2016) and Laksmana and Yang (2009) document a significant positive and complementary impact of CER activities on the value relevance and persistence of accounting information.

The scope of this chapter is to examine the impact of CER related performance on firms' financial reporting quality, proxied through the value relevance of accounting numbers (earnings and book values of equity), earnings persistence and predictability. The chapter extends previous evidence on earnings' value relevance and persistence by utilizing a multi-country sample within the EU and by considering the indirect (moderating) effect of CER on the value relevance of accounting information. We respond to the call for more research on the topic made by Reverte (2016) and Alipour et al. (2019), by considering the moderating impact of CER on value relevance and the predictability of earnings. CER performance can have an indirect impact on accounting quality because it affects the economic uncertainty inherent in corporate financial information, thus contributing to enhanced predictability of future cash flows. In the following section we provide the theoretical background on the impact of CER on stock prices and the testable hypotheses. The third section discusses the data selection procedure and the research design. The fourth section provides the discussion of empirical results and the last section concludes the chapter.

10.2 CER-Value Relevance and Testable Hypothesis

CER reporting provides non-financial information to stakeholders and mainly investors, yet its implication could have important financial dimensions for market participants. According to Gregory et al. (2016), if firms ignore the social and environmental expectations of shareholders and other stakeholders, could lead capital markets to create diminished expectations about future earnings and cash flows. This is attributed, to the fact that low CER performing firms are accompanied by enhanced information asymmetries, leading to increased perception of risk which diminishes firms' market valuation (Reverte 2016). If a firm enhances its CER disclosures and activities beyond what is dictated by laws and regulations, it will create reputational benefits either by attracting more competent employees, loyal customers and trustworthy creditors and lenders. (Brammer and Pavelin 2006; Gregory et al. 2016). As discussed in the previous chapter, CER engagement creates a reputational capital (an intangible resource which is very difficult to imitate by competitors) which helps the firm to gain competitive advantages, improve organizational capabilities and communicate those improved prospects to the market. As argued by Servaes and Tamayo (2013), the environmental expectations of customers and various stakeholders has increased over the years and the financial penalties for non-environmental responsible firms have raised significantly. So low CER performing firms tend to perceived as caring more future risk (litigation and environmentally related costs) and so CER disclosures are financially relevant for investors (Suto and Takehara 2018).

Moreover, Gregory et al. (2016), Bollen (2007) and El Ghoul et al. (2011), argue that CER activities could enhance the value of the firm even though they are not maximizing the present value of future cash flows. This is because many investors have a multi-facet utility function which dictates that their utility is not determined by simple financial measures (profits or dividends) but also from environmental responsibility dimensions of corporate activity. Following the evidence on previous chapters and Suto and Takehara (2018), CER performing firms tend to be associated with higher profitability and lower cost of capital due to lower systematic risk. The implementation of environmental friendly technology and innovations, the reduction of carbon emissions and the more effective use of resources lead firms to avoid fines and litigation costs, which further do not diminish future cash flows. Finally, Kim et al. (2012) and Alipour et al. (2019), assert that there is the ethical dimension on corporate and managerial behavior. Managers have the incentive to be ethical towards their stakeholders and so they are motivated to continue their social and environmental activities and disclosures and publish information of higher quality, transparency and reliability.

However, the opposite impact of CER on market valuation is also evident in the literature. According to Alipour et al. (2019) and Williams (1999), managers may use CER activities as a way to pursue their personal goals and interests and even to legitimize their decisions as giving the picture of a responsible and environmentally aware firm. Also, CER could also be used to mask the true economic condition of the firm (via earnings manipulation) leading to lower quality of financial reporting. Salewski and Zülch (2014) provide evidence of an opportunistic managerial behavior for firms engaging in CSR activities. Nevertheless, since the majority of the empirical findings points towards a positive impact of CER on the quality and value relevance of accounting information, we state the first research hypothesis as follows:

H1: CER performance contributes to the value relevance and persistence of accounting numbers.

In addition, CER performance can have an indirect (moderating effect) on the value relevance and predictability of earnings. Cormier and Magnan (2007), Reverte (2016), and Wang (2016) argue that voluntary environmental disclosures can improve the perceived firm transparency by the investor's, by allowing to reach in more resourceful and accurate estimations about future earnings and cash flows. Practically, the impact of CER on the value relevance of earning and book value of equity could be determined by two contradictory perspectives. The first is that CER reporting is exclusive in nature which can be used by analysts and investors to move away some of the uncertainty on firms' reported financial numbers, so improving the transparency and reliability of such numbers. In turn, increased accounting reliability can improve stock market valuation (Dimitropoulos and Asteriou 2008, 2010). Nevertheless, managers can rely on voluntary CER reporting in order to gain support and legitimacy among their stakeholders, who in turn can derive more precise estimates on future cash flows and earnings volatility. Thus, even such non-financial type disclosures can improve the value relevance of earnings and book values.

This complementary value relevance of CER towards accounting numbers can be perceived by investors as a source of additional information fulfilling the nature and trends of the classic financial accounting measures. Put it differently, CER reports and activities do not only provide information about the firms' current and past commitments and engagements, but also about their future endeavors, projects, and decisions regarding the allocation of firms' scare economic and financial resources (Reverte 2016). Consequently, our prediction is that CER will have a moderating role on enhancing the value relevance of earnings and book values. Thus, based on the above discussion the second research hypothesis is stated as follows:

H2: CER performance has an indirect impact on share prices, moderating the value relevance of accounting numbers.

10.3 Data Selection and Research Design

10.3.1 Data Selection Procedure

The current chapter utilizes a sample of listed only corporations from 23 EU-member countries over the period 2003–2018. All financial, governance and environmental responsibility data have been extracted from Datastream database. The sample collection procedure started by including all corporations that have been covered by the environmental, social and governance (ESG) scores provided by Datastream during the sample period. 2003 has been selected as the initial sample year since that year was the first year with significant coverage (taking into consideration that this database officially was initiated in 2002). From a total of 656,734 firm-year observations we excluded those firms without environmental score data for at least five consecutive years in order to avoid bias in the empirical results and also to facilitate the estimation of differenced variables. Furthermore, we excluded firms with incomplete financial data related to market value, shares outstanding and stock prices, as well as those that do not close their fiscal year on December. In addition, we winsorized the upper and down 1 per cent of the data distribution in order to mitigate the influence of significant outliers in the empirical analysis. After this procedure was finalized, we ended up with a final sample of 1288 distinct firms summing up to 115,748 firm-year observations.

10.3.2 Research Design

For examining the validity of the research hypotheses we will follow previous studies and estimate a linear regression model of firm stock prices as the dependent variable and CER performance along with other accounting numbers as independent variables (Dimitropoulos and Asteriou 2009; Laksmana and Yang 2009; Dimitropoulos et al. 2013; Reverte 2016). The approach that we will follow in this section includes the examination of the relation between earnings per share and the book value of equity per share with stock prices and how CER performance (and its main components) interact or affect accounting numbers value relevance and earnings persistence. In order to estimate a measure of share prices that remains unaffected by average differences across industries we applied the methodology proposed by Barth et al. (2008) and Dimitropoulos et al. (2013), and regressed stock price P on industry fixed effects. Then we extracted the residuals from the previous step, denoted as P* and use them as the dependent variable on the regression of book value of equity per share (BVPS) and net income per share (EPS). In order to ensure that published accounting information at fiscal year-end has been fully considered by market participants, we followed Lang et al. (2003, 2006) and estimated stock price P 6 months after the fiscal year end. Also, in order to mitigate concerns of any omitted correlated variables we will utilize a panel fixed effects regression analysis as in Reverte (2016) including year, industry and country fixed effects. The functional form of the equations includes the average CER for each firm and year estimated as the annual average of the ESG resource use score, emissions reduction score and environmental innovation score, where i denotes the firm and t the year and e is the error term:

$$\begin{aligned} P*_{it} = {} & a_0 + a_1 EPS_{it} + a_2 BVPS_{it} + a_3 CER_{it} + a_4 CER * EPS_{it} + a_5 CER \\ & * BVPS_{it} + Year\ F.E + Industry\ F.E. + Country\ F.E. + u_{it} \end{aligned} \tag{10.1a}$$

$$\begin{aligned} P*_{it} = {} & a_0 + a_1 EPS_{it} + a_2 BVPS_{it} + a_3 EN\_INNOV_{it} + a_4 EN\_INNOV \\ & * EPS_{it} + a_5 EN\_INNOV * BVPS_{it} + Year\ F.E + Industry\ F.E. \\ & + Country\ F.E. + u_{it} \end{aligned} \tag{10.1b}$$

$$\begin{aligned} P*_{it} = {} & a_0 + a_1 EPS_{it} + a_2 BVPS_{it} + a_3 EMM_{it} + a_4 EMM * EPS_{it} \\ & + a_5 EMM * BVPS_{it} + Year\ F.E + Industry\ F.E. + Country\ F.E. \\ & + u_{it} \end{aligned} \tag{10.1c}$$

$$\begin{aligned} P*_{it} = {} & a_0 + a_1 EPS_{it} + a_2 BVPS_{it} + a_3 RES\_USE_{it} + a_4 RES\_USE * EPS_{it} \\ & + a_5 RES\_USE * BVPS_{it} + Year\ F.E + Industry\ F.E. \\ & + Country\ F.E. + u_{it} \end{aligned} \tag{10.1d}$$

Where P* are the residuals from a regression of actual stock prices (P) on industry fixed effects, EPS is net income per share, BVPS is the book value of common equity per share. EN_INNOV denotes each company's environmental innovation score, which reflects the firm's ability to mitigate the environmental costs towards its customers or even creating new market and product opportunities via enhanced environmental technologies and processes. EMM denotes companies' emission reduction score estimated via their efforts and effectiveness in reducing emissions from their production and operation. RES_USE denotes the firms' resource use score and indicates firms' ability to reduce the usage of hazardous materials, energy, water and other valuable natural resources. Finally, CER is the firms' annual average of the aforementioned three scores. All environmental scores range between zero (0) and 100 so the higher the score the more environmentally responsible and efficient is the firm. If hypothesis H1 is valid we expect coefficient a_3 to be positive and statistically significant for all model estimations, indicating that CER performance has incremental value relevance beyond that of earnings and equity. In order to accept H2 we expect coefficients a_4 and a_5 to be statistically significant and have different sign of impact relative to coefficients a_1 and a_2.

Moving on to examine the impact of CER performance on earnings persistence and predictability, we will measure earnings persistence by regressing the ratio of current year net income before extraordinary items to lagged total assets (NI_TA_{it}), on the lagged NI_TA_{it-1} following the previous studies by Francis et al. (2004), Laksmana and Yang (2009) and Alipour et al. (2019) and Dimitropoulos et al. (2019). Also, in order to mitigate concerns of any omitted correlated variables we will utilize a panel fixed effects regression analysis as in Alipour et al. (2019) and Dimitropoulos et al. (2019) including year, industry and country fixed effects. This fact allows us to detect many effects that are not detectable in the simple cross-sectional OLS data analysis. The functional form of the equation is stated as follows, where i denotes the firm and t the year and e is the error term:

$$NI\_TA_{it} = \beta_0 + \beta_1 NI\_TA_{it-1} + Year\ F.E + Industry\ F.E. + Country\ F.E. + e_{it} \tag{10.2}$$

Coefficient β_1 is the earnings persistence measure and the higher its value, the higher the earnings persistence of sample firms, suggesting a more stable and predictable earnings stream. Moreover, following Laksmana and Yang (2009), Kousenidis et al. (2013) and Gaio (2010), Dimitropoulos et al. (2019) and Alipour et al. (2019) we considered the predictability of earnings as an additional favorable characteristic of accounting quality. For this reason we estimated earnings predictability as the standard deviation of residuals of model (10.2). Earning predictability measures the ability of earnings to be predicted in the future and thus be more persistent and have higher quality. Larger values of the standard deviation of residuals indicate less predictable earnings and vice versa. In order to examine the impact of CER performance on earnings persistence and predictability we will estimate model (10.2) and extract the residuals from each estimation after separating sample firms between different groups of CER performance as in Suto and Takehara (2018). Specifically we will form three portfolios (groups) by estimating the average of the CER variable per year and country and assigned each firm as low CER firm if its annual score is below the 1st quartile, high CER firms are those that their CER score is above the 3rd quartile while all other firm observations lying between the 1st and 3rd quartiles are assigned as medium CER firms. If H1 is valid we expect firms in the high-CER group to have more persistent and predictable earnings, practically higher β_1 coefficient and smaller standard deviation of residuals relative to low-CER firms.

10.4 Empirical Results

10.4.1 Descriptive Statistics and Correlations

Table 10.1 presents the descriptive statistics of the variables utilized in the regression models. First of all, the mean CER is 64.5 (with a standard deviation of 21.1) indicating that the sample firms have a satisfactory environmental performance score. The highest average score is that of resources usage (66.5) followed by emission reduction activities (65.9) while environmental innovation activities lag behind with an average of 61.3. Also, the average stock price is 9.47 euros and average EPS are positive and up to 1.35 per share outstanding. Both market numbers are far below that of single country studies in Spain, Taiwan and US (Reverte 2016; Wang 2016; Gregory et al. 2016). Thus European firms tend to have a lower stock market performance during the sample period, relative to firms in Taiwan and the US. Nevertheless, our sample firms present higher average of BVPS (24.56) which is higher that the numbers of BVPS reported by Gregory et al. (2016), Wang (2016) and Reverte (2016). So our sample firms seem to be more capitalized relative to Asian and US firms.

Table 10.2 presents the differences between the firm's average stock price, earnings pers. share and book value per share after separating firms within low, medium and high CER performance groups as in Suto and Takehara (2018). The column under the title "difference" reports the average CER difference between the high CER portfolio firms and the low CER portfolio. The "p-value" column presents the corresponding probability values from estimating the Welch two-sample t-test

Table 10.1 Descriptive statistics of the sample variables

Variables	Mean	Standard deviation	Min	Max
CER	64.597	21.167	3.161	99.429
EN_INNOV	61.333	25.962	0.202	99.919
EMM	65.915	26.189	0.161	99.919
RES_USE	66.542	25.730	0.805	99.844
P	9.469	3.122	0.013	22.08
EPS	1.357	3.854	−5.008	12.78
NI	21.28	4.825	−27.86	174.14
BVPS	24.56	49.75	−6.019	150.08

Table 10.2 Firm cash ratio and dividend payout ratio based on CER ranked portfolios

Variables	P1—High CER	P2—Medium CER	P3—Low CER	Difference (P1 − P3)	p-value
P	12.973	8.262	2.189	10.784***	0.001
EPS	1.257	2.396	1.129	0.128	0.254
BVPS	27.25	8.212	7.431	19.825***	0.001

Table 10.3 Pearson correlation coefficients of sample variables

Variables	EPS	BVPS	P	CER
1.EPS	1			
2.BVPS	**0.081**	1		
3.P	**−0.033**	**0.431**	1	
4.CER	0.019	**0.018**	**0.034**	1

Note: Correlation coefficients in bold indicate statistical significance at the 5 per cent significance level at least

and asterisks indicate statistical significance at the 1 per cent significance level (Suto and Takehara 2018). As we can see from the last column, high CER performance firms are characterized by statistically higher stock prices and BVPS relative to low CER performing firms. Both figures seem to increase as we move from the low CER performance group to the high CER performance group. The average EPS does not present and statistical significant difference between the two groups and specifically medium CER firms seem to report higher average EPS (2.39) relative to the other two groups.

Table 10.3 presents the Pearson correlation coefficients of the sample variables. Correlation coefficients in bold indicate statistical significance at the 5 per cent level at least. First, EPS is significantly and positively correlated with BVPS, yet the size of the coefficient is smaller relative to studies by Gregory et al. (2016), Reverte (2016), and Dimitropoulos et al. (2013, 2019). This suggests that more capitalized firms are more profitable and vice versa. Stock prices are positively correlated with BVPS (0.43) but negatively correlated with EPS (−0.03). This finding may suggest that during our sample period investors attach higher value relevance on equity rather than earnings for assessing firms' stock prices. Finally, CER performance is positively and significantly correlated with P and BVPS indicating that investors consider CER activities on assessing the risk of firms and adjusting their decisions on the stock market. However, EPS is not significantly associated with EPS a fact that again suggests the decreased relevance of EPS relative to BVPS in our sample of EU firms. The VIF estimation of the sample variables provided values that were far below the threshold of 10 indicating no multicollinearity. However, the correlation coefficients do not suggest causality, so we will proceed with the fixed effect regression analysis in the following sub-section.

10.4.2 Regression Results

Table 10.4 presents the fixed effects regression results from the estimation from models (10.1a)–(10.1d) using the residuals after regressing stock price P on industry fixed effects. First and foremost, the coefficient on EPS is positive and highly significant in almost all regression models (except for emissions score—EMM). Specifically in the first column using the overall CER score as explanatory variable, the EPS coefficient is 2.14 indicating that 1 per cent increase in EPS contributes 2.14 per cent increase is the stock price, ceteris paribus. This indicates that earnings have

Table 10.4 Fixed effects regression results of CER and accounting numbers value relevance

Model (10.1a–d)—Dep. Variable: P*	CER	EN_INNOV	EMM	RES_USE
EPS	2.146*** (10.78)	0.440*** (3.40)	0.203 (1.14)	2.130*** (22.57)
BVPS	−1.305*** (−22.71)	−0.111*** (−3.69)	−0.801*** (−14.21)	−2.055*** (−35.14)
CER	−0.960 (−0.37)			
EN_INNOV		−0.290 (−0.11)		
EMM			0.241 (0.10)	
RES_USE				−0.849 (−0.41)
CER*EPS	0.032*** (11.51)			
CER*BVPS	0.021*** (27.76)			
EN_INNOV*EPS		0.005*** (3.10)		
EN_INNOV*BVPS		0.006*** (14.04)		
EMM*EPS			0.004* (1.87)	
EPP*BVPS			0.014*** (19.61)	
RES_USE*EPS				0.036*** (24.74)
RES_USE*BVPS				0.025*** (39.14)
Constant	6.390 (0.35)	2.172 (0.13)	2.806 (0.02)	6.301 (0.40)
R-sq	0.505	0.298	0.413	0.556
F-Stat (p-value)	638.82*** (0.001)	265.86*** (0.001)	439.74*** (0.001)	782.55*** (0.001)

Note: *, **, *** indicate statistical significance at the 10%, 5% and 1% significance level respectively. T-statistics in the parentheses

Table 10.5 Earnings persistence and predictability between different CER performing groups

Model (10.2)—Dep. Variable: NI_TA_{t+1}	LOW CER	MEDIUM CER	HIGH CER
NI_TA_{it} (β_1)	−0.008*** (4.54)	−0.002*** (−9.75)	0.018** (2.13)
St. Dev of model (10.2) residuals	0.432	1.214	0.090

Note: *** indicate statistical significance at the 1% significance level respectively. T-statistics in the parentheses

incremental value relevance and are used by market participants in order to assess their expectations about firms' future prospects, which in turn is depicted on firms' stock prices. On the contrary, BVPS produced negative and significant coefficients indicating that book values of common equity are negatively associated with stock prices, yet they are also value relevant for investors in shaping their decisions (Dimitropoulos et al. 2013). Practically, a 1 per cent increase in BVPS lead on average on a reduction of stock price by 1.3 per cent (first column), ceteris paribus. However, none of the environmental performance variables produced significant coefficients. This CER performance and its components does not seem to affect stock prices when examined in isolation from the other variables in the models. This evidence contradicts empirical results by Reverte (2016) and Wang (2016) and lead us to reject H1 and argue that CER performance per se does not present incremental value relevance beyond that of earnings and common equity.

However, when we consider the interaction terms between the CER performance variables (and its components) with EPS and BVPS we receive a complete different picture. All regression coefficients on the interaction terms between CER (and its components) with both EPS and BVPS are positive and statistically significant. More precisely, as CER score gets higher and EPS (BVPS) increase by one percentage point, stock prices will increase by 0.3 (0.2) per cent. This result corroborates the findings by Berthelot et al. (2012), Reverte (2016), Gregory et al. (2016) and Wang (2016) and indicate that environmental responsibility plays a moderating role on the value relevance of accounting numbers. Being more specific, EPS are slightly more value relevant that BVPS as the CER performance score is increased. So for high environmentally responsible firms, both common equity and earnings provide value relevant information to investors, who assess both variables in a positive manner. These results yield support to our second hypothesis and so we can accept that financial information along with environmental and social activities disclosures both give valuable insights to investors because they lead to lower economic uncertainty, allowing them to anticipate future earnings changes and cash flows with mitigated uncertainty (Robinson et al. 2011; Reverte 2016).

In order to examine further the validity of H2 we estimated model (10.2) and extracted the coefficient β_1 as an indication of earnings persistence, following the work by Francis et al. (2004) and Laksmana and Yang (2009) and Alipour et al. (2019). Results are presented on Table 10.5 and as we can see, all coefficients β_1 (our indication of earnings persistence) are statistically significant but for the low and medium CER group of firms the coefficients are negative and become positive for the high CER group of firms. In addition, we extracted the residuals from the estimation of model (10.2) for every CER group and estimated the standard deviation as in Laksmana and Yang (2009). Smaller standard deviation suggests higher predictability of earnings. As we can see, the smallest standard deviation is evident on the high CER sub-group followed by the low CER groups, while the earnings of the medium CER group are the less predictable. Taking together these findings, we can argue that highly environmentally responsible firms are associated with enhanced earnings persistence and predictability, thus they have more favorable accounting quality attributes compared to less environmentally responsible firms. This result could assist investors to better determine the uncertainty of future earnings and cash flows and so they can reach in informed and profitable investment decisions (Alipour et al. 2019). Therefore, evidence on Table 10.5 corroborate our main inferences and lead us to accept H2.

10.4.3 Sensitivity Analysis

In order to check the robustness of the main empirical results, we performed a number of sensitivity tests regarding the research design, and various definitions of the determinant variables. At first, we followed previous studies by Wang (2016), Laksmana and Yang (2009) and Alipour et al. (2019) and included several control variables in the estimation of model (10.1a). Specifically we considered firms size (SIZE) measured as the natural logarithm of total assets. Larger firms have easier access to financial markets since they have lower likelihood of financial distress and they disclose more publicly available information to the investors. This allows market participants to better assess the uncertainty of future earnings are more probably associated positively to stock prices. So we expect a positive coefficient on the SIZE variable. LEV captures firm leverage and is estimated as the ratio of total debt to total assets. Higher leverage is associated with higher uncertainty which may affect negatively stock prices. So a negative coefficient is expected on this variable. Another significant determinant of stock prices is firms' growth opportunities (GROWTH), which are measured as the annual change in sale revenues. Firms with high growth opportunities are more probable to generate future earnings and cash flow streams, a fact that may affect their stock prices upwardly (Dimitropoulos and Asteriou 2009). So a positive coefficient is expected for the GROWTH coefficient. Finally, we considered the impact of profitability measured as the return on assets (ROA). More profitable firms tend to attract higher attention by the investors and are highly valued by the market. Thus a positive coefficient is expected on the ROA. The results are depicted on Table 10.6. As we can observe, the sign and significance of the coefficients remains unchanged relative on those presented on Table 10.4. EPS have a positive impact on P* while BVPS impacts negatively on stock prices. CER is not significant within conventional levels, but when is interacted with EPS and BVPS, it yields positive and highly significant coefficients. So the moderating effect of CER is once again verified. So overall we can accept H2

Table 10.6 Fixed effects regression results on the association between CER and accounting numbers value relevance, using additional control variables

Model (10.1a)—Dep. Variable: P	Coefficient	T-test	P-value
EPS	0.117***	2.32	0.025
BVPS	−1.794***	−20.99	0.001
CER	−0.213	−0.67	0.501
CER*EPS	0.013***	2.85	0.004
CER*BVPS	0.026***	24.62	0.001
SIZE	−0.956	−0.31	0.758
GROWTH	−0.414	−1.39	0.165
LEV	−0.252	−0.11	0.912
ROA	0.176***	10.43	0.001
Constant	0.317	0.45	0.650
R-sq	0.537		
F-Stat (p-value)	322.08*** (0.001)		

Note: **, *** indicate statistical significance at the 5% and 1% significance level respectively. T-statistics in the parentheses

but reject H1. As for the control variables, only ROA produced a positive and significant coefficient (as expected) while the rest of the controls were insignificantly impacting on P*. So we can infer that our results are robust after changing the functional form of the model and including additional controls.

Furthermore, in order to examine whether the main findings are sensitive to the estimation of the stock price we re-estimated model (10.1a) after including the actual stock price 6 months after the fiscal year end (instead of the residuals of a regression of P on industry fixed effects) and by collecting the actuals stock price at the end of the fiscal year (instead 6 months later). Results remained robust to the definition of stock prices and estimation period of stock prices and again we accepted H2 while rejecting H1. In addition, following Cai et al. (2016) we replaced the CER variable definition (which is the simple average of the three sub-scores) by estimating a principal component from the three sub-scores. So we constructed the CER_PCA variable which equals the first principal component from the abovementioned scores. The principal component was determined using eigen values greater than 1.0. The principal component analysis was used since the solutions generated from principal component analysis do not have significant differences from the factor analysis techniques. In this context, we utilized the orthogonal method of extraction and particularly Varimax which tries to maximize the dispersion of loadings within factors. Eventually, for including the factor scores as an independent variable in model (10.1a) we employed the Anderson–Rubin method. The outcome of the principal component analysis supported that data do not suffer from multicollinearity since the KMO statistic of sampling adequacy had a value up to 0.91 indicating that this kind of analysis is appropriate for our data. Eventually the extracted communalities are close to 1 proposing that the factor explains the original data adequately, and the factor loadings are above the critical value of 0.70 proposing that the estimated PCA is of statistical significance. Results remained once again robust relative to those on Table 10.4.

Furthermore, additional sensitivity testing was performed following different definitions of the dependent variables. Specifically, we estimated LEV as the ratio of long term debt to total assets, GROWTH as the ratio of market value to book value of equity (MV/BV) and ROA was replaced by ROE but the results remain unchanged after these modifications. Also EPS was estimated as earnings to lagged total assets and BVPS was replaced with book value of common equity to lagged total assets, leaving the main inferences unaffected. Finally, in order to mitigate any omitted variable bias we performed a change analysis in order to extract more efficient inferences regarding the impact of CER on the differences of share prices (returns). So we re-estimated models (10.1a) and (10.2) by substituting the dependent variables with the annual change of stock prices (return) and NI_TA from year (t-1) to year (t) (Gregory et al. 2016). Once again the results remained qualitatively unchanged compared with those on Table 10.4 verifying on the second research hypothesis.

10.5 Conclusion

Several studies on the environmental responsibility research field have focused on examining the impact of CER activities on the value relevance of accounting information and specifically earnings and common equity. Researchers have expressed different views on how a company's social and environmental performance affects the quality of its accounting information and several arguments have made stressing that more environmentally responsible firms provide more reliable, transparent and quality financial information so as to meet the ethical requirements of their stakeholders and investors (Alipour et al. 2019; Unerman and Bennett 2004). The scope of this chapter is to examine the impact of CER related performance on the value relevance of earnings and book values of equity by examining the direct and moderating effect of CER on stock prices, earnings persistence and earnings predictability. We respond to the call for more research on the topic made by Alipour et al. (2019), by covering a large multi-country sample (24 EU countries) of listed European firms (1288 corporations) over a long period of time (2003–2018).

Empirical analysis suggested that CER performance alone does not have a significant impact on stock prices, something that contradicts several studies in other countries, but CER exerts a moderating impact on the value relevance of accounting numbers. Specifically, we found that high CER performing firms have earnings which are more persistent 1 year ahead, are more predictable (show less variability) and report profits and common equity figures which are positively impacting stock prices. This finding is in line with evidence by Reverte (2016) in Spain, Wang (2016) in Taiwan and Alipour et al. (2019) in the Iranian stock market. The empirical results remain robust after several sensitivity tests regarding the definition of the dependent variable, the functional form of the model and alternative definitions of control variables. The findings of this chapter contribute to the ongoing debate about the incremental value relevance of CER disclosures and the quality of financial reporting. At first we complement previous studies by Laksmana and Yang (2009), Wang (2016) and Reverte (2016), by incorporating a multi-country instead of a single country research setting. Also, we incorporate in the analysis a decomposition of the CER variable and indicate which sub-category of CER performance contributes to value relevance, something that has not been considered extensively by previous researchers.

References

Al-Tuwaijri, S.A., T.E. Christensen, and K.E. Hughes II. 2004. The relations among environmental disclosure, environmental performance, and economic performance: A simultaneous equation approach. *Accounting, Organizations and Society* 29: 447–471.

Alipour, M., M. Ghanbari, B. Jamshidinavid, and A. Taherabari. 2019. The relationship between environmental disclosure quality and earnings quality: A panel study of an emerging market. *Journal of Asia Business Studies* 13 (2): 326–347.

Barth, M., W.R. Landsman, and M.H. Lang. 2008. International accounting standards and accounting quality. *Journal of Accounting Research* 46: 467–498.

Bernard, V.L., and J.K. Thomas. 1989. Post-earnings-announcement drift: Delayed price responseor risk premium? *Journal of Accounting Research* 27 (1): 1–36.

Berthelot, S., M. Coulmont, and V. Serret. 2012. Do investors value sustainability reports? A Canadian study. *Corporate Social Responsibility and Environmental Management* 19 (6): 355–363.

Bollen, N. 2007. Mutual fund attributes and investor behavior. *Journal of Financial and Quantitative Analysis* 42: 683–708.

Brammer, S., and S. Pavelin. 2006. Corporate reputation and corporate social performance: The importance of fit. *Journal of Management Studies* 43 (3): 35–55.

Cai, L., J. Cui, and H. Jo. 2016. Corporate environmental responsibility and firm risk. *Journal of Business Ethics* 139: 563–594.

Clarkson, P., X.H. Hua Fang, Y. Li, and G. Richardson. 2013. The relevance of environmental disclosures: Are such disclosures incrementally informative? *Journal of Accounting and Public Policy* 32: 410–431.

Cormier, D., and M. Magnan. 2007. The revisited contribution of environmental reporting to investors' valuation of a firm's earnings: An international perspective. *Ecological Economics* 62: 613–626.

Dimitropoulos, P., and D. Asteriou. 2008. Timeliness, conservatism and financial transparent firms under the Greek accounting setting. *Review of Accounting & Finance* 7 (3): 252–269.

———. 2009. The impact of accruals and cash flows on the returns-earnings relation: Evidence from Greece. *International Journal of Accounting, Auditing and Performance Evaluation* 5 (4): 384–407.

———. 2010. Accounting relevance & speculative intensity: Empirical evidence from Greece. *Journal of Applied Accounting Research* 11 (3): 195–212.

Dimitropoulos, P., and K. Koronios. 2018. Earnings persistence of European football clubs under UEFA's FFP. *International Journal of Financial Studies* 6: 43.

Dimitropoulos, P., D. Asteriou, D. Kousenidis, and S. Leventis. 2013. The impact of IFRS on accounting quality: Evidence from Greece. *Advances in Accounting: Incorporating Advances in International Accounting* 29 (1): 108–123.

Dimitropoulos, P.E., O. Vrondou, and K. Koronios. 2019. Earnings predictability of the Greek hospitality industry during the crisis. In *Smart tourism as a driver for culture and sustainability*, Springer Proceedings in Business and Economics, ed. V. Katsoni and M. Segarra-Oña, 647–658. New York: Springer.

El Ghoul, S., O. Guedhami, C. Kwok, and D. Mishra. 2011. Does corporate social responsibility affect the cost of capital? *Journal of Banking and Finance* 35: 2388–2406.

Foster, G., C. Olsen, and T. Shevlin. 1984. Earnings releases, anomalies, and the behavior of security returns. *The Accounting Review* 59: 574–603.

Francis, J., R. LaFond, P.M. Olsson, and K. Schipper. 2004. Costs of equity and earnings attributes. *The Accounting Review* 79: 967–1010.

Gaio, C. 2010. The relative importance of firm and country characteristics for earnings quality around the world. *European Accounting Review* 19: 693–738.

Gregory, A., J. Whittaker, and X. Yan. 2016. Corporate social performance, competitive advantage, earnings persistence and firm value. *Journal of Business Finance and Accounting* 43 (1–2): 3–30.

Hassel, L., H. Nilsson, and S. Nyquist. 2005. The value relevance of environmental performance. *European Accounting Review* 14 (1): 41–61.

Kang, T., G.V. Krishnan, M.C. Wolfe, and H.S. Yi. 2012. The impact of eliminating the 20-F reconciliation requirement for IFRS filers on earnings persistence and information uncertainty. *Accounting Horizons* 26: 741–765.

Kim, Y., M. Park, and B. Wier. 2012. Is earnings quality associated with corporate social responsibility? *The Accounting Review* 87: 761–796.

Kousenidis, D.V., A.C. Ladas, and C.I. Negakis. 2013. The effects of the European debt crisis on earnings quality. *International Review of Financial Analysis* 30: 351–362.

Laksmana, I., and Y.-w. Yang. 2009. Corporate citizenship and earnings attributes. *Advances in Accounting: Incorporating Advances in International Accounting* 25 (1): 40–48.

Lang, M., J. Raedy, and W. Wilson. 2006. Earnings management and cross listing: Are reconciled earnings comparable to US earnings? *Journal of Accounting and Economics* 42: 255–283.

Lang, M., J. Raedy, and M. Yetman. 2003. How representative are firms that are cross listed in the United States? An analysis of accounting quality. *Journal of Accounting Research* 41: 363–386.

Latane, H.A., and C.P. Jones. 1979. Standardized unexpected earnings 1971–1977. *Journal of Finance* 34: 717–724.

Lourenço, I.C., M.C. Branco, J.D. Curto, and T. Eugénio. 2012. How does the market value corporate sustainability performance. *Journal of Business Ethics* 108 (4): 417–428.

Moneva, J.M., and B. Cuellar. 2009. The value relevance of financial and non-financial environmental reporting. *Environmental Resource Economics* 44 (3): 441–456.

Oei, R., A. Ramsay, and P. Mather. 2008. Earnings persistence, accruals and managerial share ownership. *Accounting and Finance* 48: 475–502.

Reverte, C. 2016. Corporate social responsibility disclosure and market valuation: Evidence from Spanish listed firms. *Review of Managerial Science* 10: 411–435.

Richardson, S.A., R.G. Sloan, M.T. Soliman, and I. Tuna. 2005. Accrual reliability, earnings persistence and stock returns. *Journal of Accounting and Economics* 39: 437–485.

Robinson, M., A. Kleffner, and S. Bertels. 2011. Signaling sustainability leadership: Empirical evidence of thevalue of DJSI membership. *Journal of Business Ethics* 101: 493–505.

Salewski, M. and H. Zülch. 2014. The association between corporate social responsibility (CSR) and earnings quality – evidence from european blue chips. Accessed from https://papers.ssrn.com/sol3/papers.cfm?abstract_id=2141768

Schadewitz, H., and M. Niskala. 2010. Communication via responsibility reporting and its effect on firm value in Finland. *Corporate Social Responsibility and Environmental Management* 17: 96–106.

Servaes, H., and A. Tamayo. 2013. The impact of corporate responsibility on firm value: The role of customer awareness. *Management Science* 59 (5): 1045–1061.

Sloan, R.G. 1996. Do stock prices fully reflect information in accruals and cash flows about future earnings? *The Accounting Review* 71: 289–315.

Suto, M., and H. Takehara. 2018. *Corporate social responsibility and corporate finance in Japan*. Singapore: Springer Nature.

Unerman, J., and M. Bennett. 2004. Increased stakeholder dialogue and the internet: Towards greater corporate accountability or reinforcing capitalist hegemony? *Accounting, Organizations and Society* 29 (7): 685–707.

Wang, M.C. 2016. The relationship between environmental information disclosure and firm valuation: The role of corporate governance. *Quality and Quantity* 50 (3): 1135–1151.

Williams, S.M. 1999. Voluntary environmental and social accounting disclosure practices in the Asia-Pacific region: An international empirical test of political economy theory. *The International Journal of Accounting* 34 (2): 209–238.

Chapter 11 Corporate Environmental Responsibility and Earnings Management

Corporate environmental reporting has increased significantly over the last decades and is considered as a core information dissemination strategy of corporations to investors and other stakeholders. This fact makes the CER related activities closely connected with the quality of financial reporting practices of firms. In the literature earnings management (EM) is considered as the discretion exercised by managers and accountants in depicting the firms' financial numbers within the limits allows by Generally Accepted Accounting Principles (GAAP). Several research studies examining the relationship between corporate environmental responsibility (CER) and earnings manipulation (EM) has reached contradictory evidence due to the fact that they incorporate different measures of CER, are focused on a single country setting and utilize only on the accrual accounting process of financial reporting. The purpose of this chapter is to study the impact of CER related activities on the earnings management behavior using a multi-country setting and examining different research frameworks for estimating earnings management including accrual management, real earnings management and income smoothing activities.

11.1 Introduction

The value of accounting information is largely determined by its quality (CFA 2019). Accounting information acts as the communication medium (Obigbemi et al. 2016), has a significant impact on the decision-making behavior of stakeholders and the stability of the capital market, thus its quality has attracted widespread attention. There is no uniform definition of accounting information quality in the existing literature. For internal managers, accounting information quality is the degree to meet their decision-making needs; for investors, accounting information quality means the transparency and credibility of financial information (Katmon and

P. Dimitropoulos, K. Koronios, *Corporate Environmental Responsibility, Accounting and Corporate Finance in the EU*, CSR, Sustainability, Ethics & Governance, https://doi.org/10.1007/978-3-030-72773-4_11

Farooque 2017); for regulators, accounting information quality refers to the degree to which financial reports meet the disclosure requirements of accounting standards.

There exist several ways to measure accounting information quality in prior literature, the most popular accounting-based measure is that of earnings management (Niu 2006). From the perspective of accounting, accounting information quality is mainly affected by opportunistic earnings management (EM). Earnings management commonly refers to the act that managers influence the financial reporting process deliberately for self-interest by exercising discretion over reported earnings (Xie et al. 2003). As managers manipulate earnings primarily through discretionary accruals consisting of a great deal of discretion (Bergstresser and Philippon 2006), accounting information quality can be inversely represented by discretionary accruals level. The higher the discretionary accruals level, the lower the accounting information quality.

Under this framework, extant research studies have considered the impact of CER related activities and performance on the quality of financial information published by firms. The political cost hypothesis is one theory framing this association and it posits that firms which are a target of political scrutiny will resort to earnings decreasing manipulation in order to avoid regulatory control and monitoring. Put it differently, firms have greater incentives to reduce their earnings in order to avoid regulatory scrutiny, so that theory predicts a positive association between CER and EM (Heltzer 2011). However, institutional theory remains one of the alternative theoretical frameworks pointing that CER may be unrelated to EM because CER is affected by institutional factors not related with earnings manipulation or firms with less CER performance can be more associated with EM since environmentally responsible firms are more concerned with enhancing their stakeholder relationships so they are less probable to hide unfavorable earnings and resort to EM. So the institutional theory predicts a negative association between CER and EM.

Empirical evidence so far have provided contradictory evidence on the impact of CER on EM behavior. Prior et al. (2008) and Buertey et al. (2019) provide evidence of a positive association between CER and EM supporting the political cost hypothesis. On the contrary, Heltzer (2011), Litt et al. (2014), Yoon et al. (2019), Kim et al. (2019) and Palacios-Manzano et al. (2019) all point to a negative association between CSR and CER activities on earnings management. The negative association is justified by the fact that high CER performance firms are more profitable and their profitability is the result of real activities and not earnings manipulation. Specifically Litt et al. (2014) argue that firms engaged in pollution prevention and climate protection activities are least likely to manage their accounting numbers. The conflicting results in the literature are explained by the fact that CER performance is a multi-faceted factor that is comprised by many categories of activities and investments, each one with different association with EM and because motives for EM may differ between countries and business settings.

The scope of this chapter is to shed further light on the impact of CER performance and its components on the EM behavior by utilizing a multi-country research setting and incorporating various sub-categories of environmental performance and earnings management metrics including accrual earnings manipulation, real activities manipulation and income smoothing, thus providing more thorough evidence in the existing literature regarding the impact of CER on the quality of accounting numbers. The empirical analysis in this chapter corroborates previous arguments in the literature on the negative association between CER and EM. In the following section we provide the theoretical background on the impact of CER on EM and the testable hypothesis. The third section discusses the data selection procedure and the research design. The fourth section provides the discussion of empirical results and the last section concludes the chapter.

11.2 CER-EM Association and Testable Hypothesis

The association between CER and EM has been examined and viewed through various theoretical lenses and hypotheses. According to Chih et al. (2008), the myopia avoidance hypothesis states that an environmentally responsible firm will not resort to earnings management in order to hide unfavorable earnings so higher CER performance is associated with higher accounting quality via less earnings management. On the contrary, the predictable earnings hypothesis and the multiple objectives hypothesis (which both originate in the political cost hypothesis mentioned in the previous section) predict a positive association between CER and earnings management (Heltzer 2011). The predictable earnings hypothesis posits that firms which are less devoted to CER activities and want to increase the informational advantage of outside investors relative to firm insiders, are less prone to utilize income smoothing techniques and thus are least probable to resort to EM to manage their accounting numbers. The multiple objectives hypothesis deals with the fact that when firms try to accommodate various stakeholders' interests (via CSR and CER activities), managers are not so accountable for corporate resources' usage compared to those firm managers that are mainly focused on short-term financial goals. So CER performing firms can choose earnings management mechanisms in order to mask their sub-optimal allocation decisions (Heltzer 2011). The final theoretical hypothesis mentioned in the literature is the institutional hypothesis which argues that CER performance and earnings management behavior are unrelated due to the fact that financial malpractices and even fraud are not the outcome of a decline in ethic stances and behavior but their source is originated to other institutional factors (Litt et al. 2014).

The fact is that CER activities and investments have been increasing over the last decade and more and more firms are devoting human, capital and financial resources in various such efforts to prove their legitimacy, ethical behavior and environmental awareness to their stakeholders. Regulators, customers, investors, media and other stakeholders have increased their scrutiny over corporate activities. This high public concern has become a significant driver of CER performance and so firms are more careful on their implementation of corporate environmental behavior. According to Litt et al. (2014), this enhanced monitoring by the markets, regulators, media and the society on environmentally responsible firms can have a significant effect on the quality of published accounting information as well. Empirical evidence by Yu (2008) indicate that enhanced market coverage and monitoring serves as a substitute governance mechanism which helps to mitigate earnings management behavior. So practically, high CER performing firms are more probable to report higher quality financial information. Empirical evidence by Yoon et al. (2019), Kim et al. (2019) and Palacios-Manzano et al. (2019) verify these arguments in different country settings and economic sectors.

Moreover, firms with enhanced environmentally responsible behavior are less probable to experience self-interest and discretionary managerial behavior which may contradict corporate interests (higher managerial-corporate interest alignment). Also highly CER performing firms tend to be long-term oriented, less focused on short-term financial goals and disclose more non-financial voluntary information (Litt et al. 2014). So under this framework, CER performing firms are expected to be associated with higher transparency and thus will have established a corporate culture that would not likely resort to earnings manipulation in order artificially enhance their financial performance. Therefore, following the abovementioned discussion it is expected that enhanced CER performance will be associated with higher financial reporting quality (aka less earnings management). So the main research hypothesis is stated as follows:

H1: CER performance will have a negative impact on EM.

11.3 Data Selection and Research Design

11.3.1 Data Selection Procedure

The current chapter utilizes a sample of corporations (both listed and unlisted) from 23 EU-member countries over the period 2003–2018. All financial, governance and environmental responsibility data have been extracted from Datastream database. The sample collection procedure started by including all corporations that have been covered by the environmental, social and governance (ESG) scores provided by Datastream during the sample period. 2003 has been selected as the initial sample year since that year was the first year with significant coverage (taking into consideration that this database officially was initiated in 2002). From a total of 656,734 firm-year observations we excluded those firms without environmental score data for at least five consecutive years in order to avoid bias in the empirical results and also to facilitate the estimation of differenced variables. Furthermore, we excluded firms with incomplete financial data on operating cash flows, revenues, accounts receivables, fixed assets and those that do not close their fiscal year on December. In addition, we winsorized the upper and down 1 per cent of the data distribution in order to mitigate the influence of significant outliers in the empirical analysis. After this procedure was finalized, we ended up with a final sample of 121,154 firm-year observations. The analysis and structure of the final sample is provided on Table 4.1 on the fourth chapter of the book.

11.3.2 Research Design

In order to examine the main research hypothesis we utilized various measures of earnings management pointed in the literature. The most commonly used measure of earnings manipulation is the modified Jones (1991) model capturing the discretionary accruals of firms. The term "accruals" originates from the accrual basis of accounting which dictates that all expenses and revenues that have been incurred but not yet paid or received must be reported within the financial statements and are estimated as the difference between net income and operating cash flows (Dimitropoulos 2011). According to Jones (1991), the indication of earnings manipulation refers to the use of total accruals in such manner where private benefits can be extracted for corporate managers. The Jones (1991) model distinguished total accruals between normal (or non-discretionary accruals) and abnormal (discretionary accruals) based on their nature and origin. Normal accruals are the outcome of adjustments to cash flows due to accounting rules and regulations. On the contrary, discretionary accruals are those adjusted or manipulated by the managers according to disclosure decisions made explicitly by them. Some classic examples of discretionary accruals used by managers to manipulate their financial statements are the disclosure of increased depreciation and amortization expenses prior to initial public offerings, increasing bad debt provisions and disclosing increased provisions for deferred taxation (Dimitropoulos 2011).

The most popular empirical model for estimating discretionary accruals is the modified Jones model which has been proved to be more efficient in detecting EM relative to other accrual models in the literature (Dechow et al. 1996). For this reason in this study we estimated the cross-sectional Jones (1991) model, as being modified by Dechow et al. (1996) and Kothari et al. (2005), in order to extract the discretionary or abnormal accruals (Litt et al. 2014; Heltzer 2011). This model estimates discretionary accruals as a function of difference between change in sales and change in accounts receivables, the level of property, plant and equipment and the level of return on assets by estimating the following cross-sectional OLS equation:

$$\begin{aligned} \mathrm{ACC_i/TA_{t-1}} &= \mathrm{a_0} + \alpha(1/\mathrm{TA_{t-1}}) + \beta([\Delta\mathrm{SALES_i} - \Delta\mathrm{REC_i}]/\mathrm{TA_{t-1}}) \\ &\quad + \gamma(\mathrm{PPE_i/TA_{t-1}}) + \mathrm{gROA_i} + \mathrm{e_i} \end{aligned} \tag{11.1}$$

Where

ACC is total accruals defined as the difference between net income and operating cash flows divided by lagged total assets

ΔSALES is the change in net sales deflated by lagged total assets ($\mathrm{Sales_{it}} - \mathrm{Sales_{it\text{-}1}}$)

ΔREC is the change in accounts receivables ($REC_{it} - REC_{it-1}$)

PPE is the level of property plant and equipment for each year deflated by lagged total assets

ROA is the end of year return on assets estimated as net income over total assets

TA is firm's total assets at the end of the fiscal year

As Kothari et al. (2005) argue, the inclusion of a constant term in the Jones (1991) model provides an additional control for heteroscedasticity, it mitigates any omitted size variable problems and produces discretionary accrual measures that are more symmetric, making the power of the test comparisons more clear and overcoming model misspecifications. The discretionary accruals are defined as the residuals from estimating model (11.1). Following previous studies by Litt et al. (2014), Yoon et al. (2019) and Kim et al. (2019), we estimated the absolute value of discretionary accruals (ABS_DACC) from Eq. (11.1) as our measure of earnings management because earnings management can involve either income increasing or income decreasing accruals to meet earnings targets (Warfield et al. 1995; Reynolds and Francis 2000; Klein 2002; Wang 2006). A higher value of ABS_DACC indicates a greater level of manipulation in the financial statements thus lower earnings quality.

Moreover, firms can manipulate their accounting numbers through their daily operations and not by using the accrual process of accounting. Real earnings management is defined in the literature as the discretion exercised by managers through the boosting of sales revenue through pricing discounts or altering the credit terms offered to customers which can lead to abnormally lower operating cash flows compared with those that firm will have achieved if such activities were not in place. According to Kim et al. (2012), real earnings management has a direct impact on the actual revenue and expenses numbers and affect the firm performance more directly compared to accruals earnings management, but that characteristic makes real earnings management less detectable from auditors and regulators. One way to proxy for real earnings management activities is to estimate the discretionary portion of operating cash flows. Following, Roychowdhury (2006), Kim et al. (2019) and Yoon et al. (2019) we estimated the following cross-sectional OLS model which projects the operating cash flows as a function of the level and change of revenues:

$$
\begin{aligned}
CFO_i/TA_{t-1} = a_0 + \alpha(1/TA_{t-1}) + \beta(SALES_i/TA_{t-1}) \\
+ \gamma(\Delta SALES_i/TA_{t-1}) + e_i
\end{aligned}
\tag{11.2}
$$

Where:

CFO is operating cash flows divided by lagged total assets

SALES is the annual net sales divided by lagged total assets

ΔSALES is the change in net sales deflated by lagged total assets ($Sales_{it} - Sales_{it-1}$)

TA is firm's total assets at the end of the fiscal year

The residuals from model (11.2) represent the unexplained part of that regression and are the discretionary portion of operating cash flows (DCFO) and our proxy of real earnings manipulation. If CEOs manipulate their earnings numbers via expanding their credit policy or price discount policy, or expand their accounts receivables, then a lower cash flow figure is expected relative to its predicted level. So lower DCFO values indicate more real earnings management manipulation (Kim et al. 2012).

The final measure of earnings manipulation is income smoothing and following previous studies by Leuz et al. (2003), Barth et al. (2008) and Heltzer (2011) we estimated it via the Spearman correlation between accruals and cash flows between firms of high, medium and low CER performance. High CER firms are those that their CER score is above the 3rd quartile, low CER firms are those that their CER score is below the 1st quartile and firms with CER scores between the 1st and 3rd quartiles are denoted as medium CER performing firms. In order to capture any confounding effects of factors that are not attributable to the financial reporting setting, we adopt the methodology proposed by Barth et al. (2008) and compare the correlations of the accruals and cash flows residuals (CF* and ACC*), from the following regression models, instead of comparing the correlations between accruals and cash flows directly. The relative models have the following form:

$$CF_{it} = a_0 + a_1 SIZE_{it} + a_2 MV\_BV_{it} + a_3 LEV_{it} + a_4 ROA_{it} + a_5 DLIST_{it} + a_6 R\&D + e_{it} \tag{11.3}$$

$$ACC_{it} = a_0 + a_1 SIZE_{it} + a_2 MV\_BV_{it} + a_3 LEV_{it} + a_4 ROA_{it} + a_5 DLIST_{it} + a_6 R\&D + e_{it} \tag{11.4}$$

Where:

CF is the annual operating cash flow divided by lagged total assets

ACC is total accruals measured as the difference between net income and CF divided by lagged total assets

SIZE is the natural logarithm of end year total assets

MV_BV is the ratio of market value to book value of equity

LEV is the ratio of end year total liabilities to end year total assets

ROA is the ratio of net income before taxes to total assets

DLIST is a dummy receiving (1) for publicly listed firms and (0) otherwise

R&D is the ratio of R&D expenses to revenues

Barth et al. (2008) interpret a more negative correlation between accruals and cash flows as earning smoothing due to the fact that managers respond to poor cash flow outcomes by increasing accruals (through the increase of net income since accruals are the difference between net income and operating cash flows). If this is true we predict that firms with higher CER performance will present a less negative correlation between accruals and cash flows residuals compared with medium and low CER firms, thus will be associated with lower income smoothing. In order to test our research hypotheses we introduced the absolute value of discretionary accruals (ABS_DACC) and the discretionary operating cash flows (DCFO) as the dependent variables and the average CER and its components along with control variables in the following models where i denotes the firm and t the year and e is the error term:

$$ABS\_DACC_{it} = a_0 + a_1CER_{it} + a_2SIZE_{it} + a_3LEV_{it} + a_4MV\_BV_{it} + a_5ROA_{it} + a_6R\&D\_EXP_{it} + Year\ F.E + Industry\ F.E. + Country\ F.E. + e_{it} \quad (11.5)$$

$$ABS\_DACC_{it} = a_0 + a_1EN\_INNOV_{it} + a_2SIZE_{it} + a_3LEV_{it} + a_4MV\_BV_{it} + a_5ROA_{it} + a_6R\&D\_EXP_{it} + Year\ F.E + Industry\ F.E. + Country\ F.E. + e_{it} \quad (11.5a)$$

$$ABS\_DACC_{it} = a_0 + a_1EMM_{it} + a_2SIZE_{it} + a_3LEV_{it} + a_4MV\_BV_{it} + a_5ROA_{it} + a_6R\&D\_EXP_{it} + Year\ F.E + Industry\ F.E. + Country\ F.E. + e_{it} \quad (11.5b)$$

$$ABS\_DACC_{it} = a_0 + a_1RES\_USE_{it} + a_2SIZE_{it} + a_3LEV_{it} + a_4MV\_BV_{it} + a_5ROA_{it} + a_6R\&D\_EXP_{it} + Year\ F.E + Industry\ F.E. + Country\ F.E. + e_{it} \quad (11.5c)$$

$$DCFO_{it} = a_0 + a_1CER_{it} + a_2SIZE_{it} + a_3LEV_{it} + a_4MV\_BV_{it} + a_5ROA_{it} + a_6R\&D\_EXP_{it} + Year\ F.E + Industry\ F.E. + Country\ F.E. + u_{it} \quad (11.6)$$

$$DCFO_{it} = a_0 + a_1EN\_INNOV_{it} + a_2SIZE_{it} + a_3LEV_{it} + a_4MV\_BV_{it} + a_5ROA_{it} + a_6R\&D\_EXP_{it} + Year\ F.E + Industry\ F.E. + Country\ F.E. + u_{it} \quad (11.6a)$$

$$DCFO_{it} = a_0 + a_1EMM_{it} + a_2SIZE_{it} + a_3LEV_{it} + a_4MV\_BV_{it} + a_5ROA_{it} + a_6R\&D\_EXP_{it} + Year\ F.E + Industry\ F.E. + Country\ F.E. + u_{it} \quad (11.6b)$$

$$DCFO_{it} = a_0 + a_1RES\_USE_{it} + a_2SIZE_{it} + a_3LEV_{it} + a_4MV\_BV_{it} + a_5ROA_{it} + a_6R\&D\_EXP_{it} + Year\ F.E + Industry\ F.E. + Country\ F.E. + u_{it} \quad (11.6c)$$

All models will be estimated with panel Generalized Least Squares (GLS) random effects. The panel random effect estimation method was chosen due to the fact that panel analysis contains more information with more variability and less collinearity among the variables, providing more efficient estimates and precise parameters of model estimation, allowing us to detect many effects that are not detectable in the cross-sectional data analysis. In order to choose the random effect estimation we performed the Breusch-Pagan Lagrange multiplier test for random effect which tests the null hypothesis that the residuals variance is equal to zero. The test produced a highly significant statistic (Chi2 2546.22, $p < 0.00001$) leading us to reject the null hypothesis stating that the variance of residuals is equal to zero, and thus the random effects estimation was considered as the most appropriate method for our case.

EN_INNOV denotes each company's environmental innovation score, which reflects the firm's ability to mitigate the environmental costs towards its customers or even creating new market and product opportunities via enhanced environmental technologies and processes. EMM denotes companies' emission reduction score estimated via their efforts and effectiveness in reducing emissions from their production and operation. RES_USE denotes the firms' resource use score and indicates firms' ability to reduce the usage of hazardous materials, energy, water and other valuable natural resources. Finally, CER is the firms' annual average of the aforementioned three scores. All environmental scores range between zero (0) and 100 so the higher the score the more environmentally responsible and efficient is the firm. If our main hypothesis H1 is valid we expect a negative a_1 coefficient in model (11.5) suggesting that firms with enhanced CER performance are associated with lower values of discretionary accruals thus higher accounting quality, while for model (11.6) we expect a positive a_1 coefficient suggesting that high CER firms are related to more abnormal cash flows. This is because if CEOs try to manipulate their earnings through the price discount policy or via account receivables the predicted level of cash flows will be more compared to its actual level (Yoon et al. 2019).

In addition, we included several control variables that have proved significant determinants of earnings management in the literature. SIZE is measured as the natural logarithm of total assets (LnTA) at the end of the fiscal year. Watts and Zimmerman (1990), Van Tendeloo and Vanstraelen (2005), Dimitropoulos and Asteriou (2010) and Dimitropoulos (2011), argue, larger firms are more likely to prefer downward earnings management because they are more prone to regulatory scrutiny and since large firms are more profitable have less incentives to manipulate their accounting numbers. On the contrary, large firms may resort to higher earnings management due to more instability of their operations thus are more motivated to smooth their earnings (Palacios-Manzano et al. 2019). Therefore, due to the conflicting arguments in the literature we cannot infer any sign of association between the SIZE variable and the earnings management proxies (or alternatively to avoid earnings decreases).

Additionally, we control for the impact of firm leverage (LEV) measured as the ratio of total debt to total assets. Van Tendeloo and Vanstralen (2005) and Billings (1999) and Litt et al. (2014) document that high leveraged firms are more likely to engage in upward earnings management to avoid debt covenant violations. Therefore we believe that leverage will have a positive relation with earnings management. Also we control for firms' growth prospects (MV_BV) measured as the ratio of market value of equity to book value of equity. Lee et al. (2006) provide evidence that higher growth firms are more likely to manipulate earnings. Moreover, Abarbanell and Lehavy (2003) suggest that high growth firms have stronger incentives to meet or beat analyst's earnings forecasts resorting to earnings management (Gaver and Gaver 1993; Ittner et al. 2002; Cohen and Zarowin 2010). Consequently, we predict that firms with increased MV_BV will manipulate accounting numbers more frequently and so the respective coefficient is expected to have a positive sign.

ROA is return on assets measured as the ratio of net income before taxes to total assets. Dimitropoulos and Asteriou (2010), Palacios-Manzano et al. (2019), Yoon et al. (2019) and Kim et al. (2019) document a negative association between profitability and earnings management suggesting that firms with increased profits have less incentives to manipulate accounting numbers through accruals.

Consequently we expect a negative coefficient on the ROA variable. The last control variable is the R&D expenses to total revenue. Kim et al. (2012) argue that high R&D intensity firms are more prone to manipulate their accounting numbers in order to sustain their presence to the capital market or even gain access to various forms of financing so as to continue expanding or sustaining their R&D projects. However, Yoon et al. (2019) documented that high R&D intensity firms are associated with higher earnings quality in the Korean market. So based on the above findings we expect a positive association between R&D and earnings management.

11.4 Empirical Results

11.4.1 Descriptive Statistics, Correlations and Income Smoothing

Table 11.1 presents the descriptive statistics of the variables utilized in the regression models. As we can see, discretionary accruals have a positive average (18.93) while the average DCFO is negative (−0.085) indicating that our sample firms utilize more accrual earnings management relative to real earnings management. This findings is similar to evidence provided by Kim et al. (2019) in the Chinese market. Also, the mean CER is 64.5 (with a standard deviation of 21.1) indicating that the sample firms have a satisfactory environmental performance score. The highest average score is that of resources usage (66.5) followed by emission reduction activities (65.9) while environmental innovation activities lag behind with an average of 61.3. The sample firms are highly profitable (mean ROA 21.28) and present significant growth opportunities (mean MV_BV at 55.68). However, our sample firms are highly leveraged since the average total debt covers more than four times the sample firms' total assets and spend also the 2 per cent of their revenues in R&D investing

Table 11.1 Descriptive statistics of the sample variables

Variables	Mean	Standard deviation	Min	Max
DACC	18.933	33.833	−52.983	117.54
ABS_DACC	20.777	33.833	0.0001	117.54
DCFO	−0.085	13.604	−27.737	1.271
CER	64.597	21.167	3.161	99.429
EN_INNOV	61.333	25.962	0.202	99.919
EMM	65.915	26.189	0.161	99.919
RES_USE	66.542	25.730	0.805	99.844
SIZE	20.496	2.848	2.302	30.311
LEV	4.627	4.319	−0.682	25.691
MV_BV	55.688	43.913	−22.166	256.91
ROA	21.281	48.257	−27.861	174.17
R&D_EXP	0.018	0.017	−0.025	0.082

activities. This number however is larger compared to R&D expenses in the Korean market which averages up to 0.7 per cent (Yoon et al. 2019).

Table 11.2 presents the Pearson correlation coefficients between the sample variables. The majority of the correlation coefficients are significant with economic meaning, and are not very high indicating the absence of multicollinearity on the data. The VIF estimation of the sample variables provided values that were far below the threshold of 10 indicating no multicollinearity. As we can see, CER is not significantly associated with all earnings management measures something which does not provide support to our initial expectations. However, DACC and ABS_DACC are positively correlated with environmental innovation and emissions scores and negatively correlated with resource usage score. This evidence suggest that different environmental performance variables create different earnings management motives on the sample firms. This outcome justifies the use of decomposed environmental performance indicators instead of using an aggregate measure of CER. Additionally, SIZE is positively associated with DACC and DCFO indicating that larger firms manipulate their accounting numbers through accruals and real activities manipulation. Also, highly leveraged and more profitable firms are associated with higher earnings management. The only difference is observed in the resource use score (RES_USE) which is negatively associated with PROF and with SIZE. Finally, growth opportunities did not provide any significant correlation coefficients with the CER related and financial performance.

As an initial stage of our analysis we compared earnings manipulation via income smoothing between firms of high, medium and low CER performance. Following previous studies by Leuz et al. (2003), Barth et al. (2008) and Heltzer (2011) we estimated income smoothing via the Spearman correlation between accruals and cash flows residuals form the estimation of model (11.3) and (11.4) (Barth et al. 2008). High CER firms are those that their CER score is above the 3rd quartile, low CER firms are those that their CER score is below the 1st quartile and firms with CER scores between the 1st and 3rd quartiles are denoted as medium CER performing firms. A more negative correlation between accruals and cash flows is considered as more earning smoothing so we expect that firms with high CER performance will present a less negative correlation between accruals and cash flows residuals compared with medium and low CER firms. Results are presented on Table 11.3. As we can see, the Spearman rank correlation coefficient increases in magnitude moving from the P1 portfolio of the high CER performing firms (−0.685) to the P3 group of low CER performing firms (−0.913). This result provides an initial support to H1 indicating the firms engaging in CER activities are associated with higher accounting quality through income smoothing. Of course, correlations do not indicate causality or direction of association so we will proceed with the regression analysis results in order to extract more specific inferences regarding the validity of H1.

Table 11.2 Pearson correlation coefficients of sample variables

Variables	1	2	3	4	5	6	7	8	9	10	11	12
1. DACC	1											
2. ABS_DACC	**0.926**	1										
3. CER	0.054	0.005	1									
4. EN_INNOV	**0.015**	**0.016**	**0.728**	1								
5. EMM	**0.008**	**0.011**	**0.854**	**0.377**	1							
6. RES_USE	**−0.011**	**−0.011**	**0.863**	**0.405**	**0.710**	1						
7. SIZE	**0.135**	**0.148**	**−0.011**	**0.013**	−0.002	**−0.040**	1					
8. LEV	**−0.016**	**0.051**	0.006	**0.007**	0.001	**0.006**	**−0.046**	1				
9. MV_BV	0.001	0.001	−0.003	−0.007	0.001	−0.003	−0.021	−0.001	1			
10. ROA	**0.518**	**0.581**	**0.007**	**0.018**	**0.011**	**−0.011**	**0.127**	**−0.070**	−0.001	1		
11. R&D_EXP	**−0.028**	−0.016	−0.028	**−0.045**	−0.018	−0.005	**−0.025**	**−0.019**	−0.181	**−0.054**	1	
12. DCFO	**0.019**	**−0.018**	−0.006	−0.005	−0.003	−0.006	**0.048**	**−0.526**	0.001	**0.042**	**−0.024**	1

Note: Correlation coefficients in bold indicate statistical significance at least at the 5% significance level

Table 11.3 Income smoothing based on CER ranked portfolios

Spearman correlation	P1—High CER	P2—Medium CER	P3—Low CER
Coefficient	−0.685***	−0.723***	−0.913***
p-value	0.001	0.001	0.001

Note: *** indicate statistical significance at the 1% significance level

Table 11.4 Regression results of ABS_DACC on CER and its components

Variables	Model (11.5)	Model (11.5a)	Model (11.5b)	Model (11.5c)
Constant	−12.673*** (−3.33)	−12.120*** (−2.96)	−11.889*** (−2.82)	−14.065*** (−3.60)
CER	−0.022** (−2.40)			
EN_INNOV		−0.015* (−1.68)		
EMM			−0.014* (−1.71)	
RES_USE				−0.017** (−2.13)
SIZE	1.023*** (6.70)	0.953*** (5.62)	1.000*** (6.46)	1.067*** (6.68)
LEV	−1.087 (−0.98)	−0.657 (−0.63)	−1.405 (−1.08)	−1.047 (−0.94)
MV_BV	−0.006*** (−2.59)	−0.005** (−2.42)	−0.006** (−2.46)	−0.004*** (−2.91)
ROA	−1.035 (−1.16)	−0.648 (−0.62)	−1.112 (−1.22)	−1.339 (−1.57)
R&D_EXP	−0.645** (−2.16)	−0.607** (−2.10)	−0.655** (−2.17)	−0.622** (−2.01)
R^2-overall	0.722	0.721	0.698	0.706
Wald − x^2	6398.10***	5344.18***	1489.2***	2633.7***
p-value	0.001	0.001	0.001	0.001

Note: *, **, *** indicate statistical significance at the 10%, 5% and 1% significance level respectively, z-statistics in the parenthesis

11.4.2 Regression Results

Table 11.4 presents the regression results from the GLS random effect regression estimation of models (11.5, 11.5a, 11.5b and 11.5c) using the CER and its components as the main explanatory variables. The Wald Chi-square values are highly significant and the R^2s are very satisfactory considering the size of the sample and other relative studies published in the field. CER produced a negative and significant coefficient (−0.022) indicating that high CER performing firms are associated with lower discretionary accruals in absolute value. Moreover, CER components (environmental innovation, emissions and resource utilization) scores all produced negative and statistically significant coefficients supporting our arguments that firms investing more on environmental innovation, efficient resource utilization and emission reduction, published higher quality financial information by manipulating their earnings figures less. The finding verifies the negative association between CER and earnings management thus we can accept H1. Evidence in Table 11.4 corroborate arguments in the literature that high CER performing firms with enhanced monitoring by the markets, regulators, media and the society tend to be reporting higher quality accounting information. Our finding corroborate empirical evidence by Yu (2008), Litt et al. (2014), Heltzer (2011), Yoon et al. (2019), Kim et al. (2019) and Palacios-Manzano et al. (2019) suggesting that enhanced stakeholder oriented management (including social and environmental initiatives) serves as a substitute governance mechanism which helps to mitigate earnings management behavior.

Regarding the rest of the control variables, larger firms proved to manipulate the earnings in a higher extent (higher ABS_DACC) since SIZE produced a positive and highly significant coefficient in all models estimation, suggesting that large firms may resort to higher earnings management due to more instability of their operations thus are more motivated to smooth their earnings (Palacios-Manzano et al. 2019). MV_BV produced a negative and significant coefficient in all models indicating that higher growth opportunities within a firm are associated with less managerial discretion on reporting accounting earnings. This result contradicts previous evidence in the literature that higher growth firms are more likely to manipulate earnings (Lee et al. 2006). This result could be attributed to the fact that high growth firms with enhanced market valuation tend to be more scrutinized by regulators and investors and this may motivate them to sustain high quality financial reports and thus exercise accrual earnings manipulation less often. Finally, the coefficient on R&D expenses also provided a negative and significant coefficient verifying the findings by Yoon et al. (2019) in the Korean market who documented that high R&D intensity firms were associated with higher earnings quality.

Table 11.5 presents the regression results from the GLS random effect regression estimation of models (11.6, 11.6a, 11.6b and 11.6c) using the CER and its components as the main explanatory variables. The Wald Chi-square values are highly significant and the R^2s are very satisfactory considering the size of the sample and other relative studies published in the field. CER produced a positive and significant coefficient (0.015) as expected indicating that high CER performing firms are associated with higher discretionary cash flows and thus higher accounting quality. In addition, CER components (environmental innovation, emissions and resource utilization) scores all produced positive and statistically significant coefficients supporting our arguments that firms investing more on environmental innovation, efficient resource utilization and emission reduction, published higher quality earnings via less real activities and accruals manipulation. The findings verify arguments in the literature by Yoon et al. (2019), Kim et al. (2019) and thus corroborate evidence on Table 11.4 and thus give support to H1. Referring to the control variables only LEV and ROA produced positive and significant coefficients suggesting that highly profitable and leveraged firms manipulate less their earnings through real activities. This result is quite logical since real earnings management has consequences on firms' financial numbers (cash and other receivables) and so

Table 11.5 Regression results of DCFO on CER and its components

Variables	Model (11.6)	Model (11.6a)	Model (11.6b)	Model (11.6c)
Constant	−0.004 (−0.35)	−0.001 (−0.13)	−0.012 (−0.73)	0.004 (0.06)
CER	0.015** (2.36)			
EN_INNOV		0.011** (2.02)		
EMM			0.013** (2.51)	
RES_USE				0.011** (2.10)
SIZE	−0.004 (−0.82)	−0.001 (−0.41)	−0.002 (−0.58)	−0.001 (−1.17)
LEV	0.014*** (3.36)	0.011** (2.28)	0.019*** (4.30)	0.014*** (3.31)
MV_BV	−0.015 (−0.03)	−0.004 (−0.08)	0.013 (0.03)	0.002 (0.04)
ROA	0.022*** (4.00)	0.019*** (3.50)	0.024*** (4.44)	0.022*** (4.25)
R&D_EXP	0.128 (0.37)	0.107 (0.31)	0.165 (0.53)	0.122 (0.41)
R^2-overall	0.524	0.511	0.509	0.488
Wald – x^2	471.22***	443.68	487.14***	353.47***
p-value	0.001	0.001	0.001	0.001

Note: *, **, *** indicate statistical significance at the 10%, 5% and 1% significance level respectively, z-statistics in the parenthesis

profitable and highly leveraged firms cannot risk to mask their true economic performance by affecting their ability to serve their debt obligations.

11.4.3 Sensitivity Analysis

In order to control for the robustness of the main empirical findings we conducted several sensitivity tests for corroborating evidence presented on Tables 11.4 and 11.5. At first, we replaced the absolute value of discretionary accruals (ABS_DACC) with the signed discretionary accruals and re-estimated model (11.5). Results are presented on Table 11.6 and yield similar inferences compared to Table 11.4. Specifically, CER coefficient is still negative and significant indicating that high CER performing firms report less DACC. We received similar results when we re-estimated models (11.5a), (11.5b) and (11.5c) using DACC instead of ABS_DACC, thus H1 is again verified. In addition, models (11.5) and (11.6) were re-estimated by including all CER components in the models simultaneously. The results are presented in Table 11.7 and again provide strong support on the results on

Table 11.6 Regression results of DACC on CER (model 11.5)

Variables	Coefficient	z-test	p-value
Constant	−18.545***	−3.48	0.001
CER	−0.031**	−2.50	0.013
SIZE	1.156***	5.62	0.001
LEV	0.298	0.20	0.841
MV_BV	−0.004*	−1.67	0.096
ROA	0.012	0.01	0.993
R&D_EXP	−0.505	−1.11	0.265
R^2-overall	0.703		
Wald – x^2	1288.99***		
p-value	0.001		

Note: *, **, *** indicate statistical significance at the 10%, 5% and 1% significance level respectively

Table 11.7 Regression results of DACC and DCFO on CER components

Variables	Model (11.5)—ABS_DACC	Model (11.6)—DCFO
Constant	−19.382*** (−3.10)	0.005 (0.35)
EN_INNOV	−0.019** (−2.29)	0.010** (2.52)
EMM	−0.008* (−1.69)	0.018* (1.93)
RES_USE	−0.021** (−2.35)	0.003** (2.06)
SIZE	1.174*** (4.74)	0.001 (0.05)
LEV	0.937 (0.61)	0.005 (1.14)
MV_BV	−0.004* (−1.72)	−0.003 (−0.36)
ROA	0.047 (0.03)	0.016*** (4.46)
R&D_EXP	−0.454 (−0.97)	0.149 (0.91)
R^2-overall	0.713	0.467
Wald – x^2	7277.19***	323.72***
p-value	0.001	0.001

Note: *, **, *** indicate statistical significance at the 10%, 5% and 1% significance level respectively, z-statistics in the parenthesis

both Tables 11.4 and 11.5. In detail, all CER components produced negative and significant coefficients for the ABS_DACC dependent variable and positive and significant coefficient for the DCFO dependent variable, providing additional support to H1.

Furthermore, we performed additional sensitivity testing by estimating the classic Jones (1991) model and the Kasznik (1999) model for estimating discretionary accruals (Palacios-Manzano et al. 2019). Evidence (untabulated) produced qualitatively similar results with those on Tables 11.4 and 11.5. Also, following Heltzer (2011) we proxy for income smoothing by directly estimating the Spearman correlation between the change in accruals and change in operating cash flows instead of the residuals from models (11.3) and (11.4). Again the results did not change relative to those presented on Table 11.3 and high CER firms had smaller negative correlation relative to low CER firms. Finally, we employed an instrumental variables approach (2SLS estimation) in order to control for endogeneity as in Palacios-Manzano et al. (2019). For this reason we treated the CER variable and its components as endogenous and employed a 1 and 2 year lags as instruments. Again the results were consistent with those on Tables 11.4 and 11.5.

11.5 Conclusion

An extensive number of research studies have considered the impact of CER activities and performance on the quality of financial information published by firms and specifically the level of managerial discretion on reporting accounting numbers. Empirical evidence so far have provided contradictory evidence on the impact of CER on earnings management with some studies pointing to a positive association while others provide evidence of a negative association (Prior et al. 2008; Buertey et al. 2019; Heltzer 2011; Litt et al. 2014; Yoon et al. 2019; Kim et al. 2019; Palacios-Manzano et al. 2019). The scope of this chapter was to provide more concrete evidence on the impact of CER performance and its components on the earnings management behavior by utilizing a multi-country research setting and incorporating various sub-categories of environmental performance and earnings management measures evidenced in the accounting literature such as accrual earnings manipulation, real activities manipulation and income smoothing. The current chapter provides more thorough evidence in the existing literature regarding the impact of CER on the quality of accounting numbers. The empirical results verified evidence and arguments in the literature that CER performance improves the quality of published accounting information by mitigating earnings management behavior via accruals and real activities manipulation. This argument is corroborated after performing several sensitivity tests relating to the estimation of discretionary accruals, the functional form of the models and other robustness tests.

References

Abarbanell, J., and R. Lehavy. 2003. Biased forecasts or biased earnings? The role of reported earnings in explaining apparent bias and over/underreaction in analysts' earnings forecasts. *Journal of Accounting and Economics* 36: 105–146.

Barth, M.E., W.R. Landsman, and M.H. Lang. 2008. International accounting standards and accounting quality. *Journal of Accounting Research* 43: 467–498.

Bergstresser, D., and T. Philippon. 2006. CEO incentives and earnings management. *Journal of Financial Economics* 80: 511–530.

Billings, B.K. 1999. Revisiting the relation between the default risk of debt and the earnings response coefficient. *The Accounting Review* 74: 509–522.

Buertey, S., E.-J. Sun, J.S. Lee, and J. Hwang. 2019. Corporate social responsibility and earnings management: The moderating effect of corporate governance mechanisms. *Corporate Social Responsibility and Environmental Management*. https://doi.org/10.1002/csr.1803.

CFA. 2019. Accounting quality. Accessed September 21, 2019, from https://www.cfainstitute.org/en/advocacy/issues/accounting-quality

Chih, H.L., C.H. Shen, and F.C. Kang. 2008. Corporate social responsibility, investor protection and earnings management: Some international evidence. *Journal of Business Ethics* 79: 179–198.

Cohen, D.A., and P. Zarowin. 2010. Accrual-based and real earnings management activities around seasoned equity offerings. *Journal of Accounting and Economics* 50: 2–19.

Dechow, P., R. Sloan, and A. Sweeney. 1996. Causes and consequences of earnings manipulation: An analysis of firms subject to enforcement actions by the SEC. *Contemporary Accounting Research* 13: 1–36.

Dimitropoulos, P. 2011. Corporate governance and earnings management in the European football industry. *European Sport Management Quarterly* 11 (5): 495–523.

Dimitropoulos, P., and D. Asteriou. 2010. The effect of board composition on the informativeness and quality of annual earnings: Empirical evidence from Greece. *Research in International Business and Finance* 25: 190–205.

Gaver, J.J., and K.M. Gaver. 1993. Additional evidence on the association between the investment opportunity set and corporate financing, dividend and compensation policies. *Journal of Accounting and Economics* 16: 125–160.

Heltzer, W. 2011. The asymmetric relationship between corporate environmental responsibility and earnings management: Evidence from the United States. *Managerial Auditing Journal* 26 (1): 65–88.

Ittner, C.D., R.A. Lambert, and D.F. Larcker. 2002. The structure and performance consequences of equity grants to employees of new economy firms. *Journal of Accounting and Economics* 34: 89–127.

Jones, J. 1991. Earnings management during import relief investigations. *Journal of Accounting Research* 29: 193–228.

Kasznik, R. 1999. On the association between voluntary disclosure and earnings management. *Journal of Accounting Research* 37 (1): 57–81.

Katmon, N., and O.A. Farooque. 2017. Exploring the impact of internal corporate governance on the relation between disclosure quality and earnings management in the UK listed companies. *Journal of Business* 142 (2): 345–367.

Kim, Y., M.S. Park, and B. Weir. 2012. Is earnings quality associated with corporate social responsibility? *The Accounting Review* 87 (3): 761–796.

Kim, S.H., P. Udawatte, and J. Yin. 2019. The effects of corporate social responsibility on real and accrual-based earnings management: Evidence from China. *Australian Accounting Review* 23 (3): 580–594.

Klein, A. 2002. Audit committee, board of director characteristics, and earnings management. *Journal of Accounting and Economics* 33: 375–400.

Kothari, S.P., A.J. Leone, and C.E. Wasley. 2005. Performance matched discretionary accrual measures. *Journal of Accounting and Economics* 39: 163–197.

Lee, C.J., L.Y. Li, and H. Yue. 2006. Performance, growth and earnings management. *Review of Accounting Studies* 11 (2): 305–334.

Leuz, C., D. Nanda, and P.D. Wysocky. 2003. Earnings management and investor protection: An international comparison. *Journal of Financial Economics* 69: 505–527.

Litt, B., D. Sharma, and V. Sharma. 2014. Environmental initiatives and earnings management. *Managerial Auditing Journal* 29 (1): 76–106.

Niu, F.F. 2006. Corporate governance and the quality of accounting earnings: A Canadian perspective. *International Journal of Managerial Finance* 2 (4): 302–327.

Obigbemi, I.F., E.O. Omolehinwa, D.O. Mukoro, E. Ben-Caleb, and O.A. Olusanmi. 2016. Earnings management and board structure: Evidence from Nigeria. *SAGE Open* 6 (3): 1–15.

Palacios-Manzano, M., E. Gras-Gil, and J.M. Santos-Jaen. 2019. Corporate social responsibility and its effect on earnings management: An empirical research on Spanish firms. *Total Quality Management*. https://doi.org/10.1080/14783363.2019.1652586.

Prior, D., J. Surroca, and J.A. Tribo. 2008. Are socially responsible managers really ethical? Exploring the relationship between earnings management and corporate social responsibility. *Corporate Governance: An International Review* 16 (3): 160–177.

Reynolds, J.K., and J.R. Francis. 2000. Does size matter? The influence of large clients on office-level auditor reporting decisions. *Journal of Accounting and Economics* 30: 375–400.

Roychowdhury, S. 2006. Earnings management through real activities manipulation. *Journal of Accounting and Economics* 42 (3): 335–370.

Van Tendeloo, B., and A. Vanstrelen. 2005. Earnings management under German GAAP versus IFRS. *European Accounting Review* 14: 155–180.

Wang, D. 2006. Founding family ownership and earnings quality. *Journal of Accounting Research* 44 (3): 619–656.

Warfield, T., J. Wild, and K. Wild. 1995. Managerial ownership, accounting choices, and informativeness of earnings. *Journal of Accounting and Economics* 20 (1): 61–91.

Watts, R., and J. Zimmerman. 1990. Positive accounting theory: A ten-year perspective. *The Accounting Review* 65: 131–156.

Xie, B., W.N. Davidson, and P.J. DaDalt. 2003. Earnings management and corporate governance: The role of the board and the audit committee. *Journal of Corporate Finance* 9 (3): 295–316.

Yoon, B., B. Kim, and J.H. Lee. 2019. Is earnings quality associated with corporate social responsibility? Evidence from the Korean market. *Sustainability* 11: 4116.

Yu, F. 2008. Analyst coverage and earnings management. *Journal of Financial Economics* 88: 245–271.

Chapter 12
Corporate Environmental Responsibility and Accounting Conservatism

The emergence of the stakeholder theory, both within academia and business practice, has created a movement where corporate environmental reporting (CER) has been in the forefront of business strategy in order to create mutual benefits for the firms and society at large. CER could be used by managers with different incentives, where some may utilize CER opportunistically for private benefits or to mask the true financial situation of the firm. On the contrary, some managers may use CER based on ethical and philanthropic reasons. Either way, CER performance is expected to be associated with the quality of financial reporting. In the literature accounting conservatism is considered as the differential verifiability required for the recognition of profits versus losses, within the limits allows by Generally Accepted Accounting Principles (GAAP). Several research studies examining the relationship between corporate environmental responsibilities (CER) and accounting conservatism has reached contradictory evidence due to the fact that they incorporate different measures of CER, are focused on a single country setting and utilize only limited methodologies for measuring conditional conservatism. The purpose of this chapter is to study the impact of CER related activities on conservative accounting reporting using a multi-country setting and examining different research frameworks for estimating conservatism.

12.1 Introduction

As discussed in the previous chapters, the quality of accounting information is largely determined by several facets such as persistence, predictability and aggressiveness, measured via the level of manager's discretion on the reporting of accounting numbers like earnings manipulation and accounting conservatism. As argued by Obigbemi et al. (2016), accounting information acts as the signaling mechanism which provides acute signals to investors and market participants and so has a

P. Dimitropoulos, K. Koronios, *Corporate Environmental Responsibility, Accounting and Corporate Finance in the EU*, CSR, Sustainability, Ethics & Governance, https://doi.org/10.1007/978-3-030-72773-4_12

significant impact on the decision-making behavior of stakeholders and the stability of the capital market overall. Reliable and transparent financial reporting allows several stakeholder groups associated with any firm to make informed assessments about the firm and adjust their behavior accordingly. Stakeholder theory posits that managers have the responsibility to sustain an equitable distribution of wealth among the main stakeholders. This fact requires ethical judgements on behalf of the managers, especially in the presentation and dissemination of financial accounting information. Accounting conservatism plays a significant role on this issue since it can create the proper foundation for coordinating and satisfying the firms' multiple stakeholders interests (Francis et al. 2013).

According to Watts (2003), accounting conservatism is defined as the differential verifiability recognition of profits versus losses. Put it differently, conservatism is based on the merit that expenses are recognized faster (more timely) in financial statements relative to revenues, leading to lower net assets values and cumulative net income (Francis et al. 2013). Conservatism, as a basic accounting principle internationally, presents some advantages to contracting parties. Since conservatism requires more strict verifiability for revenues relative to expenses, it restricts the ability of managers to inflate earnings and revenues, thus does not allow them to divert wealth from the stakeholders to themselves (Leventis et al. 2013; Manganaris et al. 2017; Francis et al. 2013). Moreover, under conservative accounting reporting it is easier for shareholders and stakeholders to exert efforts towards managers to invest in positive NPV projects which ultimately benefits all stakeholder groups. Also, managers by exercising conditional accounting reporting they avoid the costs associated with litigation and tax payments. Litigation costs increase in cases were assets are overstated and evidence by Shackelford and Shevlin (2001) document that conservatism is associated with less tax payments. So empirical and theoretical evidence so far suggest that accounting conservatism contributes to firm value preservation and the reduction of the likelihood of wealth distributions to one stakeholder group at the expense of others.

Under this framework, extant research studies have considered the impact of CER related activities on the quality of financial information published by firms and specifically accounting conservatism. According to Burke et al. (2020) and Cho et al. (2020), accounting conservatism could have a negative association with CER performance for two reasons. The first is that CER signals to the market the commitment of firms' to incorporate the various stakeholders' need to business operations and even develop long-term contracting relationships with those groups. This fact enhances trust and reliability that the firm will not engage in managerial opportunism, thus mitigating related parties concerns (Lins et al. 2017). Secondly, CER activities improve firms' information environment by reducing informational asymmetries between the firm and outside stakeholders (creditors, lenders etc.) leading to less demand for conservative financial reporting (Cui et al. 2016; LaFond and Watts 2008; Burke et al. 2020). On the contrary, Francis et al. (2013) argue that CER performance could have a positive association with accounting conservatism. As conservatism improves the equitable wealth distribution between various stakeholder groups and via timely recognition and identification of negative NPV projects, firm value is preserved and protected. So under conservative financial reporting firms have the required room to mitigate conflicts of interests between stakeholders and so they are more incentivized to invest more on CER related activities. According to Cheng and Kung (2016), CER can play a complementary role for conservatism to actively improve the transparency of accounting information, so both CER and conservatism are expected to be positively associated.

Empirical evidence so far have provided contradictory evidence on the impact of CER on accounting conservatism. Burke et al. (2020), Cho et al. (2020), Hong (2020), Yang et al. (2014) and Patro and Pattanayak (2017) provide evidence of a negative association between CER and conservatism supporting the notion that high CER performing firms are more transparent and trustworthy, thus stakeholders demand less conservative accounting reports. On the contrary, Cheng and Kung (2016), Francis et al. (2013), Hong and Andersen (2011) and Pyo and Lee (2013) all point to a positive association between conservatism and CER activities on. The positive association is justified by the fact that high CER performance complements firms' governance and financial reporting practices so high CER firms could be perceived as less trustworthy by external stakeholders, who require more timely signals on firm's financial condition, thus increasing the demand for conditional conservatism. The conflicting results in the literature are explained by the fact that CER performance is a multi-faceted factor that is comprised by many categories of activities and investments, each one with different association with conservatism and because motives for conservative reporting may differ between countries and business settings.

The scope of this chapter is to shed further light on the impact of CER performance and its components on the level of conservative financial reporting by utilizing a multi-country research setting and incorporating various sub-categories of environmental performance and conservatism metrics including the verification of negative earnings, negative cash flows and a firm-specific measure of conservatism (the C-score proposed by Khan and Watts (2009)), thus providing more thorough evidence in the existing literature regarding the impact of CER on the conservatism of accounting numbers with the EU. The empirical analysis in this chapter corroborates previous arguments in the literature on the negative association between CER and conservatism. In the following section we provide the theoretical background on the impact of CER on conditional conservatism and the testable hypothesis. The third section discusses the data selection procedure and the research design. The fourth section provides the discussion of empirical results and the last section concludes the chapter.

12.2 CER-Conservatism Association and Testable Hypothesis

The association between CER and accounting conservatism has been examined and viewed through various theoretical lenses and hypotheses. According to Burke et al. (2020), any firm could be perceived as a nexus of contracting relationship between various parties and stakeholders. Those parties institute mechanisms in order to create the proper environment for managerial control and supervision. One such mechanism is accounting conservatism which reduces managerial abilities and incentives to engage in opportunistic behavior, by recognizing losses timelier than profits, thus facilitates managerial monitoring by the contracting parties of the firm (Watts 2003). Also, conservatism plays a significant informational role in the markets. Outside investors tend to require conservative financial reports in order to reduce any information asymmetries they have relative to firm insiders and specifically managers. This fact allows the flow of "softer" information in the capital marker, reducing deadweight losses and increasing firm value (LaFond and Watts 2008; Burke et al. 2020; Cho et al. 2020).

Theoretical underpinnings on the literature propose two different potential associations between CER and conservatism. The first proposal asserts that CER and conservatism are negatively associated. The reasons for that is first, CER related firms can signal lower levels of managerial opportunism to the markets and so the demand for conditional conservatism from contracting parties (stakeholders) to mitigate managerial discretion may be lower. As documented by Burke et al. (2020), Cho et al. (2020) and Hong (2020), contracting parties' conditional conservatism demands differs with the level of managerial opportunism. Empirical evidence in Chaps. 10 and 11 suggest that highly CER performing firm are associated with higher market valuation, more earnings persistence and predictability and lower earnings manipulation. This is attributed to the fact that highly CER performing firms are more likely to consider stakeholder needs (including needs for transparent financial information) and create long lasting relations with contracting parties (Lins et al. 2017; Flammer and Bansal 2017). In other words, being environmentally responsible a firm tends to uphold and fulfil implicit or explicit obligations with its contracting parties, thus contradicting any opportunistic behavior that can hurt those long lasting relations. So superior CER performance reflects enhanced stakeholder management and less short-term opportunistic behavior on behalf of managers, reducing contracting costs. Thus all else equal, CER performing firms are expected to be negatively associated with conditional conservatism. Another reason that complements that argument is the fact that CER related firms are associated with enhanced information dissemination to the public. CER related firms tend to disclose more information (Ballou et al. 2018; Ioannou and Serafeim 2017) which enhance value relevance of other financial information. Also, Cui et al. (2016) document that CER performing firms engage in a dynamic information exchange with markets and key stakeholders reducing any information asymmetries between them, leading to lower demand for conservative financial reports. Consequently, CER performance reduces the demand for conservative financial reporting by reducing stakeholders' concerns relative to managerial opportunism and by minimizing information asymmetries in the markets. Empirical evidence by Burke et al. (2020), Cho et al. (2020), Hong (2020), Yang et al. (2014) and Patro and Pattanayak (2017) support these theoretical arguments by empirically documenting a negative association between CER and conditional conservatism.

The second theoretical underpinning posits a positive association between CER and accounting conservatism. According to Fombrun and Shanley (1990), Kim et al. (2012) and Cheng and Kung (2016), the provision of transparent and reliable financial information is perceived as a social responsibility of corporations towards their stakeholders. The provision of reliable and transparent financial information allows stakeholders to effectively assess future cash flows, operating risks and ultimately the true economic condition and value of the firm. This in turn leads to informative and more efficient investment decisions with reduced information risk. Reduced information risk in the market mitigates litigation, supervisory and other social costs (Cheng and Kung 2016). Accounting conservatism is a mechanism which also can increase the precision of financial information, reduce information asymmetries and improve contracting efficiency between firms and stakeholders (Francis et al. 2013; Pyo and Lee 2013) through minimization of managerial opportunism and expropriation of corporate wealth. Since accounting conservatism sustains organizational resources, CER could play a moderating role for governance mechanisms which enhance financial transparency and reliability (Cheng and Kung 2016). Thus conservative firms could be motivated to invest more on CER related activities (due to their enhanced utilization and protection of corporate resources), so a positive association is proposed by this stream of literature. Empirical evidence by Pyo and Lee (2013), Cheng and Kung (2016). Francis et al. (2013) support the above-mentioned arguments within different markets. Therefore, following the abovementioned discussion it is expected that enhanced CER performance will be associated with accounting conservatism. So the main research hypotheses are stated as follows:

H1a: CER performance will have a negative impact on accounting conservatism.
H1b: CER performance will have a positive impact on accounting conservatism.

12.3 Data Selection and Research Design

12.3.1 Data Selection Procedure

The current chapter utilizes a sample of corporations (both listed and unlisted) from 23 EU-member countries over the period 2003–2018. All financial, governance and environmental responsibility data have been extracted from Datastream database. The sample collection procedure started by including all corporations that have been covered by the environmental, social and governance (ESG) scores provided by

Datastream during the sample period. 2003 has been selected as the initial sample year since that year was the first year with significant coverage (taking into consideration that this database officially was initiated in 2002). From a total of 656,734 firm-year observations we excluded those firms without environmental score data for at least five consecutive years in order to avoid bias in the empirical results and also to facilitate the estimation of differenced variables. Furthermore, we excluded firms with incomplete financial data on operating cash flows, revenues, accounts receivables, fixed assets and those that do not close their fiscal year on December. In addition, we winsorized the upper and down 1 per cent of the data distribution in order to mitigate the influence of significant outliers in the empirical analysis. After this procedure was finalized, we ended up with a final sample of 101,212 firm-year observations.

12.3.2 Research Design

In order to examine the main research hypotheses we utilized various measures of accounting conservatism pointed in the literature. The most commonly used measure of conditional conservatism is the model proposed by Ball and Shivakumar (2005, 2006) where they indicated that accruals could be utilized to capture potential economic gains and losses. The term "accruals" originates from the accrual basis of accounting which dictates that all expenses and revenues that have been incurred but not yet paid or received must be reported within the financial statements and are estimated as the difference between net income and operating cash flows (Dimitropoulos et al. 2012). Dechow (1994) asserted that both cash flows and accruals are information—sensitive, and in the case that cash flows are unaffected by accruals she assumed that under conservative financial reporting there is enhanced probability to recognize economic losses faster than profits (Cheng and Kung 2016). Thus negative operating cash flows increase the possibility of recognizing losses. This assumption is operationalized in the following cross-sectional OLS equation:

$$ACC_{it} = a_0 + \alpha_1 DCFO_{it} + a_2 CFO_{it} + a_3 DCFO_{it} * CFO_{it} + e_{it} \tag{12.1}$$

Where

ACC is total accruals defined as the difference between net income and operating cash flows divided by lagged total assets.

CFO is the ratio of operating cash flows to lagged total assets

DCFO is a dummy variable receiving unity (1) if CFO is negative and (0) otherwise.

As Ball and Shivakumar (2005) argue, if coefficient a_3 is positive and significant then the firm is recognizing negative operating cash flows faster than positive cash flows, so the firm is characterized by increased conservatism. Coefficient a_2 is expected to be negative and significant designating the negative associated between

CFO and accruals (Dechow 1994). In order to examine the impact of CER performance on the level of conditional conservatism we extended model (12.1) by incorporating CER performance and its main components (environmental innovation score, emissions score and resource use score) as follows where i denotes the firm and t the year and e is the error term:

$$\begin{aligned} ACC_{it-1} = {} & a_0 + \alpha_1 DCFO_{it} + a_2 CFO_{it-1} + a_3 DCFO_{it} * CFO_{it} \\ & + a_4 CER_{it} + a_5 CER_{it} * DCFO_{it} + a_6 CER_{it} * CFO_{it} \\ & + a_7 CER_{it} * DCFO_{it} * CFO_{it} + Year\ F.E + Industry\ F.E. \\ & + Country\ F.E. + e_{it} \end{aligned} \tag{12.2a}$$

$$\begin{aligned} ACC_{it-1} = {} & a_0 + \alpha_1 DCFO_{it} + a_2 CFO_{it-1} + a_3 DCFO_{it} * CFO_{it} \\ & + a_4 EN\_INNOV_{it} + a_5 EN\_INNOV_{it} * DCFO_{it} \\ & + a_6 EN\_INNOV_{it} * CFO_{it} + a_7 EN\_INNOV_{it} * DCFO_{it} \\ & * CFO_{it} + Year\ F.E + Industry\ F.E. + Country\ F.E. + e_{it} \end{aligned} \tag{12.2b}$$

$$\begin{aligned} ACC_{it-1} = {} & a_0 + \alpha_1 DCFO_{it} + a_2 CFO_{it-1} + a_3 DCFO_{it} * CFO_{it} \\ & + a_4 EMM_{it} + a_5 EMM_{it} * DCFO_{it} + a_6 EMM_{it} * CFO_{it} \\ & + a_7 EMM_{it} * DCFO_{it} * CFO_{it} + Year\ F.E + Industry\ F.E. \\ & + Country\ F.E. + e_{it} \end{aligned} \tag{12.2c}$$

$$\begin{aligned} ACC_{it-1} = {} & a_0 + \alpha_1 DCFO_{it} + a_2 CFO_{it-1} + a_3 DCFO_{it} * CFO_{it} \\ & + a_4 RES\_USE_{it} + a_5 RES\_USE_{it} * DCFO_{it} + a_6 RES\_USE_{it} \\ & * CFO_{it} + a_7 RES\_USE_{t} * DCFO_{it} * CFO_{it} + Year\ F.E \\ & + Industry\ F.E. + Country\ F.E. + e_{it} \end{aligned} \tag{12.2d}$$

According to Cheng and Kung (2016) and Hong (2020), if coefficients a_7 is positive and significant this will suggest that firms with enhanced CER performance report more conservatively so H1b will be valid. On the contrary if coefficients a_7 are found negative and significant this will lead us to accept H1a. All models will be estimated with cross-sectional OLS with robust standard errors in order to control for heteroscedasticity. EN_INNOV denotes each company's environmental innovation score, which reflects the firm's ability to mitigate the environmental costs towards its customers or even creating new market and product opportunities via enhanced environmental technologies and processes. EMM denotes companies' emission reduction score estimated via their efforts and effectiveness in reducing emissions from their production and operation. RES_USE denotes the firms' resource use score and indicates firms' ability to reduce the usage of hazardous materials, energy, water and other valuable natural resources. Finally, CER is the firms' annual average of the aforementioned three scores. All environmental scores range between zero (0) and 100 so the higher the score the more environmentally responsible and efficient is the firm.

Furthermore, in order to reach into more decisive conclusions on the association between conditional conservatism and CER, and avoid econometric problems from model misspecification, we followed Hong (2020) and applied the Basu (1997) model which captures conditional conservatism as the reaction of firm's current earnings changes relative to past negative earnings changes. Conservative accounting is larger if any reversals of changes in net losses during the previous year are larger than the reversals of changes in net profits. This model is operationalized as follows:

$$\Delta NI_{it} = \beta_0 + \beta_1 D\Delta NI_{it-1} + \beta_2 \Delta NI_{it-1} + \beta_3 D\Delta NI_{it-1} * \Delta NI_{it-1} + u_{it} \quad (12.3)$$

Where:

ΔNI is the current and lagged annual difference of net income before taxes and extraordinary items to lagged total assets.

DΔNI is a dummy variable receiving unity (1) if ΔNI is negative and (0) otherwise.

As Basu (1997) and Hong (2020) argue, if the absolute value of coefficient β_3 is small and significant then the firm is recognizing past negative earnings changes faster than positive earnings changes in the current year's earnings, so the firm is characterized by increased conservatism. In order to examine the validity of the research hypotheses we will estimate model (12.3) after separating the sample firms into three groups based on the annual CER score and estimate it for each CER group. So following Suto and Takehara (2018) we will construct three portfolios based on the average of the CER variable per year and country and assigned each firm as low CER firm if its annual score is below the 1st quartile, high CER firms are those that their CER score is above the 3rd quartile while all other firm observations lying between the 1st and 3rd quartiles are assigned as medium CER firms. Thus, if H1a is valid we expect coefficient β_3 to decrease as we move from the low CER group of firms to the high CER performing group. On the contrary if coefficient β_3 increases as we move from the low to the high CER groups then H1b will be accepted.

12.4 Empirical Results

12.4.1 Descriptive Statistics, Correlations and Income Smoothing

Table 12.1 presents the descriptive statistics of the variables utilized in the regression models. As we can see, total accruals have a positive average (21.88) while the average CFO is positive comprising the 3 per cent of total assets indicating that our sample firms report higher earnings (accruals) relative to cash flows. This findings on ACCRUALS and CFO averages are higher relative to evidence provided by Hong (2020) in the Taiwanese capital market. Moreover, DCFO has an average almost up to 0.18 indicating that 18 per cent of the sample firms reported negative CFOs. Also, the mean CER is 64.5 (with a standard deviation of 21.1) indicating that the sample

Table 12.1 Descriptive statistics of the sample variables

Variables	Mean	Standard deviation	Min	Max
ACCRUALS	21.886	48.812	−24.918	174.17
CFO	0.030	5.042	−294.25	389.94
DCFO	0.178	0.382	0	1
CER	64.597	21.167	3.161	99.429
EN_INNOV	61.333	25.962	0.202	99.919
EMM	65.915	26.189	0.161	99.919
RES_USE	66.542	25.730	0.805	99.844
ΔNI	38.589	10.45	−23.007	284.07
D_ΔNI	0.504	0.499	0	1
GROWTH	58.915	141.720	−69.711	409.071
LEV	4.627	4.319	−0.682	25.691
PPE	0.596	0.245	0.001	0.999

firms have a satisfactory environmental performance score. The highest average score is that of resources usage (66.5) followed by emission reduction activities (65.9) while environmental innovation activities lag behind with an average of 61.3. Sample firms presented an average positive change of net income (38.5) but almost half of them (0.50) experienced negative changes in earnings. The positive change in earnings is actually justified by a positive average change in revenues (GROWTH average is up to 58.91). However, our sample firms are highly leveraged since the average total debt covers more than four times the sample firms' total assets and fixed assets comprise almost the 60 per cent of total assets, an average which is again a lot higher relative to the asset tangibility of Taiwanese firms (Hong 2020).

Table 12.2 presents the Pearson correlation coefficients between the sample variables. The majority of the correlation coefficients are significant with economic meaning, and are not very high indicating the absence of multicollinearity on the data. The VIF estimation of the sample variables provided values that were far below the threshold of 10 indicating no multicollinearity. As we can see, CER is significantly and positively associated with total accruals something which indicates that environmentally responsible firms report higher earnings relative to cash flows, yet the association is very small. However, CER is negatively associated to DCFO indicating that CER performing firms are associated with lower negative CFOs. Total accruals are also positively correlated with environmental innovation and emissions scores and negatively correlated with resource usage score. This evidence suggest that different environmental performance variables create different financial reporting motives on the sample firms. This fact is also evidenced by the positive correlation between D_ΔNI with RES_USE (0.011) suggesting that firms with higher RES_USE scores are associated with more negative earnings changes. This outcome justifies the use of decomposed environmental performance indicators instead of using an aggregate measure of CER. Of course, correlations do not indicate causality or direction of association so we will proceed with the regression

Table 12.2 Pearson correlation coefficients of sample variables

Variables	1	2	3	4	5	6	7	8	9
1. CER	1								
2. EN_INNOV	**0.728**	1							
3. EMM	**0.854**	**0.377**	1						
4. RES_USE	**0.863**	**0.405**	**0.710**	1					
5. ACCRUALS	**0.008**	**0.018**	**0.011**	**−0.011**	1				
6. CFO	−0.002	0.001	−0.002	−0.005	**0.052**	1			
7. DCFO	**−0.011**	**−0.013**	**−0.011**	−0.002	**−0.024**	**−0.023**	1		
8. ΔNI	−0.001	−0.001	0.006	−0.001	**0.045**	**−0.004**	−0.002	1	
9. D_ΔNI	0.004	0.002	−0.002	**0.011**	**−0.036**	−0.001	**0.151**	**−0.019**	1

Note: Correlation coefficients in bold indicate statistical significance at least at the 5% significance level

analysis results in order to extract more specific inferences regarding the validity of our research hypotheses.

12.4.2 Regression Results

Table 12.3 presents the regression results from the OLS cross-sectional regression estimation of models (12.2a, 12.2b, 12.2c and 12.2d) using the CER and its components as the main explanatory variables. The F-stat values are highly significant and the R^2s are very satisfactory considering the size and heterogeneity of the sample and other relative studies published in the field. At first, coefficient a_3 (DCFO*CFO) is positive and statistically significant in all model estimations indicating that our sample firms utilize conservative accounting practices since they recognize negative operating cash flows faster than positive cash flows. Moreover, coefficient a_2 is negative and highly significant as expected, designating the negative associated between operating cash flows and accruals (Dechow 1994). The main variable of interest in the three-way interaction variables between CFO, DCFO and environmental performance scores. Specifically, coefficient a_7 in model (12.2a) was negative and statistically significant (−2.289) indicating that high CER performing firms are associated with less conservative financial reporting. Similar coefficients were estimated on the three-way interaction variables with EN_INNOV, EMM and RES_USE (−2.753, −2.098 and −1.692 respectively) indicating that all components of CER performance contribute to lower conditional conservatism.

The evidence on Table 12.3 provide support to H1a and the stream of literature documenting a negative association between CER and accounting conservatism. So our study complements arguments by Cui et al. (2016), Burke et al. (2020), Cho et al. (2020), Hong (2020), Yang et al. (2014) and Patro and Pattanayak (2017) who document that CER performing firms engage in a dynamic information exchange with markets and key stakeholders reducing any information asymmetries between them, leading to lower demand for conservative financial reports. Consequently, CER performance reduces the demand for conservative financial reporting by reducing stakeholders' concerns relative to managerial opportunism and by minimizing information asymmetries in the markets.

Table 12.4, presents the empirical results from the estimation of model (12.3) (following the studies by Basu (1997) and Hong (2020), after separating the sample firms into three groups based on the annual CER score and estimate it for each CER group (Suto and Takehara 2018). As we can see, coefficient β_3 (ΔNI*DΔNI) is positive and significant in all groups but as we move from the low-CER group to the high-CER group of firms the coefficient is increasing in size and statistical significance. This results suggests that high CER performing firms are recognizing past negative earnings changes slower than positive earnings changes in the current year's earnings, so the firm is characterized by lower levels of conditional conservatism, relative to low CER performing firms. This results yields support to H1a and

Table 12.3 Regression results of CER and its components on conservatism

Variables	Model (12.2a)	Model (12.2b)	Model (12.2c)	Model (12.2d)
Constant	13.66*** (4.08)	5.708 (1.37)	10.07*** (4.48)	40.54*** (11.19)
DCFO	−12.71*** (−3.70)	−5.308 (−1.27)	−9.708*** (−4.24)	−39.75*** (−10.84)
CFO	−29.25** (−2.39)	−63.19* (−1.80)	−11.80* (−1.86)	−16.33*** (−3.13)
DCFO*CFO	28.73** (2.32)	62.74* (1.78)	11.32* (1.73)	16.57** (3.12)
CER	0.002 (0.02)			
CFO*CER	2.293** (2.49)			
DCFO*CER	−0.023 (−0.27)			
DCFO*CFO*CER	−2.289** (−2.49)			
EN_INNOV		0.148 (1.11)		
CFO*EN_INNOV		2.757* (1.83)		
DCFO*EN_INNOV		−0.161 (−1.21)		
DCFO*CFO*EN_INNOV		−2.753* (−1.83)		
EMM			0.045 (0.78)	
CFO*EMM			2.101*** (3.22)	
DCFO*EMM			−0.057 (−0.98)	
DCFO*CFO*EMM			−2.098*** (−3.22)	
RES_USE				−0.371*** (−6.03)
CFO*RES_USE				1.685*** (2.99)
DCFO*RES_USE				0.353*** (5.63)
DCFO*CFO*RES_USE				−1.692*** (−3.00)
R^2	0.123	0.168	0.139	0.104
F-stat	29.69***	29.51***	30.12***	30.64***
p-value	0.001	0.001	0.001	0.001

Note: *, **, *** indicate statistical significance at the 10%, 5% and 1% significance level respectively, t-statistics in the parenthesis

Table 12.4 Regression results of Basu (1997) model based on CER groups (model 12.3)

Variables	Low—CER	Medium—CER	High—CER
Constant	3.757*** (3.17)	2.245** (2.47)	3.602*** (6.81)
ΔNI	−0.003 (−1.19)	−0.001 (−0.44)	−0.005** (−2.09)
D_ΔNI	4.233 (0.10)	7.254 (0.77)	−3.981 (−0.02)
ΔNI*DΔNI	0.006* (1.70)	0.009* (1.77)	0.019** (2.36)
R^2	0.098	0.087	0.018
F-stat (p-value)	3.62** (0.012)	4.04*** (0.001)	3.84*** (0.009)

Note: *, **, *** indicate statistical significance at the 10%, 5% and 1% significance level respectively, t-statistics in the parenthesis

corroborates evidence on Table 12.3, so the negative association between CER and accounting conservatism is verified in the EU sample firms.

12.4.3 Sensitivity Analysis

In order to examine the robustness of the main findings we performed several sensitivity tests regarding the functional forms of the main models and the methodologies for measuring conditional conservatism. At first, we re-estimated model (12.2a) by applying panel regressions instead of cross-sectional OLS. This estimation method was chosen due to the fact that panel analysis contains more information with more variability and less collinearity among the variables, providing more efficient estimates and precise parameters of model estimation, allowing us to detect many effects that are not detectable in the cross-sectional data analysis. In order to extract more precise inferences we re-estimated model (12.2a) with random and fixed effects despite the fact that the Breusch-Pagan Lagrange multiplier test for random effect produced a highly significant statistic (Chi2 354.98, $p < 0.00001$) indicating that the random effects estimation is the most appropriate method. In addition, we included several control variables that have proved significant determinants of conservatism in the literature. So we control for the impact of firm leverage (LEV) measured as the ratio of total debt to total assets, growth opportunities (GROWTH) measures as the annual percentage change of sales revenues and asset tangibility (PPE) measured as the ratio of property, plant and equipment to total assets as in Hong (2020), Cheng and Kung (2016) and Pyo and Lee (2013). The relative results are presented on Table 12.5.

As we can see on the first and second column of Table 12.5, the F-stat and Wald Chi-square values are highly significant and the R^2s are very satisfactory considering the size of the sample and other relative studies published in the field. CER produced

Table 12.5 Panel regression results of CER on conservatism, including control variables

Variables	Model (12.2a)—Fixed effects	Model (12.2a)—Random effects	Model (12.2a)—Random effects with controls
Constant	25.55*** (6.44)	26.22*** (4.57)	19.56*** (3.50)
DCFO	−4.490 (−1.58)	−7.267 (−0.75)	−7.455** (−2.46)
CFO	−12.20** (−2.33)	−15.25*** (−3.93)	−21.03* (−1.79)
DCFO*CFO	11.82** (2.24)	14.48** (2.53)	21.80** (1.91)
CER	−0.155** (−2.13)	−0.111 (−1.45)	−0.134* (−1.68)
CFO*CER	0.914** (2.15)	1.159*** (4.48)	1.494** (2.07)
DCFO*CER	0.119** (2.12)	0.103 (0.72)	0.140** (2.00)
DCFO*CFO*CER	−0.913** (−2.14)	−1.155*** (−4.29)	−1.505** (−2.09)
GROWTH			−0.731*** (−5.30)
PPE			8.977** (2.25)
LEV			0.002 (1.55)
R^2	0.111	0.131	0.143
F-stat/Wald X^2	20.21***	22.42***	32.97***
p-value	0.001	0.001	0.001

Note: *, **, *** indicate statistical significance at the 10%, 5% and 1% significance level respectively, t-statistics in the parenthesis (for first column) and z-statistics in the parentheses (second and third column)

a negative and significant coefficient (−0.155) indicating that high CER performing firms are associated with lower total accruals. The three-way interaction variable (DCFO*CFO*CER) produced negative and significant coefficients both with fixed and random effects estimations, verifying our existing findings and again providing support to H1a. Also similar findings were extracted on the third column of Table 12.5 were we introduced additional control variables on model (12.2a). The findings verify arguments in the literature by Burke et al. (2020) and Hong (2020) and thus corroborate evidence on Table 12.3 and thus give support to H1a. Referring to the control variables only GROWTH and PPE produced statistically significant coefficients suggesting that firms with enhanced growth opportunities and lower asset tangibility (PPE) report lower accruals and so are associated with more conservative financial reporting.

Despite the fact that the Basu (1997) and Ball and Shivakumar (2005) models are extremely recognized in the literature, their main disadvantage is that they estimate the overall conservatism for a given year for all firms. On the contrary, Khan and

Table 12.6 Average C-Score (Khan and Watts 2009) based on CER groups

	Low—CER	Medium—CER	High—CER	Difference High—Low CER
C-SCORE	0.498	0.248	0.053	−0.445*** (0.001)

Note: *** indicate statistical significance at the 1% significance level respectively, p-value in the parenthesis

Watts (2009) proposed a method that allows us to estimate a firm-specific measure of conservatism based on the Basu (1997) model. So Khan and Watts (2009) incorporated additional controls affecting conservatism like firm size (SIZE), market to book ratio (MV_BV) and leverage (LEV) and constructed a firm-specific measure of conservatism termed the C-score from the following model which estimation is restricted to listed firms only due to the requirements for stock market data (Cheng and Kung 2016; Burke et al. 2020; Hong 2020):

$$\begin{aligned} EARN_{it} = \beta_0 + \beta_1 DR_{it} + (\mu_1 + \mu_2 SIZE_{it} + \mu_3 MV\_BV_{it} + \mu_4 LEV_{it})R_{it} \\ + (\lambda_1 + \lambda_2 SIZE_{it} + \lambda_3 MV\_BV_{it} + \lambda_4 LEV_{it}) DR_{it} * R_{it} + \varepsilon_{it} \end{aligned} \tag{12.4}$$

Where:

EARN is the earnings yield measured as the ratio of net income before extraordinary items divided to the beginning of year market value of equity.

R is the annual return estimated from compounding monthly stock returns starting 5 months after the fiscal year end on order to be sure that the information published in the financial statements will have been incorporate on stock returns (Cheng and Kung 2016).

DR is a dummy variable receiving unity if the R is negative and (0) otherwise.

SIZE is the natural logarithm of total assets at the end of the fiscal year.

MV_BV is the market to book ratio at the same period as stock returns.

LEV is leverage measured as the ratio of total debt to total assets.

Equation (12.4) is expanded by including the estimates of firm-specific coefficients μ_1 to μ_4 and λ_1 to λ_4 and then those coefficients are used to estimate the responsiveness of firms' earnings to positive returns (G-score) and the responsiveness of firms' earnings to negative returns (C-score) which is the main variable of interest. The estimation of the C-score and G-Score is based on the following model:

$$\text{G-score} = \beta_2 = \mu_1 + \mu_2 SIZE_{it} + \mu_3 MV\_BV_{it} + \mu_4 LEV_{it} \tag{12.5}$$

$$\text{C-score} = \beta_3 = \lambda_1 + \lambda_2 SIZE_{it} + \lambda_3 MV\_BV_{it} + \lambda_4 LEV_{it} \tag{12.6}$$

The higher the C-score the more conservative is the firm's financial reporting. So in order to examine the validity of the research hypothesis we will estimate the C-score for each firm after assigning it to portfolios based on the average CER performance as in Table 12.4. The results are presented on the following Table 12.6. The "difference" column presents the corresponding probability values from estimating the Welch two-sample t-test and asterisks indicate statistical significance at the 1 per cent significance level (Suto and Takehara 2018). As we can see, the average C-score decreases as we move from the low-CER to the high-CER performing group of firms and the difference between the two C-scores is statistically significant at the 1 per cent level. This suggests that firms with enhanced environmental performance (and responsibility) report less conservative financial numbers, so our main findings remain robust and thus H1a is again verified.

12.5 Conclusion

An extensive number of research studies have considered the impact of CER activities and performance on the quality of financial information published by firms and specifically the level of conditional conservatism that firms are reporting their accounting numbers. Empirical evidence so far have provided contradictory evidence on the impact of CER on conditional conservatism with some studies pointing to a positive association while others provide evidence of a negative association (Burke et al. 2020; Cho et al. 2020; Hong 2020; Yang et al. 2014; Patro and Pattanayak 2017; Cheng and Kung 2016; Francis et al. 2013; Pyo and Lee 2013). The scope of this chapter was to provide more concrete evidence on the impact of CER performance and its components on the accounting conservatism by utilizing a multi-country research setting and incorporating various sub-categories of environmental performance and conditional conservatism methodologies evidenced in the accounting literature by Ball and Shivakumar (2005), Basu (1997) and Khan and Watts (2009).

The current chapter provides more thorough evidence in the existing literature regarding the impact of CER on the quality of accounting numbers. The empirical results verified evidence and arguments in the literature that CER performance impacts negatively on accounting conservatism providing support to arguments regarding the enhanced transparency of CER performing organizations which leads to reduced demand for conservative financial reporting by contracting stakeholders. This conclusion is corroborated after performing several sensitivity tests relating to the estimation of conditional conservatism at firm level, the functional form of the models and other robustness tests.

References

Ball, R., and L. Shivakumar. 2005. Earnings quality in UK private firms: Comparative loss recognition timeliness. *Journal of Accounting and Economics* 39 (1): 83–128.

———. 2006. The role of accruals in asymmetrically timely gain and loss recognition. *Journal of Accounting Research* 44 (2): 207–255.

Ballou, B., P. Chen, J.H. Grenier, and D.L. Heitger. 2018. Corporate social responsibility assurance and reporting quality: Evidence from restatements. *Journal of Accounting and Public Policy* 37 (2): 167–188.

Basu, S. 1997. The conservatism principle and the asymmetric timeliness of earnings. *Journal of Accounting and Economics* 24 (1): 3–37.

Burke, Q., P.-C. Chen, and G.L. Lobo. 2020. Is corporate social responsibility performance related to conditional accounting conservatism? *Accounting Horizons*. https://doi.org/10.2308/horizons-18-111.

Cheng, C.-L., and F.-H. Kung. 2016. The effects of mandatory corporate social responsibility policy on accounting conservatism. *Review of Accounting and Finance* 15 (1): 2–20.

Cho, S.Y., P.K. Kang, C. Lee, and C.K. Park. 2020. Financial reporting conservatism and voluntary CSR disclosure. *Accounting Horizons*. https://doi.org/10.2308/hotizons-17-093.

Cui, J., H. Jo, and H. Na. 2016. Does corporate social responsibility affect information asymmetry? *Journal of Business Ethics* 148 (3): 1–24.

Dechow, P.M. 1994. Accounting earnings and cash flows as measures of firm performance: The role of accounting accruals. *Journal of Accounting and Economics* 18 (1): 3–42.

Dimitropoulos, P., D. Asteriou, and C. Sıriopoulos. 2012. Euro adoption and the quality of accounting information. *Managerial Auditing Journal* 27 (3): 299–328.

Flammer, C., and P. Bansal. 2017. Does a long-term orientation create value? Evidence from a regression discontinuity. *Strategic Management Journal* 38 (9): 1827–1847.

Fombrun, C., and M. Shanley. 1990. What's in a name? Reputation building and corporate strategy. *Strategic Management Journal* 33 (2): 233–258.

Francis, R.N., S. Harrast, J. Mattingly, and L. Olsen. 2013. The relation between accounting conservatism and corporate social performance: An empirical investigation. *Business and Society Review* 118 (2): 193–222.

Hong, S. 2020. Corporate social responsibility and accounting conservatism. *International Journal of Economics and Business Research* 19 (1): 1–18.

Hong, Y., and M.L. Andersen. 2011. The relationship between corporate social responsibility and earnings management: An exploratory study. *Journal of Business Ethics* 104 (4): 461–471.

Ioannou, I., and G. Serafeim. 2017. The consequences of mandatory corporate sustainability reporting. Harvard Business School Research Working Paper No. 11-100. Retrieved from https://ssrn.com/abstract=1799589

Khan, M., and R.L. Watts. 2009. Estimation and empirical properties of a firm-year measure of accounting conservatism. *Journal of Accounting and Economics* 48 (2-3): 132–150.

Kim, Y., M.S. Park, and B. Wier. 2012. Is earnings quality associated with corporate social responsibility? *The Accounting Review* 87 (3): 761–796.

LaFond, R., and R.L. Watts. 2008. The information role of conservatism. *The Accounting Review* 83 (2): 447–478.

Leventis, S., P. Dimitropoulos, and S. Owusu-Ansah. 2013. Corporate governance and accounting conservatism: Evidence from the banking industry. *Corporate Governance: An International Review* 21 (3): 264–286.

Lins, K.V., H. Servaes, and A. Tamayo. 2017. Social capital, trust, and firm performance: The value of corporate social responsibility during the financial crisis. *The Journal of Finance* 72 (4): 1785–1824.

Manganaris, P., E. Beccalli, and P. Dimitropoulos. 2017. Bank transparency and the crisis. *The British Accounting Review* 49 (2): 121–137.

Obigbemi, I.F., E.O. Omolehinwa, D.O. Mukoro, E. Ben-Caleb, and O.A. Olusanmi. 2016. Earnings management and board structure: Evidence from Nigeria. *SAGE Open* 6 (3): 1–15.

Patro, B., and J.K. Pattanayak. 2017. Association of CSR disclosure with accounting conservatism: A study of select Indian mining firms. *Indian Journal of Finance* 11 (3): 7–25.

Pyo, G., and H.-Y. Lee. 2013. The association between corporate social responsibility activities and earnings quality: Evidence from donations and voluntary issuance of CSR reports. *The Journal of Applied Business Research* 29 (3): 945–962.

Shackelford, D.A., and T. Shevlin. 2001. Empirical tax research in accounting. *Journal of Accounting and Economics* 31: 321–387.
Suto, M., and H. Takehara. 2018. *Corporate social responsibility and corporate finance in Japan*. Singapore: Springer Nature.
Watts, R.L. 2003. Conservatism in accounting - Part I: Explanations and implications. *Accounting Horizons* 17 (3): 207–222.
Yang, J.S., Y.J. Yeo, and O.J. Kwon. 2014. The effect of excellence in corporate social responsibility activities on accounting conservatism. *Korean Management Review* 43 (6): 1931–1961.

Chapter 13
Conclusions and Implications

As has been documented in the previous chapters, Corporate Environmental Responsibility (CER) has been on the forefront of the academic and business community for over two decades, especially because of the importance of environmental awareness for the functioning and operation of the organizations. Despite the increased interest of CER on various business sectors the concept has recently entered the accounting and finance disciplines, since many corporations have evolved as a growing economic force within the national and international markets, and are more intensively engaging in socially and environmentally responsible activities. The book has built on the stakeholder and institutional theories proving that CER is a significant dimension of corporate responsibility and addressed key implications of CER related to corporate accounting and finance and specifically on financial performance, capital structure, firm risk, innovating activities, and the quality of accounting reporting. The aim of this book was to provide answers on several research questions referring to CER and corporate finance and accounting by utilizing data from a sample of EU listed and unlisted firms for the period 2003–2018. The final chapter of the book is dedicated to summarizing the main findings and conclusions in response to our research questions stated on the first chapter, and also provide useful policy suggestions and directions for future research.

13.1 Corporate Environmental Responsibility in Action

Corporate environmental responsibility (CER) has gained significant momentum and became an important component of corporate strategies and decisions. Since the last decade executives are positive towards CER since it contributes to firm's profitability and has been a permanent addition to the managerial agenda ever since. The trend is expected to persist in the future because CER and sustainability

P. Dimitropoulos, K. Koronios, *Corporate Environmental Responsibility, Accounting and Corporate Finance in the EU*, CSR, Sustainability, Ethics & Governance, https://doi.org/10.1007/978-3-030-72773-4_13

in general have been introduced themselves as a core theme on the board discussions and executive agendas.

Following the discussion in previous chapters of the book, CER refers to the implementation and design of business strategies and activities that deal with the interests of the firms but simultaneously sustaining natural resources for future utilization and exploitation. The benefits of CER activities on organizations have include among others: the reduction of waste, addressing environmental issues that matter to the public, increase operation efficiency, reduce litigation and reputational risks, and even discourage regulatory scrutiny and interventions. CER activities and investments give corporations the necessary tools and mechanisms to produce innovative products and services based on environmental standards, take actions to promote environmental awareness and responsibility and add value to shareholders and other stakeholders while abiding to laws and regulations at the same time.

Environmental responsibility programs and regulations have been shaped by the interventions of several international organizations and global initiatives. At first, the United Nations Global Compact (UNGC) was established in 2000 and is considered as the first type of intervention from a global non-governmental organization on the matter. Specifically, the UNGC adopted three main principles on environmental responsibility relating to supporting a precautionary strategy on environmental issues, to undertake specific initiatives to promote greater environmental responsibility and to incentivize the creation and dissemination of environmentally friendly technologies.

In the same direction, OECD has also developed practices and regulations on environmental responsibility in order to create new forces of economic growth via environmental friendly processes, green growth programs and innovations. Specifically, in 2008 OECD reported that there are several environmental issues that are not effectively managed by corporations including water scarcity, hazardous waste management, air quality and ecosystem fragmentation. Since then several corporations exerted efforts to cope with the proposed directions and resolve environmental problems created by their operation. Among the most significant environmental practices adopted by corporations are energy efficiency policies along with environmental innovation activities, incorporation of emission reduction programs on business operations and the implementation of waste reduction policies and efficient utilization of resources. Those three issues have been examined extensively on this book not only descriptively, but also empirically by considering their association with corporate finance and accounting quality.

The rest of the chapter is organized as follows: In the following section we provide a brief discussion on environmental responsibility issues faced by corporations followed by a section on several emerging issues on environmental responsibility. The fourth section will discuss the answers to the main research questions that have been raised on the book. The fifth section will provide useful policy implications based on the empirical findings of the book and will also offer fruitful directions for future research. The last section will conclude the chapter with some parting thoughts on environmental and sustainability reporting for firms and corporations around the globe.

13.2 CER Issues Faced by Corporations

Nowadays, the corporate world seems more vulnerable than even to environmental changes and challenges. Corporations within various sectors are potentially sensitive to shortages in resources from the natural environment and has become clearer now that the environment cannot longer support business operations that degrade the ecosystem and become environmentally hazardous for the society and the reputation and brand-loyalty of the firm. Consequently, societies today cannot add value to products and services (and so firms cannot yield profits) that are the outcome of hazardous and polluting processes and operations. This fact has been recognized by the majority of economic sectors, where managers realize that their core business operations will be significantly affected by natural resources shortages in the following 5 years or more. Business sectors with a significant environmental footprint (chemicals, industrials, telecommunications and manufacturing) have started to adjust their operations around the abovementioned concerns and proceed to production operations that are more environmentally friendly and protect natural resources via enhanced waste management.

Business environmental responsibility together with the dissemination of voluntary (or mandatory) reports, documents, and practices can further contribute to the protection of the environment, while simultaneously improving the financial results (revenues, profits and productivity) of corporations. Another current issue that has emerged at the top of the corporate agenda is the report of greenhouse gas emissions (GHG). This trend has increased over the years, not only from the awareness on behalf of managers and corporations on the significance of air pollution to resources and social wellbeing, but also for reputation management, efficiency strategies and goals, and customer awareness and expectations.

Since independent organizations and database firms started to rate firms based on their environmental responsibility activities and achievements, the reputation issue has been on the forefront of corporate decisions. As years go by, more and more firms realize that greenhouse gas emissions are at the bottom line a form of waste which adds no value to the firm or its customers and stakeholders and is also being considered as a proxy for risk and inefficiency by several investors as well. It is not a surprise that almost three out of four corporations worldwide have set emission reduction goals on an annual basis and tend to report on the relative outcomes publicly. Thus, the reduction of GHG has been considered as an efficiency mechanism and as a way to mitigate risk and future liabilities of the company.

Finally, water management is another issue that has gained attention by firms. Most firms in EU and US consider water scarcity as a substantial risk for their business operations and simultaneously identify opportunities on this realm of activity along with profit potentials on the production of innovative products with minimum water consumption and waste. In the same vein, firms internationally have started to set goals on the reduction of water waste and also report extensively on the water usage along with future targets on water management. This tendency is even clearer in water-intensive sectors such as mining, oil and gas, chemicals, energy, and food and beverage where the majority of the corporations have set goals on water footprint reduction.

13.3 Emerging Issues on CER

As CER is evolving over the years, there are some emerging issues that capture the public eye with more interest than other. For example, the most important audiences of environmental responsibility reports are customers, employees, shareholders (investors), analysts and various national and international NGOs. This fact indicates that various types of stakeholders have evolved into crucial players on the environmental agenda and their choices and decisions have significant impact on corporate activities. Among the most prominent group that is expected to impact firms' CER behavior and strategies the following decade are investors and lenders. The fact is that the increased interest of investors for the issue of climate change and degradation of environmental resources is expected to shift investors' focus and assessment of corporate risks associated with economic, environmental and social activities. This in turn is expected to have a significant effect on corporate performance and overall business strategies.

Moreover, the interest and focus of investors on environmental reporting and performance is extremely important and has been enhanced over the years, since we have experienced an important growth on social and environmental responsible investments (SERI) mounting up to trillion of dollars. Additionally as we shown on the previous chapters, there is a positive association between environmental performance and financial performance in terms of profitability, lower systematic risk, better accounting quality and improved capital structure. Also, banks and financial services firms have already established environmental and sustainability research departments which provide appropriate information to analysts and investors, increasing the scrutiny, coverage and comparability of corporate environmental information internationally. This fact has increased the investors' awareness on the value relevance of environmental information and incorporate such data on their investment decisions and risk appraisals. Finally, institutional investors (banks, hedge funds, pension funds, mutual funds, insurance companies, foundations and asset managers) have also indicated a growing interest for environmental information, CER performance and even GHG and climate-change risk disclosures. All those facts lead to the logical conclusion that CER related information and disclosures will enhance over the next 10 years not only in quantity, but also in quality since there are several stakeholders who consider CER reporting as a proxy of corporate risk, leading to enhanced reporting and assurance of such information.

13.4 Research Questions and Answers

On this section we will provide a summary of the main empirical conclusions of the book by connecting them to the eight research questions that were stated on the first chapter of the book. Specifically:

1. Which are the main financial determinants of CER performance?

   Empirical analysis on Chap. 4 suggested that profitability, leverage, size and growth opportunities have a positive impact on the CER and its components (environmental innovation, emission reduction and resource utilization), while on the contrary CEO duality had a negative impact on CER performance. Moreover, the level of CER engagement and performance is associated with the existence of a CSR committee within the firm which provides strategic direction and focus on corporate activities. Finally, in order to take into consideration the impact of firms' legal framework and environment, we proceed with the separation of the sample firms between code law and common law firms. Results indicated that the aforementioned associations are more significant for code law firms suggesting that countries' legal system has an impact on corporate decisions regarding CER.
2. How CER performance contributes to corporate innovative activities and how it affects firms' decisions to invest in R&D, and the impact of that decision on their financial performance.

   Empirical analysis on Chap. 5 suggested that CER generates competitive resources leading to competitive advantages and enhanced corporate financial performance (CFP). Also, the specification of our empirical design indicated that there is a bidirectional association between CER and CFP. Our evidence confirm that highly profitable firms are associated with higher CER performance due to their enhanced ability to finance such activities. In turn higher CER performance contributes to enhance CFP which contributes towards higher environmental investments and CER performance. However, when we considered the innovative activities of firms on Chap. 6, we found that the R&D indicator variable (R&D_D) and its interaction with profitability (ROE) yield insignificant coefficients in all model estimations. This result corroborates arguments in the literature that less innovative firms resort to increased CER activities in order to gain market differentiation (or competitive advantages) and enhance their financial performance. Consequently, R&D investments in our EU sample plays a substitution role on the CER-CFP relation and so it has an insignificant impact on CSR and corporate financial performance. Also, we examined the components of the CER variable and how they interact with profitability for innovative and non-innovative firms. All CER components produced positive and significant coefficients in almost all tables indicating that higher environmental innovation score, emissions score and resource use score lead to higher profitability. However R&D investments produced insignificant coefficients. Specifically, R&D intensive firms receive lower emissions score or invest less in emission reduction activities relative to their less R&D intensive counterparts.
3. What is the association between CER and corporate capital structure decisions?

Empirical analysis on Chap. 7 suggested that CER performance has a negative impact on long term debt but on the contrary it has a positive impact on short term debt and shareholders' equity indicating a differential impact of CER on capital structure decisions. Practically, we can argue the CER performing firms rely more on their own shareholders' equity and short term debt to finance their activities, and less on long term debt. This finding provides support to the arguments about the different motives behind short term and long term financing decisions and contribute to the ongoing debate about the association between CER and capital structure of EU corporations.

4. How CER performance is associated to corporate systematic risk?

   Empirical analysis on Chap. 8 suggested that CER performance provide a negative and significant impact on firm systematic risk (beta) indicating that CER investments reduce the firm's systematic risk, so leading us to accept the hypothesis of risk reduction impact of environmental responsibility. In addition, firms with more stakeholder oriented CER activities are associated with more risk aversion and ethical behavior, especially towards its employees and the environment, thus perceived as having less uncertainty and risk. This is interpreted as investors requiring a smaller risk premium to invest in such corporations. Moreover, after separating sample firms based on their country's legal origin we found that CER performing firms in common law countries are associated with higher systematic risk. This result indicates that CER activities in common law countries are considered as a waste of corporate resources contributing to increased financial risk, since investors and shareholders could ask for a higher premium in order to safeguard their investments from the risk that CER activities bear.

5. What is the impact of CER performance on working capital management decisions, and specifically on cash holdings and cash dividend payments?

   Empirical analysis on Chap. 9 indicated that CER performance has a positive and significant impact on cash holdings, supporting evidence provided on the literature that CER performance lead firms to hold more cash. This result is attributed to the fact that managers of high CER performing firms focus more on maximizing stakeholder value leading to enhanced cash holdings to achieve such goals. Furthermore, we found firms with higher emission (EMM) and resource use scores (RES_USE) are significantly associated with higher cash holdings. Furthermore, all environmental performance subscores (except for the emission score—EMM) found to impact positively on dividends payments. Practically, empirical evidence suggest that a 1 percentage point increase in CER performance score contributes to a 0.2 per cent increase to dividend to asset ratio, ceteris paribus. These evidence verify the existing arguments in the literature that a high dividend payout ratio can signal to market participants that CER related activities are not a waste of firms resources and conversely could lead to a better allocation of those resources and to enhanced shareholders' satisfaction.

6. How CER performance contributes to the value relevance of accounting information, the persistence and predictability of earnings and the firms' stock prices?

Is CER performance useful for investors and analysts to project future earnings figures?

Empirical analysis on Chap. 10 suggested that CER performance alone does not have a significant impact on stock prices, something that contradicts several studies in other countries. On the contrary, it seems that CER performance has a moderating impact on the value relevance of accounting numbers. Specifically, we found that high CER performing firms have earnings which are more persistent 1 year ahead and more predictable (have lower levels of variability), and report profits and common equity figures which have a positive impact on stock prices. The findings of this chapter contribute to the ongoing debate about the incremental value relevance of CER disclosures and the quality of financial reporting, by establishing a direct link between CER performance and the value relevance of accounting numbers.

7. What is the contribution of CER engagement and performance in mitigating earnings management behavior?

   Chapter 11 was dedicated on the examination of CER performance on the quality of accounting earnings and specifically earnings management behavior. The empirical results verified evidence and arguments in the literature that CER performance improves the quality of published accounting information by mitigating earnings management behavior via accruals and real activities manipulation. This argument is corroborated after performing several sensitivity tests relating to the estimation of discretionary accruals, the functional form of the models and other robustness tests. CER components (environmental innovation, emissions and resource utilization) scores all produced negative and statistically significant coefficients supporting our hypotheses that firms investing more on environmental innovation, efficient resource utilization and emission reduction, published higher quality financial information by manipulating their earnings figures less. Evidence in Chap. 11 corroborate arguments in the literature that high CER performing firms with enhanced monitoring by the markets, regulators, media and the society tend to report accounting information of enhanced quality.

8. How CER performance impacts on the corporate decisions regarding accounting conservatism?

   The goal of Chap. 12 was to provide more concrete evidence on the impact of CER performance and its components on the accounting conservatism of financial numbers. The empirical results verified evidence and arguments in the literature that CER performance impacts negatively on accounting conservatism providing support to our hypothesis regarding the enhanced transparency of CER performing organizations which leads to reduced demand for conservative financial reporting by contracting stakeholders. This practically means that high CER performing firms not only provide financial statements which are closer to the true economic condition of the firm (characterized by less earnings management), but also recognize bad news (losses) in a more timely manner allowing investors to more effectively assess firms' future prospects and risks. This fact indicated that environmentally responsible firms designate enhanced responsibility on financial reporting as well.

13.5 Policy Implications and Future Research Recommendations

The empirical findings of this book could be proved useful to investors, academics, lenders, regulators and various other stakeholders since they provide useful policy implications. At first, evidence on Chap. 4 could be proved useful to existing shareholders and potential investors for reaching on efficient investment decisions. Existing shareholders via their voting power in their general meetings can shift managerial focus towards CER related issues and even exert pressure to CEOs to engage in more CER related activities and disclosures in the future. Such activities could be considered as potential signals to the market for the dedication of the firm to social and environmental issues and the devotion of the firm to corporate transparency. Also, potential investors should consider the association between firm's governance structure and CER performance since firms which have not established a sustainability committee might indicate less involvement or awareness for environmental related disclosure, which may affect the perceived levels of firm risk by future investors. This fact could have implications for firms' corporate financing, future profitability and viability.

The findings of Chap. 5 could be proved useful to both managers and regulators. Regulators should seriously consider fostering their guidance and support towards firms' to follow environmental responsibility activities and investments (with various direct and indirect policies such as financing, tax exemptions etc.). Also regulators can promote the internationalization of CER activities by local firms, which in turn can help gain access to international financial markets and global resources. Moreover, the results could be proved useful to managers, since the implementation of CER related investments and activities could yield significant consequences on their financial performance. Specifically, the investments in environmental innovation, emissions reduction and more efficient resource usage are noteworthy contributors to managerial efficiency and profitability. This fact not only will allow firms to gain legitimacy for their operations but also to access various resources which could be proved significant for their longevity and viability.

Also, evidence of Chap. 6 have important managerial implications for corporate managers because if they decide to devote resources on socially and environmentally responsible activities based on their values or beliefs, our evidence point that such decisions could have a significant economic reasoning under certain conditions. The substituting impact of R&D on the CER-CFP relation suggests that managers should consider or balance both benefits and costs of each decision and channel funds towards those activities with greater value or that which can benefit shareholders and help the firm achieve its strategic goals. So at the bottom line the decision between R&D and CER investments resolves into a trade-off between costs and benefits. Alternatively, when managers are deciding to allocate limited resources to CER activities they have to consider CER and other types of long term investments such as R&D, simultaneously and not independently.

Moreover, findings of Chap. 7 have important implications for lenders and investors. Lenders of firms with high CER performance should seriously consider re-financing their activities since are firms with less information asymmetries, long term debt obligations and risk, so they have greater ability to re-pay interest and principal easier compared to their low CER counterparts. In addition, existing and potential investors could use the findings of this chapter useful in fostering their guidance and support towards firms' to follow environmental responsibility activities and investments since the implementation of CER related investments and activities could yield significant results on firms' risk exposure, and sustainability. Specifically, the investments in environmental innovation, emissions reduction and more efficient resource usage are significant contributors to managerial control (agency theory perspective). This fact not only will allow firms to gain external financing for their operations but also to access various other resources which could be proved significant for their longevity and viability.

Furthermore, the empirical findings of Chap. 8 have also implications for investors, especially those who invest within firms with high CER performance. They should seriously consider re-financing their activities since they are firms with less information asymmetries and risk, so they have greater ability to reduce the overall riskiness of their portfolio and sustain their future returns. Also, investors need to consider each firms' legal origin as a differentiating factor that impacts on the association between CER and risk. In addition, financial managers, stakeholders and policy makers dealing with CER creation guidelines and regulations could use the findings of this chapter useful in fostering their guidance and support towards firms' to follow environmental responsibility activities and investments since the implementation of CER related investments and activities could yield significant results on firms' risk exposure, and sustainability. Specifically, the investments in environmental innovation, emissions reduction and more efficient resource usage are significant contributors to risk reduction, which can not only assist firms to gain external financing for their operations but also to access various other resources which could be proved significant for their longevity and viability, this contributing to a more efficient allocation of resources within the economy.

In addition, findings of Chap. 9 could be proved useful to both regulators and investors. Investors who are interested in receiving annual cash dividends instead of uncertain future capital gains (according to the bird-in-the-hand theory of dividend policy) can focus on high CER performing firms as their environmental awareness could provide indications of an enhanced payout policy in the future. In addition, existing and potential investors could use the findings of this chapter in fostering their guidance and support towards firms' to follow environmental responsibility activities and investments since the implementation of CER related investments and activities could yield significant results on firms' cash holdings decisions and their ability to repay obligations, finance daily operations and even take advantage of future growth opportunities. Also, considering the benefits of environmental responsibility in improving stakeholders interests and needs, regulators and governments must try to sustain at first place and later on to improve the reporting framework and the motives for firms to get involved in socially and environmentally responsible activities.

Moving on to the issue of accounting quality and CER performance, the findings of Chap. 10 provide useful theoretical implications as well as policy implications for managers, investors and regulators. At first, the current study adds to the growing literature on the implementation of CER as a tool for signaling the firms' ethical stances towards society and at the same time as a mechanism for wealth creation. So under this framework, corporations must incorporate CER within their operational and strategic decisions, not only for gaining competitive advantages, but also as a moral obligation towards their stakeholders. Consistent with this moral obligation, the findings of this chapter indicate the importance of CER for enhancing the quality of financial information. Motivated by different goals and aspirations, corporate managers may engage within CER for different reasons, yet the improvement of accounting quality surpasses all market participants, investors, creditors and various other stakeholders. Practically the results of this chapter, argue that there is a win-win status for the corporations and their stakeholders since the transparency and reliability of financial information could provide a sound basis for efficient decision making for all stakeholders. Also, the findings of the chapter have also important managerial implications. Since the issue of environmental protection (climate change etc.) has gain increased attention from the public the last decade, it is comprehensible that the more managers exert efforts to enhance environmental reporting and related activities, the more transparent will be the disclosed financial information. Thus, managers may consider the engagement and publication of CER related activities and investments not only for gaining reputation and recognition among the society and their stakeholders, but also for improving the quality of their firm's persistence, predictability and value relevance of accounting information. In order to achieve such goals, managers must develop and incorporate widespread procedures for measuring, disclosing and circulating financial and environmental information to the public and even improving firm's environmental performance each year. Also, this chapter provides useful implication for investors as well. Investors need to take into consideration both the reliability and transparency of environmental and financial information in order to reach in efficient investment decisions, since CER is tightly connected and even moderates positively the value relevance of published accounting numbers. Moreover, our findings are also useful for regulators who are the official advocates and standard setters for CER related activities and disclosures. So regulators can further set more detailed rules on improving the quality of CER disclosures (such as the GRI standards which are globally renowned and accepted) and force firms to disseminate those information to the public. This can have a dual outcome, once by spreading the word about the necessity for firms to incorporate CER as a main pillar of corporate decisions and activities (thus enhancing stakeholder engagement and firm legitimacy) and secondly by improving the quality of financial information to the markets, leading to better risk assessment, better investment decisions and overall to efficient allocation of scarce resources within the economy. This last effect could be even more important to led developed countries and markets within the EU.

Additionally, evidence of Chap. 11 have important implications for regulatory agencies in the EU. Since the majority of the European countries have adopted common accounting standards and financial reporting rules, the improvement of the accounting framework (which is the ultimate goal of supranational accounting regulatory bodies such as IASB) could be achieved via enhancing the use and publications of social and environmental responsibility information and reports. Also, the current chapter could be of interest not only to academic researchers but also to potential and existing investors in European corporations. The negative association between CER performance and earnings management could be used by investors in assessing the risk of firms and the quality and reliability of financial information. As we have provided evidence of a lower systematic risk for firms engaging and reporting CSR activities and responsible investments since investors perceive such firms as less risky. The fact that CER performance impacts positively on the quality of accounting information provides additional assistance to investors for making value relevant investment decisions.

Finally, empirical evidence of Chap. 12 contribute to the ongoing debate on conservatism by providing evidence that European firms' managers utilize conservatism in the financial reports and that choice is inherent on non-traditional business obligations and activities like environmental responsibility. Also, the findings on this chapter have important implications for regulatory agencies in the EU. Since the majority of the European countries have adopted common accounting standards and financial reporting rules, our evidence could be relevant to regulators when they assess the extended permutations of CER activities within their analysis of the advantages and disadvantages of potential CER reporting regulations. Also, the current chapter could be of interest not only to academic researchers but also to potential and existing investors in European corporations. The negative association between CER performance and conditional conservatism could be used by investors in assessing the risk of firms and the quality and reliability of financial information. The fact that CER performance impacts negatively on the conservatism of accounting information provides additional guidance to investors for assessing the information quality of non-financial reports and incorporate such information on their investment decisions. Consequently, the empirical contribution of this book supports the position that the engagement on CER related activities and disclosures creates direct implications for financial decision making, the quality of published financial information and decision making ability of practitioners since environmental and sustainable behaviors can bring important advances in operational efficiency, innovation, economic performance and competitiveness.

Nevertheless, there are several other issues and challenges that future research needs to examine further. At first, the current study was focused on listed and unlisted corporations that have published financial statements and were covered by international agents on their CER behavior and performance. Future research can extend findings on this book by examining smaller firms (not having the legal form of corporation) and even family controlled firms in order to assess the impact of CER performance on finance and accounting issues. Also, a potential avenue for future research is to examine the concurrent impact of environmental and social responsibility and performance on firms' accounting and financial management. Several arguments are made on the literature on corporate sustainability which incorporates governance, social and environmental management. So it will be interesting to consider all dimensions of sustainability on corporate finance and accounting. Finally, since this study is focused on EU countries which are mainly well developed economies and follow similar regulations and laws, it will be interesting for future researchers to consider corporate sustainability within developing economies where such environmental and social regulations and directions have a clear voluntary nature and are in their initial stage of development. The potential extension of this study to developing economies can provide comparative evidence and useful policy implications for the implementation of a global environmental and sustainability reporting framework.

13.6 Parting Thoughts

Closing the final chapter of the book, it seems logical to leave the reader with some parting thoughts on the importance of environmental reporting. As discussed earlier, the existence of several corporate scandals over the last two decades and the cases of financial turmoil made evident the fact that there were lapses in the ethical behavior and decisions of corporations. Regulators and standard-setters tried to promote ethical behavior through enhanced monitoring and by introducing mechanisms for the transparent operation of corporations. Thus, ethical behavior and regulatory compliance became an integral part of corporate reporting for improving the information environment and even preventing further crises from happening.

The enhanced focus on environmental reporting and long lasting value adding activities has created a need for a new form of corporate reporting that goes beyond the classic financial statement type of reports. This new enhanced type of corporate reporting incorporates several non-financial information including environmental key performance indicators (KPIs). The utilization of both financial and non-financial KPIs may provide details on future earnings and cash flows and allow investors to assess the full risk of business activities, which do not always come from the core economic investments of the firm. So corporations can incorporate forward looking information on their annual reports in order to make them more relevant to the stakeholders.

This tendency has created a new "battleground" for accounting standard setters who need to develop an integrated reporting framework which will aim at the intersection among environmental responsibility and economic value through environmental management and disclosure of relevant environmental information. Under this framework, several stock market exchange commissions require from their member firms to disclose information (social and environmental) on their operations and the accountability of such activities. As years pass by and most firms are disclosing environmentally related information, there will be a need to converge the different types of disclosures and reports under a globally accepted framework. The framework can cover the needs of corporations to use such reporting for both internal and external purposes. Internally, CER reports can be used in order to assess the achievement of specific goals and strategies, for benchmarking purposes and evaluation of managerial effectiveness. Externally CER disclosures could assist stakeholders for decision making reasons and to validate their compliance with existing laws, rules and regulations.

Consequently, standard-setters and regulators need to face some challenges on their journey to global standardization of sustainable and environmental disclosures. At first, environmental disclosures must be centered on the key business stakeholders (creditors, investors, shareholders, customers, suppliers, employees, government and the society in general) and be value relevant for their needs and interests. Also, information on those disclosures must be forward-looking and include both financial and non-financial information and KPIs relative to the environment and other crucial social issues. Furthermore, those disclosures must be based on principal standards and not code of law, since principle-based reports define the conceptual framework of environmental reporting and propose integrated thinking by corporations. Also, environmental disclosures must be easily presented and simplistic in their language and avoid unnecessary complexities which may confuse the user. The simplicity of presentation should be made without mitigating the transparency that must characterize such disclosures and in order firms to achieve that they will have to keep a balanced stance on the disclosed information by providing both favorable and unfavorable performance data. The assurance of the disclosed information from an independent auditor can further enhance the transparency and reliability of such disclosures. Finally, there must be a relative consistency on the environmental disclosures through uniformity of presentation, measurement, recognition and identification of environmental performance. All those challenges may seem a rather vast amount of work that needs to be done for a rather long period of time, yet the first steps for a reliable and transparent implementation of CER strategies and disclosures has already been made (though under a voluntary nature) and definitely the future will unveil all the above mentioned issues under a more systematic manner and will be in the agendas of corporations and international organizations thereafter.